W9-CCX-024

Mysticism

Mysticism

HOLINESS EAST AND WEST

Denise Lardner Carmody
John Tully Carmody

New York Oxford
OXFORD UNIVERSITY PRESS
1996

Oxford University Press

Oxford New York
Athens Auckland Bangkok
Calcutta Cape Town Dar es Salaam Delhi
Florence Hong Kong Istanbul Karachi
Kuala Lumpur Madras Madrid Melbourne
Mexico City Nairobi Paris Singapore
Taipei Tokyo Toronto

and associated companies in
Berlin Ibadan

Published by Oxford University Press, Inc.
198 Madison Avenue, New York, New York 10016

Oxford is a registered trademark of Oxford University Press, Inc.

Library of Congress Cataloging-in-Publication Data
Carmody, Denise Lardner, 1935–
Mysticism : holiness east and west /
Denise Lardner Carmody, John Tully Carmody.
p. cm. Includes bibliographical references and index.
ISBN 0-19-508818-2 (cloth).—ISBN 0-19-508819-0 (pbk.)
1. Mysticism—Comparative studies. I. Carmody, John, 1939– .
II. Title.
BL625.C33 1996
291.4'22—dc20 95-14411

1 3 5 7 9 8 6 4 2

Printed in the United States of America
on acid-free paper

PREFACE

This book is an introductory study of mysticism in the world religions. It should be suitable for both general readers and college undergraduates. We provide both a theory of mysticism and surveys of its main contours in Hinduism, Buddhism, Chinese and Japanese religions, Judaism, Christianity, Islam, and the traditional religious cultures of oral peoples in North America, Latin America, Australia, and Africa. We also suggest how readers may think best about what mysticism implies for their own lives.

Our thanks to Helen McInnis and Cynthia Read of Oxford University Press for sponsoring the book, and to the many people at Oxford who worked on producing it.

Santa Clara, Calif.
February 1995

D. L. C.
J. T. C.

CONTENTS

Mysticism

1

Introduction

Scenes from an Album of Mysticism

Following the age-old traditions of his tribe, a young Native American boy on the verge of puberty receives his final instructions from the elder who has been his teacher and departs for several days of solitude in the deep forest. He is in quest of a vision, a dramatic experience that will reveal who the holy powers want him to be as an adult. Fasting, letting the solitude take over his soul, and trying to keep his spirit pure, he waits for an all-important revelation. On the fourth day it comes. A black bear crashes into his sacred space. The boy watches the bear intently. The bear pauses and seems to consider him at leisure. As the bear turns to leave, the boy hears in the depths of his mind the words, "Fear nothing. Walk through the world unhurried, with dignity." Returning to his teacher, the boy tells what he has experienced. The teacher is pleased: "Black bear will be your new name. Pray mightily that the spirit of this great brother will stand by you. Remember his message. Live without fear, as he does. Walk like a chief." Presenting the boy to the rest of the tribe, the teacher is aglow with pride. The chief asks him to lead the ceremonies celebrating Black Bear's coming of age, which he does with great dignity. The vision-quest was a crucial experience for many Native American boys. Not to receive a vision could mean that one lived among the women, one's manhood indistinct, compromised.

The passage to adulthood, which came more naturally to young women when they began to menstruate, could be the occasion for solitude and reflection that deepened their sense of themselves, but that was less dramatic than the struggle through which boys had to go. Young Native American women learned about the mysteries of fertility. Young men learned about the mysteries of killing. The two powers, to conceive life and to end life, were polar and had to stay separate. Thus women withdrew into solitude during their monthly periods and men went into isolation after hunting or warfare, and each power had its mystique, its aura of being able to take sensitive people into the heart of the sacredness, the primal reality, responsible for the world. Could the coming of each power, or the adult experience, therefore be the occasion of a mystical experience?

As we shall see, much in any discussion of mysticism depends on one's definition. Our position is that a vision-quest or an intense experience of female fertility or a soul-shaping endurance of a great storm or a passionate love that reveals the structures of the world or a dialogue with a personal divinity in deep prayer can turn into a mystical moment if the person having the experience feels drawn into a direct encounter with ultimate reality, the very foundation of everything that is. In our view, mysticism depends less on the particulars of the given experience—vision-quest, ritual dance, Zen meditation, perception of the utter perfection of a child's ear—than on what happens through it, where it "takes" the person experiencing it.

Imagine another scene. A Hasidic rabbi is lost in prayer on behalf of his suffering people. He feels himself taken to the gates of heaven, where the Master of the Universe hears all of the petitions sent up by His people. The rabbi is present in body in the synagogue. His congregation can see him bowed low. However, his spirit has taken wing. He has gone out of himself through the intensity of his prayer. At the gates of heaven the guardian angel looks him over and then lets him enter. He moves toward the throne of heaven, getting close enough to see some of the rays of the divine glory. Dazzled, he hands over his petition: "Take care of my people, your people." Suddenly, he is back on earth, his back aching from his having been bent low for so long. Listening to his soul, he tells his people, "Somehow, I don't know how exactly, the Lord will take care of us. He knows of our plight. It pains His heart. Keep faith, my brothers and sisters. Hang on. Remember our ancestors in the desert, awaiting their entrance into the Promised Land. Remember how God saved Isaac from the knife. The Lord gives, and the Lord takes away. Blessed be the Lord. All flesh is grass. All our lives are fleeting. Keep faith. Do not despair. Let the Lord be your Lord, the one who cares for you far better than you could care for yourself."

In the story, the Hasidic rabbi is doing what his famous predecessors have done: he is hoisting the sufferings of his people onto his own back and taking them before the Lord. He is making sure that God knows how much this small portion of the chosen people is suffering. The rabbi's prayer is mystical inasmuch as it takes him to the center of the universe, the source of all creation, the divine place where all that happens gets its sense; he enters into the great quilt that God is weaving from history. There, the rabbi is transformed. Call his vision fantasy or a symbol for the journey of his soul or whatever else you wish, but do not belittle it. Do not patronize the rabbi, as though your late twentieth-century view of the world were clearly superior to his. Where do you place the center of the universe? Where do you go to cry when the pain has become unbearable? If you do not know, if you have no place to go, are you not poorer than the rabbi? Who lives in the tract house and who in the palace of memory?

Lastly, imagine a traditional Chinese Buddhist nun meditating. See her sitting calmly in the lotus position, her back straight, her weight balanced perfectly, her eyes half-open and gazing without focus. She has left her *koan* (meditational word) far behind. She is just sitting, paying the stuff that floats down the stream of her consciousness no mind, letting the intrinsic knowledge, the innate Buddhahood, of her mind come closer to manifesting itself. She is lost to the world, yet not traveling imaginatively as the rabbi was. She is focused, collected, "one-pointed." So, when the small green worm of which she has been aware pushes a drop of water off the end of the brown leaf it was eating, the fall of the drop registers like a minor earthquake. Everything in existence could be in that drop. Nothing need separate the drop from the leaf, the leaf from the worm, the worm from the nun herself. The nun does not experience this intuition with the force of enlightenment. She does not feel her whole spirit tip over and place her in an outpost of wisdom. However, she does know, instinctively, that she has moved a little closer to the day when that may happen. She does leave her meditation a little more convinced that the Buddha's precept about stopping desire points to a place of utter peace.

If the nun had directly experienced ultimate reality, sensing the inner light that Buddhist teaching says shines within all things, we might call her meditation mystical. As described, what happened to her was still preparatory, preliminary, though quite advanced. Most Buddhist accounts of enlightenment describe ultimate reality in impersonal terms. There is no "Master of the Universe" hearing prayers in a heavenly palace. Equally, there is no black bear coming as a "familiar" to make the kinship between human beings and animals so vivid that thereafter the person could never doubt it. The configuration of most Buddhist peak experiences is simpler, the tone quieter, the temperature cooler; but the transformation can be just as profound. Through the transformation the world of the person granted the peak or mystical experience changes markedly, often for good, never to return to what it was before the experience, so that the person finds more beauty, more joy, than she had suspected was possible.

Theories of Mysticism

There are more theories of what happen to people such as the three whom we have described, grounding more definitions of the term "mysticism," than any general treatment can handle adequately. Thus Louis Dupré, developing an overview for *The Encyclopedia of Religion,* begins with an instructive disclaimer:

> No definition could be both meaningful and sufficiently comprehensive to include all experiences that, at some point or other, have been described as mystical. In 1899 Dean W. R. Inge listed twenty-five definitions. Since then the study of world religions has considerably expanded, and new, allegedly mystical cults have sprung up everywhere.[1]

Although the range of the term "mysticism" does indeed stretch forth endlessly, it seems fair to group the theories of mysticism into two broad camps: essentialist and empiricist. *Essentialist* theories stress the sameness of the peak experiences that human beings are likely to nominate as instances of mysticism. Further, such theories tend to assume that human nature does not vary drastically from culture to culture or from historical period to historical period. Also, essentialist theories often assume or argue that the ultimate reality with which human beings are always engaged (but perhaps never so intensely as in mystical moments) is the same always and everywhere. One may call this ultimate reality "God" or "the ground of being" or some equivalent term, but under any name it is the same for Westerners and Easterners, for those who call themselves *theists,* believers in God, and those who call themselves *atheists,* nonbelievers.

Those who hold an essentialist theory of mysticism can vary considerably in how they ground or apply it.[2] They may give historical and cultural variances more or less importance. In the final analysis, though, they reject an empiricist view of reality, according to which there can be no essences, no substantial samenesses found always and everywhere. Essentialists think that a Hindu mystic has substantially the same experience as a Jewish or Christian mystic and vice versa.

Empiricist theories of mysticism[3] comprise the second camp. This view of what happens in mystical moments, of the influence of local culture on human nature, and of reality, whether proximate or ultimate, is more concerned with the particulars of the experience than is the essentialist view. Empiricists gave more weight to the language, the historical period, the other aspects of an individual's culture, and the individual's own idiosyncrasies than to any core or essential sameness that individuals might share. Similarly, empiricists are leery of claims that in their mystical moments Hindus, Christians, Jews, and others have essentially the same experience. For those in this second camp, what essentialists would consider somewhat accidental bulks larger and constitutes more of the meaning of the experience in question. Therefore, empiricists are slow to make generalizations across the history of religions. They are more inclined than essentialists to individualize Buddhist, Muslim, Taoist, and Native American mystical moments.

In an essay titled "The Mystical Illusion," Hans H. Penner has put what we are calling the empiricist position strongly:

> The basic assumption of this essay is that there are no direct experiences of the world, or "between individuals except *through* the social relations which 'mediate' them." Once this principle has been granted, it does not make much sense to speak of states of "pure consciousness" or experiences not constituted from within a linguistic framework. The principle is not revolutionary or new. It is, however, often forgotten when scholars focus their attention on religion and mysticism. If the principle is firmly held it will follow that, *if* mystical experiences have any significance, in order to explain their significance it will be necessary to locate and explicate the set of relations which mediate them. The thesis of this essay is that "mysticism" is an illusion, unreal, a false category which has distorted an important aspect of religion.[4]

Penner goes on to say that what yogis or Western mystics, such as Meister Eckhart and John of the Cross, report is not necessarily unreal or illusory. His quarrel is not with the facts or reports that generate studies of mysticism so much as with the theoretical categories, the working methodologies, that scholars of mysticism have created. Inasmuch as these have been essentialist, he finds them illusory.

Few of the theorists who interest us in this book are naive, simple-minded, or unaware of the complexities and nuances that any adequate view of religious experience, human nature, or ultimate reality has to admit and reconcile. To do justice to any reputable, representative theorist, essentialist or empiricist, would require lengthy reporting and analysis. Sketching in broad strokes, though, one can hold to what we have described thus far: one group of theorists stresses sameness, another group stresses difference. For essentialists cultural differences are less significant than a core human likeness. For empiricists such cultural differences are the key, and there may not be a core human likeness or essence worth discussing.

Neither group can deny the multitude of reports coming from the world religions to the effect that human beings can gain or be drawn into exceptional states, nonordinary plateaus of awareness, that often they consider extremely valuable. The argument concerns the interpretation of these reports. Indeed, sophisticated forms of the argument between theorists grapple with the apparent fact that no "report" comes without a theory. The people making the reports tend to couch them in terms of their own theologies and cultures. Whether this fact wins the game for the empiricists or merely forces the essentialists to speak more carefully depends on who is serving as the umpire, what his or her own understanding of mystical experience, human nature, and ultimate reality is.

Inasmuch as we are the umpires in the present instance, let us say frankly that we find merit in both camps and that our own efforts will attempt to set in motion a dialectic, a back-and-forth discussion. Shortly, we shall propose a working description of mysticism with enough adjectival and adverbial material to put in play the experience implied by the noun "mysticism" and by the actions and passions implied in the verbs that usually accompany it. Then, as we deal with some of the innumerable data that each of the world religions supplies, we shall let Hindu, Buddhist, and other mystics color our working description, adding qualifications, challenges, and overtones.

In all of this, however, we shall be honoring the intuitions or convictions of both camps of theorists. We shall keep open the possibility that Chinese and Jewish mystics have enough in common to warrant our subsuming them under essentially the same conceptual, categorical roof. We shall also note the differences in their experiences that come from their variances in language, theology, geography, or other contextual influences. We do not want to deny that mortality or contending with the cycles of nature or other factors that seem to be universally human cut across historical periods and geographical areas or point toward a basic sameness in both human nature and human experience of ultimate reality, of the sacred or divine. However, neither do we want to deny that each mystical experience is unique inasmuch as the person who gains or receives it is unlike any other human being, in ways either small or great.

Much of the significance of the debate between essentialists and empiricists depends on the eye of the beholder. If the beholder is more interested in and more impressed by the transcendent character of mystical experiences, the regular refrain that the mystic cannot render adequately what happened, the other-worldly zone, or dimension, of reality that appeared and who or what came into play and to what end, then cultural differences are apt to seem secondary. However, if the beholder is more interested in what rituals formed the person in question—what views of nature, God, the self, and the religious community the person held—then these particular features are apt to receive more attention than any transcendence.

Our view is that transcendence is important, marking a crucial component or aspect of any experience that we would admit as mystical.

If people think that they can describe quite well what happened to them and what that happening implies, then they have not had what we would consider a mystical experience. If there is no mysteriousness, in the profound sense of an encounter with something or someone greater than human comprehension, there is in our view no mysticism.

Certainly, mystics can rattle on at great length, describing or praising or expressing their fear of what at the outset and in the middle and at the end they may say is transcendent and passionately claim cannot be expressed in human terms. Certainly, the culture in which a mystic lives colors his or her sense of such ineffability, perhaps even shapes it decisively. The mystical experience itself is in part a function of what the mystic thinks can happen. No experience of transcendence ever happens to a person who inhabits no culture or thinks in no particular language.

All mystical experiences, therefore, vary somewhat. Each is unique. In our view the uniqueness supported by the data on mystical experiences does not negate the claim for transcendence or ultimacy or perhaps other characteristics that would free the experience from the reductionist potential in empiricism, that is, the power of empiricism to imply or state that there is only the culturally conditioned component, that nothing transcending culture is involved. That is not the way that many, perhaps the majority of, mystics speak. Many say that speech breaks down, that silence is more appropriate, and that even silence is not adequate. In this way, many mystics point to a dimension that, although ultimately inseparable from the culture and personality of the given mystic nonetheless cuts across cultures, applying to Christians as much as to Aboriginal Australians.

The practical effect of our view of how the theories of mysticism differentiate themselves should remain sensitive and fair to both the inside and the outside of mystical experience—both the data that support an essentialist reading and the particularizing data that remind us how much mystics remain creatures of their own place and time, their own religious tradition. This means that our study invites comparisons of Hindu and Jewish mystics because the data legitimate such comparisons. It also means that our study is cautious, advising skepticism about claims that Hindu and Jewish mystics are more like one another than they are like fellow Hindus and Jews who are not mystics.

We could go on to speak of the legitimacy of looking for special characteristics that single out Jewish mystics, differentiating them from Muslim or Christian mystics, but we could also support the hypothesis that some Jewish mystics may be more akin to some Muslim mystics than to other Jewish mystics, let alone to nonmystical Jews. Regularly, the standards for what theoretical viewpoints are fruitful will be what the data support and what illumination a given theoretical viewpoint offers. The data are various and many of the theoretical viewpoints are interesting, intelligent, provocative, and at least potentially quite illuminating. So, we advise a certain open-mindedness as well as a large-heartedness so that the reader remains not credulous or unskeptical but also not afraid to admit the richness of both the data on mysticism and the minds that have explored mysticism. Finally, we encourage an attitude that is not afraid to admit that at present scholars of the world religions have a lot to learn about mysticism and so ought not to foreclose any reasonable possibility of illuminating it.

A Working Description of Mysticism

Having bent over backward to be balanced and open-minded, what do we suggest as a working description of mysticism? We suggest: "direct experience of ultimate reality." [5] "Ultimate reality" can connote "God," "the Tao," "*nirvana*," "the sacred," or any of the other terms that religious people have coined to indicate what is unconditioned, independent of anything else, most existent, dependable, valuable. In the Western religious traditions, "God" has been the privileged term, and "God" has carried personal overtones. The Lord worshiped by Jews, the Father worshiped by Jesus, and the Allah worshiped by Muslims have all been unlimited centers of knowledge, love, awareness, and will.

In the Eastern religious traditions, the personal has often been less ultimate than the impersonal. Thus the Tao, important in Chinese religions, has connoted the way of nature (the "Way") more than the way (customs) of the ancestors. The *Brahman*, important in Hinduism, has been the ground of nature, and although many Hindu thinkers have related it to such gods as Vishnu and Shiva, it has remained more impersonal than personal. Buddhists speak paradoxically about ultimate reality, not wanting to imply a ground beyond the network of existent beings and not wanting to make a choice between being and nonbeing. Still, our sense of "ultimate reality" can include whatever Buddhists who have come to enlightenment say they are dealing with: *nirvana*, Buddha nature, or Suchness.

This is a working definition. It is a model, a template, a hypothesis to get us going. As the data come in, we shall have to adjust our model, trim our template here and extend it there. We are not saying, at this juncture, that "ultimate reality" is going to please all Jews, Christians, and Muslims. We are not saying that Native Americans, Native Africans, or Native Australians would be happy with it as a substitute for, or an explication of, the spirits or gods that they think made the world and are the deepest part of the world, what the shaman or diviner is dealing with in ecstasy. Finally, we are not saying that Chinese or Japanese, Buddhists or Hindus, will be overjoyed with the term "ultimate reality", willing without any complaints to plug it in at all mystical junctures. We are simply saying that, for students or general readers setting out to survey mysticism, "ultimate reality" is not a bad description of what mystics say has taken them over.

The second part of our working description of mysticism, "direct experience," is similarly modest and generalized. Mystics meet ultimate reality (God, the holy, *Brahman*). They become aware of it, drawn into it, raised up and transformed by it. Whatever the stress or accent of their description of their mysticism, they leave no doubt that it is experiential. They are not speaking about something they have only read about or heard from the elders. They are not simply reasoning from premises to conclusions, as philosophical theologians might (for example, when analyzing the relations between God and the human spirit or how the ground of nature relates to various natural phenomena).

Similarly, even in traditions that depend on a scripture, the reports of mystics clearly are not just creative exegeses of privileged texts. The world over, the reports of mystics stress what happened to them personally, what they themselves saw or heard or felt. Indeed, any special authority that mystics claim comes from this experiential or immediately personal quality. "God" has become alive for the theistic mystic. "God" is something, someone, living—speaking, moving, commanding, consoling, threatening, whatever. Analogously, *nirvana* is not just something that the Buddha talked about long ago. *Nirvana* has become the name for what the mystic experienced in enlightenment, what Buddhist meditational practice offers.

Lastly, what about the word "direct"? As the cautionary, skeptical quotation from Hans Penner suggested, this is a tricky word. Here, at the outset, we mean it fairly straightforwardly, commonsensically. The reports of the mystics tend to stress that ultimate reality bypassed intermediaries. It shone in their minds or warmed their hearts directly. If they began with a book or by holding hands with another person, the book or other person receded from the center. In the center, of both the experience and the mystics themselves, ultimate reality gave them a light, created a sense of union, that revealed their very constitution—the legitimacy of saying that it (God, *Brahman*) was the inmost being, the realest reality, of the mystic himself or herself (and of the world, of everything).

This does not mean that the mystic lost all sense of separation from ultimate reality or was so united with ultimate reality as to feel dissolved into it. Some mystics have spoken in this way, claiming that all difference vanished; but other mystics have not spoken in this way, have not claimed that they became ultimate reality or that they lost all partiality or uniqueness.

Now, although the mystics themselves tend to stress directness or immediacy, those (such as ourselves) who are analyzing their reports have to add some qualifications. First, virtually all reports by mystics are shaped by their given culture and the particular convictions of their religious tradition. Thus Western mystics are far more likely to maintain that their experience did not dissolve them into the ultimate reality than are Eastern mystics because Western theism (the cultural matrix for the majority of Western mystics) insists that human beings never become God literally. That is the orthodox Western theological position, and while some Western mystics risk violating it, the majority do not.

Second, we outside analysts must further note that some mystical traditions use images to mediate their experiences, at least in the early stages of their training and sometimes through to the very end. For example, Christian mystics often use images or icons, and some argue that one should never depart from the central Christian image, Christ himself, no matter how advanced one's prayer. Other Christian mystics stress an imageless prayer and union with God.

Jewish and Muslim mystics, formed by theologies suspicious of images, usually want to avoid at all costs misrepresenting the unrepresentable God and have worked out different variations on the matter of directness. The mediation that one finds in the accounts of their mystics is less that of a personal image than of a scriptural text such as the Koran. Some Asian mystics speak as though imagelessness ought to be taken for granted, but others urge the use of mantras, mandalas, and other concrete, mediating forms.

Therefore, in putting "direct experience" into our working description of mysticism, we are not necessarily denying either mediation by the mystic's cultural tradition or a mediational use of sights, sounds, bodily postures, or other aids. Rather, we are saying that the core of the experience, what the mystic stresses when describing the moment, is a vivid presence of ultimate reality (however named) that makes any intermediary transparent and secondary. One could say, in fact, that in mystical consummation the mediator (tradition or icon) has done its job so well that it becomes a translucent lens; or, in some cases, one could say that the mystic feels as though everything intermediary has fallen away so that darkness or a wraparound presence renders ideas, words, images—everything particular or partial—inoperative, beside the point.

The key import of "direct" is the sense that ultimate reality, God or *Brahman* or *Tao,* is present singularly and is so real that the mystic feels that "ordinary" perception gives only a cardboard version, a stand-in or simulacrum barely valid enough to avoid being called a fraud.

When Blaise Pascal distinguished between the God of the patriarchs (Abraham, Isaac, and Jacob) and the God of the (antiseptic) philosophers, he pointed to this difference. God, or ultimate reality, had become for him like a blazing fire, the "jealous" God depicted in the Old

Testament. No longer was God just a notion, however lovely or powerful. No longer was God a reality in which he believed only at a distance, without having it dominate his awareness, take his breath away, or ravish his heart. In Pascal's mystical moment, God became alive, real, present directly as never before. So, although Pascal used an ancient word, "God," it had acquired a completely new energy. What it called to mind was much more powerful than the remote "God" favored by many philosophers.

We are working with a description of mysticism that stresses the ultimacy of what the mystic meets or is touched by. Our description also stresses the experiential character and the directness of mysticism. In the mystical experience, mystics are aware of ultimate reality at first hand with such vividness and such vitality that there is no room for doubt. Afterward, when "ordinary," nonmystical consciousness has returned, the mystic may reason about the experience, may even doubt what happened, but during the experience itself, there is no doubt. Ultimate reality is experienced as indubitably present.

Parmenides, the mystical Greek philosopher, spoke for many other mystics when he said, "Is!" To the fore in his experience was pure existence, simple being as opposed to being this or that. Such a formulation may be more Indo-European than East Asian or Semitic, but a little foraging shows that it has cognates in other systems—for example, with Lao-tzu, the Taoist master, as well as with some Jewish and Christian mystics, who depend on biblical texts such as Exodus 3:14, where God tells Moses how to think about the divine name, implying that the best clue to the divine nature is how God acts.

At any rate, the difficulty of clarifying the key ingredients of our working description underscores the ineffable aspect of mystical experience. Although mystics may say many things about what happens to them, they confess regularly that what they can say falls far short of what happens. The happening, the oneness or being they experience, is far more real and far richer than any words they can summon to represent it. Whatever they do—draw a picture, dance a dance, sing a song—falls short. They dance and point and sing so that they may give witness to their experience inasmuch as it seems something that would benefit others, but those others have to experience ultimate reality for themselves if they are to gain its full value.

Mystics also express themselves because expression comes naturally to people who ordinarily (apart from their mystical transports) think, feel, and know through their bodies, using images or symbols that mediate between body and spirit. When mystics are expressing their experiences, however, they are usually not in the midst of the experience itself, which tends to take them out of themselves, their prayer, their healing, their dancing, whatever they are thinking and doing. Those activities were vehicles to carry them toward their goal. In the mystical experience they reached their goal, or their goal moved to meet them, overcoming the distance that separated them at the outset.

It is fitting, then, that our working description be modest and flexible enough to adapt should we come across something to show us more directly, more experientially, the ultimate reality of mysticism, the who or what the mystics we are studying merged with, came to know and love and say was their center, the center of all of us, creation's depth and height and wholeness.[6]

The World Religions

We are focusing primarily on the mystics who have been nourished by the great systems that words such as "Hinduism," "Confucianism," and "Islam" connote.[7] The majority of mystics have grown up in such systems because the majority of human beings have. One can speak meaningfully of mystics who have lived apart from these systems, but they are a minority. It behooves us, then, to clarify the relationship we see at the outset of our study between mysticism and the world religion in which it arises.

The mysticism that occurs under the aegis of a certain religion is not the whole of that religion, and arguably it is not the center or most valuable part of it (many mystics, however, would say that it is). Mysticism is not the whole because the vast majority of the adherents of a world religion are not mystics; that is, they do not, with any regularity, experience ultimate reality directly. Moreover, the main occupations of any world religion include more than mystical communion with God or self-realization. Religions have to provide doctrines through which their people can understand what they believe. They have to provide ethical guidance—teaching, exemplification, reward, and punishment. Religions have to form communities for worship, social service, and simple mutual support. They generate myths, rituals, stories, and ceremonies through which their people can work what they believe into their imaginations, their nerves, their marrow. Although each of these activities can bear a mystical connection or dimension, none of them need do that. So, these activities indicate how the world religions are more than mystical.

Traditionally, many peoples had no special word for religion. What they believed was the core of their culture. They were Indians or Chinese or Native Americans of a particular sort, a given tribe, because of where they lived and how their myths and rituals helped them to survive, even to flourish there. Their myths and rituals both expressed who they were and situated them in the world, formed them to their particular identity. Most traditional (premodern) peoples did not separate their worship from their mores. Equally, they did not distinguish sharply between what they believed about the world and how they expected their fellow human beings to act. No legal system defined what was sacred in contrast to what was secular. Indeed, there was no secular zone in our modern sense. Everything was religious in that everything was capable of expressing ultimate reality or holiness.

Just as few traditional peoples had a special language separating the religious zone from the secular, so few located mysticism in a special, separate sphere. Certainly, people who wanted to pursue intense experiences of ultimate reality, visions by which to guide their lives, might separate themselves from the general community. Certainly, a few people might make such a separation quasi-permanent, loving solitude enough to become hermits. However, the more prominent and crucial pattern was for all members of the religious community to think that experiences of ultimate reality would illuminate human existence as such. Farming, hunting, cooking, sewing, sex, birth, death, coming of age, marrying, taking over leadership of the tribe—any significant event spotlighted the omnipresent influence of ultimate reality, of the gods, of the sacred.

Mystical experience—special insight, rapture, transport, realization—bore on the ordinary. It was extraordinary precisely because it revealed the structures, the depths, the potential of everyday, ordinary events that people normally missed. People went through their routines fairly dully. They did not experience eating, drinking, working, or having sex as dazzling revelations of the full meaning of life, of the transcendent depths holding all that is in being. What we, looking backward, would now call mystical moments or preoccupations of the traditional world religions were the times and ways in which members of those religions actualized their beliefs, became fully aware of the mysteries they were convinced were the most crucial aspects of human existence.

Western religious traditions have not equated mystical insight, extraordinary vision or experience, with what is necessary to be a member in good standing. Down the historical line, the prophet, priest, sage, or king has not been the only genuine Israelite, Jew, Christian, or Muslim. However, the people filling such roles have stood out from the generality of community members, and one of the reasons why was the likelihood that they had had extraordinary experiences.

For instance, either on the way to becoming a prophet or in the exercise of the prophet's role a Western religionist might have extraordinary experiences of ultimate reality, but such experiences would not make the prophet the only full member of the community. Others who had not had such experiences would continue to possess rights, probably including the right to disagree with the prophet, who would not have exclusive authority.

Although there have been prophets in the Eastern traditions, in Asia sages have held greater sway. The sage is a person revered for wisdom. Wisdom is more than learning, though typically the most revered sages have been masters of the traditions that the given people held sacred. In many cases, therefore, the sage has been a master of particular writings, ritual lore, and even medicines, but the mastery in question has not been what we now call academic. The sage knows from the heart, for the sake of the people's overall education and healing, the meaning of the material in question. Learning is connected with an ability to read people's hearts and know what they need to be healed physically or to escape from illusion and vice.

To gain this kind of wisdom, sages have to ruminate, chew the relevant traditional materials over and digest them, so that such materials enter their metabolisms, become part of their selves. Often some kind of meditation facilitates this process, and in meditation the sage might have extraordinary mystical experiences.

Thus the classical Eastern sage could well be a mystic, someone viewing the world on the basis of direct experiences of ultimate reality. The wisdom of the Hindu or Buddhist sage could well have come from meditational, yogic perceptions of the truth that the *Dharma,* the holy teaching, had laid out. The case is different, culturally, for the typical Chinese sage, Confucian, Taoist or Buddhist, but not wholly so. The master there would be able to epitomize the tradition for the disciple and bring it to bear practically. This would involve knowing the tradition from the heart, having penetrated its core. It would also involve knowing the individual disciple or petitioner and just what he or she required. In studying the tradition from the heart, the master might have been moved by direct experiences of ultimate reality. Such mystical experiences would not have violated the social expectation shaping the sage's work. Indeed, they would have fulfilled it. To know the Way (*Tao*), for example, one had to be moved by the Way, have the Way seduce one's heart, direct one's will, illumine one's mind. All this could be mystical.

Nonetheless, the religious complexes that we associate with Asia—Hinduism, Buddhism, Confucianism, and Taoism—sponsored rituals, myths, doctrinal concerns, ethical concerns, and community matters that collectively made their scope broader than "mysticism." In this study we are not concerned with the full span of any of the world religions. Here, our scope and focus are narrower. We are interested only in the direct experience of ultimate reality, and while this experience certainly is shaped by the other concerns of the tradition in question (and, in turn, shapes those other concerns), it is not coextensive with them.

For example, although we may speculate about connections between the *Laws of Manu,* an influential Hindu text bearing on social and ethical matters, and the mystical moments that traditionally have shaped

Hindu yogis, we need not consider Manu a mystical author nor think that the people using his laws needed direct experiences of ultimate reality to obey them. Ordinary common sense would have sufficed.

The world religions are wonderfully complex cultural entities, full of historical developments and spilling into every nook and cranny of their adherents' lives. The mystical dimension or potential in each of the world religions (each of the traditions that has lasted for centuries and formed millions of people) has been a major reason for its perennial vitality (by holding out the possibility of meeting God, encountering the sacred), but mysticism has not been coequal with any tradition. The particular world religion in question has been the whole, its mysticism a part of that whole. When we study a given mysticism we gain a lens onto the whole, but we need the whole to appreciate the part.

Dialectical Understanding

A *dialectical* understanding is one that speaks or reads or studies the reason (*logos*), or the meaning, of something by going across (*dia*) it, or working it through. Here, the "something" is the mysticism of the world religions. The assumption we make is that the understanding we seek will emerge best by working through representative data from the different traditions. More specifically, it will emerge best through our setting up and putting in play an alternation of viewpoints, an intellectual movement back and forth, from the general to the particular.

We have proposed a working definition of "mysticism" and are beginning with the thesis that mystics are people who experience ultimate reality directly. We have noted that, historically, most mystics have been people formed by one of the world religions. Their Hinduism or Islam has shaped how they prayed, what they thought about the world, and where they set the limits of what the human spirit might know. Certainly, there has been no Hinduism or Islam as such, apart from colorings from the part of India in which a particular adherent lived or the historical phase of Indonesian history in which a particular adherent participated. The shorthand usage we employ (Hinduism, Islam) to suggest the overall religious heritage common to Hindus or Muslims is nearly as abstract, or as much in danger of committing an essentialist fallacy, as is our key term "mysticism." Nonetheless, we have found it useful to begin with such a shorthand usage. Indeed, we have found it necessary to provide an orientational sense of mysticism and to assume that "Hinduism" or "Judaism" made enough sense and conveyed enough meaning to point the general reader toward the areas we intend to survey.

Part of the intrigue in any intellectual survey is the experience of having terms first clarify themselves, then show their inadequacy, and finally be taken up again with a new understanding. The new understanding does not deny the inadequacy that the survey has revealed, but it does acknowledge the inevitability of using general terms, even though they can abuse reality.

For example, Hindu mystics could well object to the definition of mysticism that we have proposed, wanting something that squared better with the discovery of deepest selfhood, which is crucially important in some Hindu schools. That such a discovery can include, paradoxically, the realization that what most people mean by the "self" is illusory does not negate the possibility of expressing the quintessence of mystical experience, of key moments of realization, in terms of self-discovery.

Similarly, a Muslim mystic might well find our generic, introductory description of mysticism inadequate, compared to the resources offered in Islamic tradition. From the Koran one can draw an understanding of God that makes plain how mysticism is bound to be insight into the priority and soleness of God and into how God's claim on whatever is real or good in any creature is greater than that creature's own claim. Compared to the Koranic language in which a Muslim could couch this realization, "direct experience of ultimate reality" could seem pale indeed. Allah is closer to us than the vein pulsing at our throats. We exist only because of the divine "let them be," and the divine "let them be" has to keep reverberating, if we are to continue in existence. For the believer, language such as this is far more satisfying than generalizations drawn from across the board of mystical experience.

When we move back and forth between general principles and specific cases, between the universal and the particular, we can generate a helical understanding (a helix being a spiral that moves upward). Of course, understanding exceeds spatial metaphors, but something (probably the location of natural light in the heavens) seems to incline human beings to picture progress in understanding as moving upward, toward the sun. Plato's famous parable of the cave pictures progress toward the love of wisdom as the turning of the brutish soul from a cavernous existence to one emerging into the light. The line of progress is an ascent, symbolizing a shift from concern with lower things, things of the belly or the loins, to things of the mind.

One does not have to accept a Platonic sense of bodily geography to want dialectical understanding to generate a helix rather than a spiral. With a spiral, we might bore ahead, but we also might go around the same point. In contrast, the helical understanding that we hope to generate moves the student and the teacher to higher ground. From higher ground one can survey more data, oversee more of reality, and so perhaps reason, estimate, and act with a better perspective.

Interestingly, some descriptions of wisdom equate it with the long view, the perspective which is most heavenly. If we could see as the gods see, we would be immeasurably wiser. Thus after probing the problem of evil, the biblical book of Job finally suggests that we human beings shall never solve it because we were not present at creation and so shall never know the proportions of the universe. It is not given to us to pontificate about what is owed us. We did not make ourselves, and we do not know how we fit into the total scheme of things.

Much of the wisdom that one comes across in the world religions is of this negative, agnostic character, as is much of the articulation of mystical experience. The mystic, like the sage, tends to realize that we need the whole if we are to describe or estimate any part correctly. The mystic also tends to realize that the whole is simply beyond us. We are finite in mind and heart as well as in body. What we meet in mystical experience, as in sagacious reflection at the end of life, is too much for us; it is infinite, without borders or endings. It stretches on and on, even though it may also condense into utter simplicity. We cannot get our minds, our imaginations, our feelings, even our radically simple selves around it. It is the measure; we are the measured. So, we die not knowing where we came from or where we are going. Not even the most vivid mystical experience solves this problem because we seek an articulation of the experience that lets us imagine it and feel it. However, no articulation can be infinite. Only ultimate reality, radically inarticulable, can "answer" our most profound questions.

There is a Jewish spin to this bottle, and also a different one congenial to Taoists. There is a Buddhist way to "sense" the nonduality of ultimate reality, how it is not other than we are, and a Christian way. In fact, there are many Jewish and Taoist spins, many Buddhist and Christian ways of "sensing," but they all fit the dialectic that we are starting. We have grounds, from within the traditions themselves, for thinking that Jews and Taoists spin similar bottles. The ten thousand things that Taoists contemplate are cousins to the sparks set off by the God of the Kabbalists and the Hasidim. Similarly, the nonduality prominent in much Buddhist meditation and metaphysics can summon a Christian effort to think again about the experience of union with a creator—indeed, with precisely the Christian Creator who spoke the world forth in the same Word that took flesh from Mary.

The difference between studying the philosophy or theology of a given tradition's (or a given traditional thinker's) approach to this issue and studying its mysticism is the stress of the latter on direct experience. The mystic's encounter with ultimate reality is vivid, holistic, and as such much more than intellectual. The mystic is not just thinking about ultimate reality. The mystic is seeking the face of ultimate reality (to use a biblical metaphor), trying to gain, or be transformed by, *nirvana* (to use Buddhist diction).

Yes, the mystic has a mind and thinks, both before the mystical experience and afterward, but in the precise mystical moment, in the exact center of the direct experience, thought (discursive, step-by-step, reasoning) is not the issue or the focus or the presiding activity. Intuition, visionary perception without steps, may be an important part of the mystical experience, but the more usual description by mystics themselves turns away from analysis of any sort, insisting on simplicity and wholeness.

Therefore, in mystical experience the heart and spirit are as important, indeed are usually more important, than the mind or understanding. Even when the mystic uses imagery (for example, Julian of Norwich describing how on her deathbed she saw the crucified Christ), the mind does not rule the experience. The whole person is involved, seized, transformed.

With pleasure, we anticipate taking our initial description of mysticism and seeing what happens when texts from the Vedas or the Koran come up for study. Such texts will not assume precisely the same things that we have, nor will their assumptions match one another. Back and forth, we can let them interact with our general descriptions and with one another. Doing that—letting them agree, disagree, fight, reconcile—we hope to generate a full, dialectical understanding.

Our minds are made to grow and learn and in this way to move forward and upward toward higher ground and fuller light. The pleasure that we feel is the pleasure of using our minds well, letting persistence with a good problem teach us new things, even change us substantially. Both the best academic scholarship and the best traditional religious wisdom honor this pleasure. Both confide in it, trusting that to pursue the light, to try to move toward higher ground, is among the most human things we can do. Neither the best academic scholarship nor the best of traditional wisdom is fixed, dogmatic, immobile. Each moves as a dialectical interaction between the general and the specific, between what we closed with yesterday and what comes to us afresh today.

Because we are finite, sure to die, we have to be masters of the dialectic and not exalt it as an idol. Because we are finite and the dialectic is potentially infinite, we sometimes have to act before the dialectic has come to term, lest people depending on us starve physically or spiritually. Nonetheless, if we participate in the dialectic honestly and generously, it will teach us a great deal about when to persevere and when to let go. In other words, we can learn valuable things about both how to think and how thinking is quite limited. We can even learn, again experimentally, why Qoheleth said, "Of the making of books there is no end," and so why some exemplary people (for example, al-Ghazali) have for longer or shorter periods dropped books to seek wisdom through other, more mystical ways.

The Academic and the Personal

As we have intimated, mysticism is a topic soliciting both academic and personal attention. We can learn a great deal from scholars, but textual studies have their limits. Admittedly, all studies are personal, calling for acts of imagination from their readers, humanistic studies most of all. With mysticism, however, the call includes a summons to go beyond imagination, to picture stopping picturing, starting to locate dark, voided places in the self like the places where the mystics say they have experienced ultimate reality directly. Because ultimate reality is not picturable, as proximate realities are, those places lie below or above or to the side of the normal imagination. Certainly, the mystic has had to return to the normal imagination to speak about mystical experience, yet such speech has had to deny itself, or try to defeat itself, to communicate the core truth.

The core truth about ultimate reality, as one gathers it from the traditional reports, is that only ultimate reality itself can render ultimate reality adequately: ultimate reality must be its own revelation and exegesis. The only way to know God is to experience God directly.

True, one can say the same about love or shooting a round of golf below par, but with mysticism and ultimate reality the matter is quite different. It is not just the full flavor of the experience that is in question or just the existential difference between the intercourse (or the birdies or the tonsillectomy or the being fired) as imagined and the experience as it is actually happening. With ultimate reality and the mysticism that experiences it directly, the very substance, the bare minimum necessary for speaking significantly at all, cannot be captured adequately in human terms. Until we can locate in ourselves the ineffable quality that stands at the mystical core, we have not found what the mystics themselves suggest is the Archimedean lever.

Indeed, mystics say that, even with such a location, we understand nothing of which we can be proud. No matter what we say about ultimate reality, it remains more unlike than like our descriptions, and any possession we gain of it is not an achievement of ours but a gift from it, a gift it need not have made.

These are not simply playful paradoxes gleaned from Zen masters. They are canonical Christian doctrines, gleaned from a passel of Church fathers. They are also staples of Jewish and Muslim theology, related to a prohibition against idols, attempts to represent the unrepresentable God. Obviously, they echo in the Hindu *neti, neti* (not this, not that), in the Mahayana Buddhist interest in the wisdom that has gone beyond, and in the Taoist conviction that those who speak do not know. Wherever one finds direct experiences of ultimate reality, the primacy of that reality and the impotence of any lesser reality to render it adequately come to the fore.

The academic enterprise that has developed in the modern West does not know what to make of this mystical wisdom. Nescience, unknowing, negativity, apophaticism, using reason to defeat reason—choose what term you will—is unlikely to appear in a university catalogue. It occurs in the minds of some professors and some students, but few of them know how to correlate mysticism with scholarship, their academic reason to be.

Unknowing, like the iconography that can seem to be its antithesis but in fact often is its cousin, is a more-than-academic matter. The standard academic program cannot handle it. Once experienced, known personally, ultimate reality dwarfs all of our studies, even our studies of ultimacy. It is whole and complete; we and our academic studies are all too partial. It never dies. We fill up cemeteries and the basements of libraries. The academic world is grass and flowers, and grass withers, flowers fade.

The personal world may be more than grass and flowers. If taken to heart, the experience of the mystics may mediate our at least dim or thin encounter with a deathless, truly ultimate reality, and so with a deathless potential in ourselves. If we read slowly and contemplatively what the saints have said has happened to them at prayer or in meditation or while walking in a wood or binding up the wounds of a leper, it may happen to us. If we pray or meditate or hie ourselves off to the woods or serve the poor and suffering, our worlds too may tip over, split apart, leave us high, dry, and transformed.

Consider al-Ghazali, once a prominent academic. His law and philosophy did not fulfill him, so he took himself to the Sufis. There, he found a new Islam, a new Prophet, and a new Koran. Of course, it was also the old Islam, Muhammad, and the same book from the beginning; but now it had a soul, had come alive, and the force of it was newly compelling. So the life he had led withdrew from him, faded, and lost its allure. The understanding he had had, so impressive to outsiders, now seemed to him a tiny thing, no cause at all for boasting.

Consider also Thomas Aquinas toward the end of his illustrious career. When he met his God in a consummating vision, all of the volumes on his shelf turned to straw. He had delighted in pure reason and had loved to pore over Aristotle and the Bible. For twenty-five years he had studied, taught, preached, and argued, but in a mystical moment all of that fell away.

Finally, consider Lao-tzu, reputed author of the *Tao te Ching*. Grappling with the Way, he found his reason clouding. All around him moved bright, busy, and certain people. They seemed clear about what they were doing, about who they were and what was happening to them. He alone seemed to feel overcast, dull, and not at all certain. The more he searched, the less he found. The longer he studied, the less he knew. It is easy to picture him trekking off into silence: the *Tao* that could be told was not the real *Tao*. However, painful though his dissociation was, hard as his alienation struck him, he was in love with the *Tao* and so was willing to suffer for it. Life without the *Tao* would have been no life. Clarity without reality and depth would have been horrible.

These are the accents in which mystics often speak, and few of them resound in academic journals. The closest we can come in disciplinary terms is the study of poetry. Even there, though, the experience is the key, and the experience requires a more-than-academic investment. The poetry that can be told is not the real poetry.

Analogues abound in creative art and science, but these too cross the bounds of what we connote by "academic." We may study books on the sources of painting or where groundbreaking physics comes from, but what we can learn from books is partial. Certainly, human beings, always partial themselves, should never despise what is partial. Nonetheless, neither should we forget the greater value of what is whole.

From the perspective of the mystics whom we find most trustworthy, the academic approach is not to be despised, as long as it confesses its partiality. When it presents itself as a wholeness, which is rarely, the mystic has to smile. Often, however, it gets so preoccupied, so absorbed with its studies, that it implies that it is the best, perhaps the only real, game in town. Of course, if the only local options are making money, getting drunk, or fornicating, this suggestion will seem to have great merit. The best and brightest will be attracted, and for a while they will stay intrigued. However, this best-and-brightest appeal itself will eventually undo them, for it will not let them forget they have been made for more.

The Buddha said that all life is suffering, but he thought that human beings had been made for more. Augustine said, "You have made us for yourself, O Lord, and our hearts are restless till they rest in Thee." Augustine was one of the best and the brightest.

If we use the criterion of direct experience of ultimate reality, both Gautama the Buddha and Augustine the saint were mystics of a high order. Both came into moments of enlightenment that changed their lives, reordering their grasp of the whole and the partial. As their words have come down to us, both spoke a poetic language, rhythmic and measured. Both formed moving sentences, chains of thought anchored in nothing partial. Moving on and on, their logical lines curve like Einsteinian space, bending round a whole experiential no-thing-ness. Thus the studies of both are endless, enduring generation upon generation.

We authors take some of our livelihood from academic institutions, so it would be churlish for us to denigrate them. However, we take our love of life from the ultimate reality described by the mystics, so it would be deceptive for us not to delimit, or circumscribe, the academy. Because the method and tone of a proper study derive from its subject matter, which in this case both admits of profitable academic study and escapes it, we think the approach proper to this book is to blend the academic and the personal. Dialectically, therefore, we shall draw on *(1)* what others better informed than we have said about sources such as the *Vedas* and the New Testament and *(2)* matters that go beyond what can be said in scholarly prose, pointing poetically to what is transcendent. That is what our mystical subject matter seems to require and, therefore, what we find most conscientious. That is also what we imagine our readers to desire: both information and encouragement.

This Book

We have introduced the subjective side of mystical studies at some length. What about the objective content with which we shall engage in this book? We shall begin in the East, more because one has to begin someplace than because the East has proprietary rights to "mysticism." Once it was the vogue to contrast the mystical East with the pragmatic West. Indians such as Tagore and Gandhi played profitably with such a contrast. Nowadays, however, commentators are more apt to note the pragmatism that drives politics everywhere, often to bloody ends. As well, at least a few commentators note the hunger for ultimate reality that occurs everywhere, from Marxist China to Castroist Cuba, from innermost Africa to the California coast.

The pragmatic mentality, arising from the need to get things done, can flourish anywhere, just as anywhere it can get out of hand and ruin the earth's ecological balance. The mystical mentality, arising from the need to ground the pragmatic (and everything else in this world that is passing) in ultimacy, also can flourish anywhere. It becomes a potential concern, a live option, as soon as the human spirit lingers with either the starry heavens above or the moral law within. Our beginning with the East is therefore no championing or stigmatizing of the East as more mystical than the West and no evaluative judgment on either family of cultures.

Hinduism represents the millennial spiritual traditions of India, the umbrella of religious concerns erected on the subcontinent. To suggest what mystical experience has been in India and to begin the dialectic between our working definition and the data it implies, we consider representative moments in Hindu history and culture. After a general review of Hinduism as a whole, we consider first some specimens of Vedic experience. The *Vedas* are the texts most revered, considered scriptural, in mainstream Hinduism. The suggestions of direct experience of ultimate reality that we find there provide solid foundations for later developments. One capital development was the rise of the Upanishadic literature at the end of the Vedic period. People wanting to probe the unchanging basis for the changing world of both natural phenomena and psychological phenomena produced analyses both poetic and profound.

The cast of mind and style of life integral to the *Upanishads* could be highly disciplined. *Yoga* is a general Indian term for discipline, especially that suitable for people in search of enlightenment. So the fourth topic in our study of Hindu mysticism is *yoga,* with particular emphasis on how stilling the reasoning mind seems to relate to both experiences and conceptions of ultimate reality. Our fifth topic, the *Vedanta,* takes us to the philosophical school that most fully developed the guiding intuition of the *Upanishads,* namely, that ultimate reality is not dual. Our last topic, *bhakti,* draws us into a consideration of the Hindu devotional love, usually theistic, that has been the most popular mode through which Indians have sought direct experience of, even union with, ultimate reality.

Our treatment of Hinduism exemplifies how we shall approach all of the world religions. Thus in Chapter 3, dealing with Buddhism, we take up stations on the Buddhist mystical journey, representative moments and texts that we consider roughly parallel to what we have taken up for Hinduism. Naturally, the tradition itself determines, or at least suggests, what examples are most representative and illuminating. Still, our general method remains the same. Each tradition has drawn the attention of numerous scholars. In each case, the mysticism in question deserves a scholarly attention that could fill a whole library. Our work here deals with all the major traditions, but it is confessedly introductory. Thus we shall have done well simply to have alerted the reader who wants to move into a more leisurely and thorough treatment of any of the traditions.

We are much concerned not to lose sight of the forest for the trees, not to forget the dialectical conversation we have promised. Our focus is the interaction of mysticism (a trans-traditional phenomenon, a generality of sorts) and the apparently direct experiences of ultimate reality that scholars tend to single out when speaking of a particular tradition. Our goal is the fire (the light and warmth) that comes from such a rubbing together. Our method is to focus on specimens that illustrate the distinctive mystical experiences of the tradition in question.

It will be more than legitimate for specialists to criticize what we offer as specimens of Buddhist, Chinese and Japanese, Jewish, Christian, and Muslim mysticism. Similarly, it will be more than legitimate for them to criticize the sense of mysticism with which we begin, the dialectic between the general and the particular that we develop, and the conclusions that we reach. There is no orthodoxy, no single consensus, on any of these points. The choice of examples, the general understanding of mysticism, the dialectical interaction, and the conclusions are only as good as the data upon which they depend and the light they shed. Our aim is to introduce the mysticism of the world religions fairly and interestingly. As long as critics accept that this has been our aim and do not substitute what they think our aim ought to have been, their reactions will be welcome.

The same aim that has shaped our work in Chapters 2 through 7, where we deal with the major literate traditions, holds for what we offer in Chapter 8, where we deal with the mysticism of oral peoples (those not shaped decisively by written, scriptural texts). What we can say about Native Americans of both continents, Native Africans, and Native Australians has to be painfully brief because of our spatial limitations, but it is an effort at least to suggest how the mystical venture has survived in the many impressive cultures that have relied on singing and dancing more than on reading.

In conclusion, we shall first review the journey that we have completed, the dialectic that we have followed, asking about its larger patterns. Thus we shall inquire how the traditions align themselves—what likenesses and differences they manifest and what principal ways they color or contest our working description. Second, we shall ask about the major methods that the traditions reveal and that recur regularly—ways to ecstasy, self-possession ("enstasis"), disenchantment with the world, acceptance of the world, and the like. Third, with an eye toward mystical practice today, we shall inquire about the major pitfalls that the traditions suggest lie in wait for anyone either setting out for a direct experience of ultimate reality or feeling drawn into it. Fourth, balancing this is the question of the major profits that, worldwide and throughout history, the mystics say they have reaped. Fifth, we reflect on the roots of mysticism in human nature—what it is that may explain why, generation after generation, human beings have felt drawn out of the simply factitious and into a holy ultimacy. Sixth and lastly, we reflect on the roots of mysticism itself, the causes of the phenomenon, that seem to lie in ultimate reality itself, raising the possibility that the fullest explanation for mysticism is an active divine allure. This final inquiry will take us to the point where we can go no further. The reader will find further help in the notes accompanying each chapter, but the rest of the dialectic will be out of our hands.

NOTES

1. Louis Dupré, "Mysticism," in *The Encyclopedia of Religion*, ed. Mircea Eliade (New York: Macmillan, 1987), 10:245. Compare this to the beginning of Ninian Smart, "Mysticism, History of," in *The Encyclopedia of Philosophy*, vol. 5, ed. Paul Edwards (New York: Macmillan, 1967), 419–420: "Mystical experience is a major form of religious experience, but it is hard to delineate by a simple def-

inition for two main reasons: First, mystics often describe their experiences partly in terms of doctrines presupposed to be true, and there is no one set of doctrines invariably associated with mysticism. . . . Second, there is quite a difference between mystical experience and prophetic and, more generally, numinous experience, but it is not easy to bring out this phenomenological fact in a short definition." There is no substantial ("macropaediac") article on mysticism in the latest (15th) edition of *The New Encyclopaedia Britannica.*

2. Huston Smith represents this theory with learning and grace. See his books *Forgotten Truth: The Primordial Tradition* (New York: Harper & Row, 1976) and *Beyond the Postmodern Mind* (New York: Crossroad, 1982). Smith is indebted to the corpus of Frithjof Schuon. See the debate about Schuon between Smith and Richard C. Bush in the *Journal of the American Academy of Religion* 44 (1976): 715–724. Our own appropriations of what might be called "transcendental" factors in human consciousness, including a dynamic orientation to ultimate reality, come from the philosophical work of Bernard Lonergan and Eric Voegelin. See Lonergan's *Insight: A Study of Human Understanding* (New York: Philosophical Library, 1958) and Voegelin's *Order and History,* esp. vol. 4 (Baton Rouge: Louisiana State University Press, 1974). Presently, the complete works of both authors are being reedited and reissued by, respectively, the University of Toronto Press and the Louisiana State University Press.

3. Two volumes of essays edited by Steven T. Katz include representative contributions from the empiricist camp. See *Mysticism and Philosophical Analysis* (New York: Oxford University Press, 1978) and *Mysticism and Religious Traditions* (New York: Oxford University Press, 1983).

4. Hans H. Penner, "The Mystical Illusion," in *Mysticism and Religious Traditions,* ed. Katz, 89.

5. This description crystallized while we were reflecting on how our enterprise here differs from that begun by Bernard McGinn in *The Presence of God: A History of Western Christian Mysticism, vol. 1, The Foundations of Mysticism* (New York: Crossroad, 1991), especially pp. xii–xx, 265–343. Note McGinn's predilection for Lonergan (especially as mediated through the work of James R. Price; see p. 473). McGinn's references in these pages and their accompanying notes amount to a fine bibliography of recent theoretical works on mysticism.

6. Arthur Green's *Seek My Face, Speak My Name* (Northfield, N.J.: Jason Aronson, 1992) is a good example of how a mystical orientation can shape a contemporary view of a religious tradition, in this case Judaism.

7. We have dealt with the world religions as wholes in our textbook *Ways to the Center,* 4th ed. (Belmont, Calif.: Wadsworth, 1993). The study of religious experience by Wayne Proudfoot, *Religious Experience* (Berkeley: University of California Press, 1985), deals with mysticism on occasion, updating the theories of Schleiermacher and William James. For our purpose, Proudfoot works at too great a remove from the texts of nonchristian mystics to be of great practical use.

2

Hinduism

General Orientation

As we have indicated in the Introduction, to locate well a given religious phenomenon, such as mysticism, it is useful to have a good sense of the whole of which it is a part. This is the function of the first sections of each of our central chapters. "General Orientation" means a sketch of the history and worldview, the flow of time and orientation of the representative minds, which provides the given mysticism its context. Here, the given mysticism is that of Hinduism, so the history is that of the Indian subcontinent and the worldview is one dominated by such ideas as *dharma, karma, moksha, yoga,* and *maya.*[1]

Periodizations of Indian history vary, as do periodizations of all other histories, due to the varying criteria that historians employ for determining separable epochs. A representative history of India, the two-volume study by Romila Thapar and Percival Spear,[2] varies even from volume to volume. Thus volume 1 covers in less than 400 pages the huge temporal span from the establishment of Aryan culture around 1000 B.C.E. to the coming of the Mughals in C.E. 1526 (2,500 years), while volume 2 devotes slightly less than 300 pages to the 450 or so years from the Mughals to the period after Nehru in the mid-1970s, when his daughter, Indira Gandhi, became the Indian leader. Volume 1 has only fourteen chapters, principally because it works with large organizing concepts: "Republics and Kingdoms c. 600–321 B.C.E.," "The Emergence of Empire c. 321–185 B.C.E.," "The Disintegration of Empire c. 200 B.C.E.–C.E. 300." Three chapters, requiring about sixty pages, cover 900 years of Indian history. Volume 2 has twenty-two chapters, principally because it works with smaller topics: "The Coming of the Mughals," "Akbar," "The Mughal Empire," "The Great Mughals," "Europeans and Mughal India," "Decline, Collapse, and Confusion." Six chapters, requiring about sixty-five pages, cover about 225 years of Indian history.

The point of this example is simply to remind us that historical context is not a simple notion and that the lengthy history in which Hindu mysticism is embedded can be understood in many different schematizations. In our own general textbook on world religions, where the historical interest naturally was the evolution of religious institutions, understanding, and cultural styles, we found useful a sixfold periodization of the entire span of the perhaps 5,000 years from before the coming of the Aryans to the development of modern Indian culture: "Pre-Vedic India," "Vedic India," "The Period of Native Challenge," "The Period of Reform and Elaboration," "The Period of Foreign Challenge," and "Contemporary Hinduism." For our purposes in that general textbook, it seemed well to assume that, at least until the period of foreign challenge (political occupation by Muslims and Christians beginning from roughly the middle of the two millennia C.E.), Hindu religious rituals, institutions, ideas about the construction of reality, and mores were the principal articulation, the clearest foci, of the general culture of the people sharing the subcontinent.

Nonetheless, even that general culture was remarkably varied. Foreign rule added a new element, complicating what previously had been a fairly autogenous, consistent development from within; but in fact the vastness of traditional India, the great diversity of languages and local customs, meant that Hinduism had always been more an umbrella of ideas and rituals than anything tidy or uniform. So, Hindu mystics have always had a rich range of influences both working on them and available for their exploitation.

The culture that existed prior to the steady immigration of the Aryans from the northwest from 2000 or so B.C.E. on was centered on the earth and fertility. As suggested by excavations at Harappa and Mohenjo-daro, two sites in the Indus Valley where pre-Aryan communities large enough to merit the name cities flourished around 2500 to 1700 B.C.E., this civilization may have been quite stable, prosperous, and peaceful. In contrast, the Aryans were warlike and nomadic, oriented to gods of the sky more than the earth, and stratified socially into priests, rulers, and workers. Taken crudely, traditional Indian culture, and so Hinduism, comes from the mergers and interactions among these two cultural streams: the pre-Aryan (sometimes called Dravidian) and the Aryan. Although the Aryans became dominant culturally, and so religiously, the pre-Aryans always maintained much cultural influence, especially among the ordinary people. Thus the dynamic Aryan orientation toward the sky found a balance or complement in the more patient Dravidian orientation toward the earth, the body, the stable local community. Stereotypically, the Aryan cultural contribution was the more masculine, the Dravidian the more feminine. The Aryans contributed the strong heavenly gods like Agni (fire) and Varuna (sky ruler), while the Dravidians contributed the Mahadevi (many-formed Great Mother). These are simplistic summaries, but they offer what we believe to be a reliable orientation.

The third and fourth epochs of Indian history, according to our sixfold scheme, when native challenges and then reform and elaboration were the key happenings, brought on the scene Jainism, Buddhism, and other heterodox versions of the system involved in the *Vedas*, the Aryan scriptures. As well, these epochs saw the development of the Upanishadic literature, which technically is closely linked with the *Vedas* and so is considered semiscriptural, and of a new Hindu dynamism, focused on ardent devotion to such personalized gods as Vishnu (Krishna) and Shiva, that overcame the challenge of Buddhism and has shaped popular Hinduism down to the present.

Epochs five and six, dominated by Muslim and British rulers, saw the introduction of the Muslim–Hindu conflicts that have riven India in the past millennium, culminating in 1947 and 1971 in the dissociation from India, the nation-state, of the areas we now call Pakistan and Bangladesh. These last periods also saw the influential *raj* (rule) of the British (1818–1947), through which Western political notions came to prevail and Christian ideals to challenge traditional Hindu ones. Presently, India continues to suffer terrible conflicts between Hindus and Muslims, as well as uncertain relations between the non-Western, non-Christian social and metaphysical ideas that have dominated its long history and the new political ideas on which its modern constitution as a nation-state is based. (For example, as a secular state, India is supposedly neutral in religious matters.)

The leading ideas that we mentioned at the outset of this chapter suggest the overall philosophical context, or worldview, within which most Hindu mystics have lived and thought.

Dharma means "teaching" or "law," and connotes authority for the culture at large because it is based on revelations given to ancient seers, responsible for the *Vedas*, by the gods who made the world. *Dharma* is teaching that expresses how the world is composed and the objective reality that human beings have to honor if they are to live sane, wise, whole, holy, happy lives. The assumption is that human prosperity comes from being in harmony with the physical cosmos, which the gods run. Religious rituals aim at restoring such harmony when it has been lost by keeping healthy the human–divine relationship that the physical cosmos mediates.

Karma is the distinctively Indian notion of metaphysical cause and effect. We act as we do because of our prior actions. We are who and what we are because at birth we have a prehistory. The entire force of our prior actions, good and bad, shapes who we are, where we stand in the social scale, and our chances for happiness. If we are female, this bad fortune is because of our prior lives. If we are born into a priestly family, this good fortune comes from virtuous acts we performed in prior lives. *Karma*, therefore, has enormous implications: reincarnation, great limits on human freedom, direct ties to a stable social system (caste).

Moksha suggests liberation from the bondage of *karma*. Hindu mystics have drawn great energy from the general cultural conviction that it is possible to escape from the cycle of reincarnation, the system of constant death and rebirth. The more Buddhist terminology for such escape is *nirvana*, while both Hinduism and Buddhism speak of *samsara* when they want to indicate the painful karmic realm where one cannot escape rebirth and so further suffering. *Moksha* takes one out of *samsara*, though into what is less clear, but certainly the Hindu assumption is that it takes one into an opposite state, a complete fulfillment: being, awareness, and bliss. If *karma* and *samsara* connote existence that is conditioned, limited, painful, unfree, then *moksha* connotes the opposite: overcoming this existence and achieving the height of human potential. (In sophisticated discussions of the relation between *moksha* and *samsara* Hindu philosophers have noted the dependence of *samsara*, the conditioned, on *moksha*, the unconditioned, and so have softened the distinction between the two realms.)

The basic connotation of *yoga* is "discipline," and it indicates the sort of physical and spiritual work through which people may accomplish liberation from *karma* to *moksha*. By disciplining the body, the mind, the emotions, work, love, and so on, those seeking liberation, the fulfillment of being, awareness, and bliss can make headway against karmic conditioning and begin their spiritual unfolding toward the unconditioned, divine All. The yogic ideal is an autonomy, a self-possession and self-control, stronger than *karma*. Hindu *dharma* provides for this yogic ideal inasmuch as in addition to legislating for the ordinary people, whose bondage to *karma* requires guiding them through such institutions as social castes, it acknowledges the extraordinary people who long for a fully authentic existence outside all distortions and imperfections of caste and the driven, karmic personality.

Lastly, *maya* refers to both the illusion and the play of the world in which ordinary people live. As karmic, that world is conditioned and so unreal. It is limited and distorted in ways that ultimate reality, or being in itself, or the cosmos in its divine foundations and core constitution, is not. However, the unreality of ordinary existence, *maya*, is not completely pernicious. Seen from the perspective of *moksha*, by those who see aright because they are not enslaved to *karma*, *maya* is sportive, seductive, not completely serious, playful.

For Hindu mystics influenced by a culture accrediting *maya*, what happens in any given life is not crucial or decisive as it would be in a linear Western conception of either history in general or the significance of a particular human life. What happens in any given lifetime may have happened previously many times and may be bound to happen many more times in the future. Wise people learn to move lightly to the music of this therefore somewhat provisional, illusionary play, usually gaining their lightness, their detachment, through yogic activities. By yogic disciplines they detach themselves from ambition, desire, striving so that they let being itself, the way the world is when we do not push and pull at it, dictate more and more of what they think and do and are.

For our purposes, it is enough to note that the people who have succeeded best in this Hindu project of detachment from *karma* and *maya* often merit the name mystics. Often their ability to let being itself dictate more and more of what they think and do, indeed of who and what they are, comes from experiences in which they believe they are dealing with ultimate reality directly, seeing and understanding and being shaped by the divine structures of the world with little if any distorting influence from *karma*, *maya*, and the other expressions of enslaved, warped social consciousness. Hindu mystics have had close ties to *yoga* and *moksha*. Both in and out of the cultural mainstream created through the historical interactions of the Dravidian and Aryan heritages, they have sought and often found relations with ultimate reality that have given them at least a foretaste of being, bliss, and awareness. Theirs are the stories within the overall Indian history and worldview to which we now turn.

The *Vedas*

We begin with Vedic mystics and what we think are legitimate indications in the most revered Aryan scriptures of direct encounters with ultimate reality. The *Vedas* are not a small, simple bible (although the Christian Bible is hardly simple, it is small when compared to the *Vedas*). As one representative general description of them begins:

> Specifically, the Vedas are often understood to comprise four collections of hymns and sacrificial formulas. In a more general sense, however, the term *Veda* does not denote only these four books, or any single book, but a whole literary complex, including the Samhitas, the Brahmanas, the

> Aranyakas, the Upanishads, the Sutras, and the Vedangas [all collections of ancient Indian religious texts]. The many texts, varied in form and content, that make up the Veda were composed over several centuries, in different localities, and by many generations of poets, priests, and philosophers. Tradition, however, will not admit the use of the word *compose* in this context, for the Veda is believed to be *apauruseva,* "not produced by human agency." It is eternal. Its so-called authors have merely "seen" or discovered it, and they are thus appropriately called *rsis,* or seers.[3]

On its own terms, then, Vedic Hindu culture thinks of *Veda* as eternal wisdom. *Veda* is objective, the reason or order by which the world stands. It is divine in the sense of holy and most ultimate. Human reception of it is due to revelation. *Rsis* "see" it because it presents itself to them, makes itself available and known. Clearly, many assumptions about human consciousness and knowing lie in the background, most of them probably inchoate or little elaborated. Later Indian philosophy was nothing if not sophisticated about spiritual knowledge, but in the Vedic period the working notions of what the *Veda* is and how human beings gain light from it usually seem more poetic and commonsensical than critical, disciplined, refined, or differentiated.

Raimundo Panikkar, a prolific scholar born of an Indian father and a European mother, has labored for years at the front lines of recent scholarship in the history of religions concerned with cross-cultural understanding. He has tried to appropriate both the Christian and the Hindu spiritual traditions, both the European and the Indian aspects of his heritage, in the process asking creative questions about consciousness, culture, and a host of other extremely difficult issues. If untreated, such issues bar the way to any adequate understanding of either *(1)* how the major religious traditions are alike and different or *(2)* how we may gain any adequate understanding of the relations among their mysticisms. Since our overall theoretical ambitions are modest, we need not follow all of Panikkar's convolutions in his mammoth project to open the way to an adequate understanding, the large volume entitled *The Vedic Experience.*[4] Let us first indicate how Panikkar orients his readers to the Vedic experience and then deal with some representative Vedic texts that suggest what "mysticism" ought mainly to connote for Vedic Hinduism.

In the midst of a fairly dense effort to locate his anthology amid recent scholarly debates (about how best to conceive of the nature of the *Vedas* both for ancient Indians and for anyone today), Panikkar offers a challenging description of Vedic revelation (whose compatibility with our working definition of mysticism as direct experience of ultimate reality should be plain):

> The Vedic Revelation is not primarily a thematic communication of esoteric facts, although a few of its sayings, as, for example, certain passages

of the Upanishads, disclose some truth that is unknown to the normal range of human experience. But for the most part the Vedic Revelation is the discrete illumination of a veil, which was not seen as a veil but as a layer, one might almost say a skin, of Man himself. The Vedic Revelation unfolds the process of Man's "becoming conscious," of discovering himself along with the three worlds [heaven, earth, underearth] and their mutual relationships. It is not the message of another party speaking through a medium, but the very illumination of the "medium," itself the progressive enlightenment of reality. It is not a beam of light coming from a lighthouse or a powerful reflector; it is dawn. It is the revelation of the Word, of the primordial Word, of the Word that is not an instrument, or even a sign, as if it were handing on or pointing to something else. It is the revelation of the Word as symbol, as the sound-and-meaning aspect of reality itself. If there were somebody who had spoken the Word first, by what other word could he communicate the meaning of the original to me? I must assume that the Word speaks directly to me, for the Vedas reveal in an emphatic manner the character of reality.

In short, the fact that the Vedas have no author and thus no anterior authority, the fact that they possess only the value contained in the actual existential act of really hearing them, imparts to them a universality that makes them peculiarly relevant today. They dispose us to listen and then we hear what we hear, trusting that it is also what was to be heard. (13–14)

Pannikar's description of Vedic revelation as dawn suggests looking at some texts from the *Rg Veda*, perhaps the most poetic and influential part of the Vedic corpus, that deal with dawn, which is sometimes personified as the goddess Usas. Let us first simply quote the texts that Panikkar provides under the heading "dawn" and then comment on the coloring they give to Vedic mysticism:

Now Dawn with her earliest light shines forth, beloved of the sky.

Just as a young man follows his beloved, so does the Sun the Dawn, that shining Goddess.

Fair as a bride adorned by her mother, you show your beauty for all to see.

Happy are you, O Dawn. Shine ever more widely, surpassing every dawn that went before.

Fresh from her toilet, conscious of her beauty, she emerges visible for all to see. Dawn, Daughter of Heaven, lends us her luster, dispersing all shadows of malignity.

Like a swift warrior she repulses darkness.

She drives off wicked spirits and dread darkness.

Usas comes carefully, fostering all creatures, stirring to life all winged and creeping things.

Bright Usas, when your rays appear, all living creatures start to stir, both four-footed and two.

Arousing from deep slumber all that lives, stirring to motion man and beast and bird.

This maiden infringes not the Eternal Law, day after day coming to the place appointed. (164–165)

In our impression, the Vedic revelation occurring in these Rg Vedic texts on dawn is oriented more outwardly than inwardly. That is to say, the truth, the light communicated in the experience that most of the verses express, is collocated in the physical appearance of the sun in the morning, "dawn" in the sense of the beginning of nature's day. However, nature's day is not the separable entity that contemporary Western consciousness and technology can make it. The *Vedas* come from a time when most of humanity rose at dawn and went to bed fairly soon after dark. Certainly, fire allowed human beings to socialize well into the night, but the need to work in the light, and so begin at dawn, tended to foreshorten night life. No electricity blurred the distinction between night and day. Dawn brought an opportunity, the beginning of the time for work in the light, that one missed at one's peril.

Dawn is the beloved of the sky, one of the most privileged moments of the day—bright, fresh, full of the promise of new beginnings. She is so lustrous and alluring that the sun follows her, smitten like a young male lover. (Here, we find a prefiguring of the *bhakti*—erotic religious love—that played an important role in later theistic Hindu mysticism.) She is beautiful, a bride adorned for her wedding (and the consummations the day may bring), a creature at the flush and flower of her promise. She is happy and shining, and those beholding her want her only to shine on, to surpass all her previous splendor and attain new sweeps of illumination. She disperses all shadows of malignity (evil hunkers in the dark, dawn is its overcoming). This makes her like a warrior, a foe of wickedness. Darkness is the abode of wicked spirits, ignorance, confusion, what ruins life and rapes it of its beauty and splendor. Dawn conquers darkness.

Still, although her approach is that of a warrior opposed to darkness, the goddess is not violent but careful, controlled, wise, manifesting nothing wanton or chaotic. She comes to nourish creation; her light gives life. All sorts of creatures quicken at her coming; animals and human beings are one in their dependence on her. She takes them from their slumber into vitality, helping them realize themselves, be themselves with energy, and her coming, her influence, are regular, reliable, indicating that she serves the law of the divine cosmos itself, happy to take her place and play her role in the eternal order of things, according to the wisdom of ultimate reality, by the decrees and designs of what simply is.

These are overtones that we find in the texts on dawn. Meditating on them, chanting them ritualistically so that they become sacred sounds, a traditional Hindu seeking *moksha* might find much nourishment. Admittedly, they are not as provocative as other texts, for example, from the *Upanishads,* if the specific issue is direct experience of ultimate reality. They do not say that all reality is one in such wise that the seeker is essentially the same as the sought. Rather, they suggest obliquely an admiring spirit that loves to contemplate the beginning of light in the heavens and in its contemplations moves through a range of appreciations. It loves the physical beauty of the dawn, which seems to it graceful and feminine. It loves the power of the light, which drives off wickedness and nourishes all life.

How deeply the ordinary user of the Rg Vedic hymns to Usas pondered the dawn of illumination in the human spirit is hard to say. Beyond doubt, though, a contemplative personality can easily find in these texts an invitation to move inward. Darkness, wickedness, and slumber admit of more than simply physical connotations. A mind and heart uplifted to the physical dawn, so in love with its beauty as to be centered and galvanized at its appearance, could be quick to sense the symbolic power of these images, how the actions of the outer natural world point to processes of the inner human spirit.

Mystics are sensitive to both the symbolic actions of the outer natural world and the processes of the inner human spirit. To any significant natural phenomenon (dawn, sunset, storm, drought), they bring a keen eye and a will to find in the phenomenon a provocation to come closer to the real, to join themselves more intimately with the divine.

So, admiration of the dawn may well have brought many Vedic mystics out of themselves, in ecstasy at the beauty of the world. Returning from that ecstasy, some may have lingered with the light of their own spirits exercised by their ecstasy in the glow of the transformation that beauty had brought to their souls. We can only conjecture this, having few historical details, but it seems likely psychologically. If so, such a lingering may also have fostered an appreciation, more tasted or intuited than controlled critically, of the attunement of the free (from *karma*) human spirit to the light, the life, the freedom that Usas represented. All of this is hypothetical, or interpretational, but we believe it can claim a toehold in the poetry of these texts.

Limitations of space forbid our bringing forward texts and reflections on the rest of the full range of topics, parallel to "dawn," that Panikkar includes in his anthology, let alone those that make up the whole Vedic corpus itself. Worth quoting, though, as a line on which to exit from what Panikkar calls "the Vedic epiphany," is the following verse from *Rg Veda* 1, 164, 39:

> He who knows not the eternal syllable of the Veda, the highest point upon which all the Gods repose, what business has he with the Veda? Only its knowers sit here in peace and concord. (3)

The "eternal syllable" may be the *Om* on which many Hindu mystics have focused their sense of the direct experience of ultimate reality or it may be any word that comes alive and burns with light, for example, "dawn." When it brings people to sit (probably yogically) in "peace and concord," however, it has achieved the fruits for which most Hindu mystics, both Vedic and non-Vedic, have labored. Eternal wisdom brings peace and joy. It is tranquility and bliss. That is why it is the great treasure worth so much toil.

The *Upanishads*

Beginning a brief overview of the *Upanishads*, William K. Mahohy has written:

> The Upanishads are codified Sanskrit philosophical speculations of varying lengths in both prose and verse form, composed orally and set to memory mostly by anonymous South Asian sages, primarily in the classical and medieval periods. While the most important and influential Vedic Upanishads date from the eighth to the fourth centuries B.C.E., some lesser-known sectarian Upanishads appear as late as the sixteenth century C.E. Individually and as a whole, the Upanishads present insights and doctrines that serve as the foundation for much of India's philosophical thought.
>
> Traditional South Asian teachings based on the Upanishads have been called the Vedanta, the "end of the Veda," for the Upanishads chronologically and formally set the closure of the Vedic canon. Perhaps more to the point, Upanishadic lessons are said to be the end of the Veda in that they purport to present the "hidden meaning" or the "real message" of religious practice and thought.[5]

Although the implications of the word "*Veda*" include a sense of ultimate reality's giving itself in a privileged "word" (revelation, disclosure of the crucial message that the divine cosmos has to give), traditional Indians who have concentrated on the *Upanishads* have tended to think that these texts are further privileged. The *Upanishads* are esoteric in the sense that hidden in them is wisdom deeper, barer, more condensed than what is spread through the rest of the Vedic literature.

In principle, any human being can aspire to attain or receive Upanishadic wisdom, but in practice these texts draw intellectuals, people of a poetic and philosophical bent, people capable of considerable spiritual sophistication.

The *Upanishads* are not uniform in their teachings, but their major impact on Indian culture has been monistic. That is to say, the excitement that one senses in these texts comes from a perception that the user, or seeker, is getting to the heart of the matter, boiling away the many accidentals, reducing things to the essence.

In *Vedas* such as the *Rg* nature is prominent: dawn, fire, storm. The vitality and diversity of the world dance, rage, beguile, draw the hymnist to worship. In the *Upanishads* a harder spirit, dissatisfied with the plurality, complexity, and confusion of the rich, pulsing world probes for simplification. What holds the plurality together? What is the core, the gist, the foundation, the supreme point of leverage? The passion of the most ardent Upanishadists was for unity. They wanted their spirits to fly like arrows to the target. What thoughts and practices could strip, or streamline, them so that *moksha,* liberation from plurality and its discontents, might carry them into freedom, or fully authentic existence?

The *Isa Upanishad,* blessedly brief, offers a few verses both representative of the Upanishadic temper at large and provocative for our inquiry. First, concerning the attitude that any successful seeker has to maintain, it makes a strong but paradoxical claim for knowledge: "Into blind darkness enter they that worship ignorance. Into darkness greater than that, as it were, they that delight in knowledge." [6]

What can we make of this text? Typical in its brevity, it assumes a context in which a disciple has a teacher who explains the paradox. On the whole, the use of Vedic texts was oral. People memorized them, turned them over this way and that, were familiar with all the nuances of their words. The teacher initiated the disciple into this pattern of oral assimilation. The commentary of the teacher set up the disciple's probings. Naturally, none of the inquiry was hard and fast, but generally a significant text came to a seeker pregnant with interpretations, sleek with savor. So, the typical serious, regular hearer of this verse would not be startled by its apparent paradox. He or she would be provoked to think again about the nature of real, liberating knowledge.

Does anyone worship ignorance? In answer, one could note the mass mind, those who pursue only karmic goods and dwell in *maya* gladly, or a deeper, more reasoned, agnostic option. In the second case, the argument would be something like this: There being no certitude in this multitudinous world, I shall concentrate on ignorance—how much cannot be known, how unknowable the world is. That is where wisdom seems to beckon. For the author of this text, either reference leads to blind darkness. Curiously, we do not know whether such blind darkness is baleful or a boon. Instinctively, we probably react negatively, but then we have to think more deeply because the second assertion of this verse draws us up short. The darkness into which those who delight in knowledge go is greater than that of those who worship ignorance.

Therefore, the following question arises: Should one delight in knowledge? It is difficult to answer no. Inasmuch as the whole slant of the Vedic literature is toward illumination, epiphany, realization, knowledge is coin of the realm and enlightenment is the key process. Answering yes brings us to the following conclusion: delighting in knowledge is the heart of the quest for authenticity; we find that according to the *Isa Upanishad* darkness spreads until it covers ultimate reality. If what we go into through delighting in knowledge is a darkness greater than what the worship of ignorance generates, then there must be something dark about reality itself, the divine cosmos in its foundations, being in its essence—and so there is.

However much we can know and be illumined by reality, it is always greater than we. Our grasp of it is always overshadowed by its limitlessness, and so our knowledge is always dark. We know mainly that we do not know the whole or the gist adequately and that we cannot. So a nescience, an unknowing, becomes blessed to seekers after ultimate reality. The darkness becomes as intrinsic as the light.

The next verse from the *Isa Upanishad* moves this sort of consideration forward a step: "Other, indeed, they say, than knowledge! Other, they say, than non-knowledge! Thus we have heard from the wise, who to us have explained it" (v. 10).

The wisdom at the end of the *Veda* is not dualistic. One cannot describe it adequately in categories of knowledge or nonknowledge. By implication, one cannot describe it adequately in categories of light and darkness. Like reality, Vedic wisdom is more basic, more original or underived.

Now, if one cannot deal with the Upanishadic *Veda* adequately using categories of knowledge and nonknowledge or light and darkness, one cannot deal with it through ordinary reason, which uses such categories. Ordinary reason is binary, geared to either/or. Vedic wisdom is unitary, whole, geared to not-this/not-that. The site of Vedic wisdom is therefore not ordinary reason but must be something more intuitive, more the bare human spirit itself in the face of ultimate reality, the revelation of what simply is.

One can call this the soul or the spirit. Any term raises difficulties, but the main point remains clear. As these verses from the *Isa Upanishad* suggest, grappling with the nature of true wisdom brought Indian reflectivity deeply into paradox. Close to paradox were silence and abnegation. Close to intellectual inquiry, therefore, came *asceticism*, the discipline to move better along wisdom's way. More and more the task became a readying of the spirit as a site sufficiently deep, central, and pure to stand beyond both sullying desire and such dichotomies as light and darkness or ignorance and knowledge, at the place where being simply was, existing in a primordial firstness.

A third verse from the *Isa Upanishad* hints at the nature of this primordial existence that drew the serious user's quest and riveted it to paradox: "It moves. It moves not. It is far, and it is near. It is within all this, and it is outside of all this" (v. 5).

The "it" is spirit and being and wisdom in one. The text embraces both the seeker, trying to know and unknown, and the sought. As well, it embraces the relationship that binds them together. All this moves, changes, drives forward, gives life to mind and heart and body. Yet it also does not move. What *is* continues to be in the same ways that make variation dubious. Similarly, spiritual subject, spiritual object, and their relationship are both far and near. Being goes beyond everything that we control, yet nothing is any closer to reality, nothing a better candidate for being the first thing one must say about something: it is. Finally, what Vedic wisdom deals with, in Upanishadic terms, is always both within and without, covering both realms but in its simplicity escaping their dualism.

Verses such as these from the *Isa Upanishad* remind us what sort of nuggets the early Hindu mystics held in their mind's eyes and considered the gold they ought to mine for wisdom. The actual activities of what we might call "working mystics" have not usually been academic or speculative in a detached sense. They have been intense probings of traditional words believed and hoped to be pregnant with a force of liberation, a power to dispel karmic ignorance and move the seeker into mokshic being, awareness, and bliss. Nonetheless, doctrines developed to put the basic convictions of the Upanishadic venture into somewhat detached intellectual formats, and among the most influential was that which equated *atman* and *Brahman*.

The term *atman* admits of several meanings, but for our present, limited purposes the one most relevant is the soul or self, the summary of the seeker that is the site of both his or her being and the revelation being sought. *Brahman* is the objective correlative, the site of ultimate being revealed, the deep other with whom the self of the seeker is engaged through mutual illumination. (In full illumination the sage may discover that *atman* and *Brahman* have never been separate.)

The Upanishadic literature represents a valiant struggle to clarify the relationship of *atman* and *Brahman*, not because it was an interesting intellectual puzzle but because it appeared crucial to the mystical quest, the drive for direct experience of ultimate reality. A high point in this struggle occurs when Yajnavalkya, the hero of the longest and most important *Upanishad*, the *Brhadaranyaka*, answers the question of a serious seeker:

> Then Usasta Cakrayana questioned him. "Yajnavalkya," said he, "explain to me who is the *Brahman* present and not beyond our ken, him who is the Self in all things. . . ." [Yajnavalkya answered] "He is your self (*atman*), which is in all things. . . ." (3:4:1)[7]

The spiritual hero says that the objective ground, or most ultimate reality, is the same as the self, the subjective ultimate. One cannot then simply conclude that all reality is one or pantheistic or diverse only by convention or accommodation, but one can realize that the most revered viewpoint that emerged from the Upanishadic period was that which claimed to have found in the identity of *Brahman* and *atman* the simple center of the panoplied, dawning reality celebrated in the *Vedas*. One can realize that through the Upanishadic quest Hindu mysticism became profoundly idealistic and monistic, attracted to the idea that spirit is the universal reality that makes the most sense of the world and is the key to liberation.

Yoga

In classical Indian thought the first name associated with *yoga*, taken as a philosophical system (a coherent explanation of both natural reality and human wisdom) has been that of Patanjali (perhaps third century C.E.). In recent Western analysis of *yoga* perhaps the name that stands out most has been that of Mircea Eliade. Here is Eliade on Patanjali:

> Patanjali's *Yoga-sutras* are the result of an enormous effort not only to bring together and classify a series of ascetic practices and contemplative formulas that India had known from time immemorial, but also to validate them from a theoretical point of view by establishing their bases, justifying them, and incorporating them into a philosophy.
>
> But Patanjali is not the creator of the Yoga "philosophy," just as he is not—and could not be—the inventor of yogic techniques. He admits himself that he is merely publishing and correcting . . . the doctrinal and technical traditions of Yoga. And in fact yogic practices were known in the esoteric circles of Indian ascetics and mystics long before Patanjali. Among the technical formulas preserved by tradition, he retained those which an experience of centuries had sufficiently tested. As to the theoretical framework and the metaphysical foundation that Patanjali provides for these practices, his personal contribution is of the smallest. He merely rehandles the [atheistic and metaphysical, matter/spirit] Samkhya philosophy in its broad outlines, adapting it to a rather superficial theism in which he exalts the practical value of meditation.[8]

Patanjali's *Yoga-sutras* collected traditional theses about meditation and the requirements and discipline of purifying the human spirit and achieving enlightenment. The *sutras* themselves are bare statements, beads on a string, that Patanjali assumes readers and disciples will ponder deeply, most likely under the direction of a teacher, or *guru*. Let us consider some of his *sutras* and ask how they orient the Hindu quest for direct experience of ultimate reality.

Interesting in light of what we have heard from Eliade, and perhaps fairly easily understood, is the following block of seven consecutive sutras:

> (1) Concentration may also be attained through devotion to Ishwara.
> (2) Ishwara is a special kind of Being, untouched by ignorance and the products of ignorance, not subject to karmas or samskaras or the results of action.
> (3) In Him, knowledge is infinite; in others it is only a germ.
> (4) He was the teacher even of the earliest teachers, since He is not limited by time.
> (5) The word which expresses Him is OM.
> (6) This word must be repeated with meditation upon its meaning.
> (7) Hence comes knowledge of the Atman and destruction of the obstacles to that knowledge.[9]

The first goal of yogic practice is concentration: collecting and focusing the mind and spirit. People begin the yogic enterprise scattered, distracted, the slaves rather than the masters of their minds. The first truly ambitious aim of yogic discipline is to gain control of the imagination, the feelings, and the operational (practical) reason itself. Yogic disciples want to bring their spiritual energies or capacities to bear on the task of knowing what is real as acutely and as deeply as possible. That is why they go to a guru and enter upon the discipline: to get control and make progress toward liberation—*moksha*, being, awareness, and bliss. Concentration is the first high ambition and victory, assuming a reformation of the moral life and moving beyond it into spiritual reform, a reconstitution of the mind. When they are not scattered but focused, not spilled out variously into wild images and desires but collected and condensed, yogis are well under way. If devotion to Ishwara facilitates the attainment of concentration, yogic disciples are well advised to become devoted. (As Eliade hints, Patanjali's theism seems quite pragmatic.)

The description makes Ishwara divine, correlated with the realm beyond ignorance. He therefore can stand for the ultimacy, the full reality, that the disciple is pursuing. Ignorance and the products of ignorance (false conceptions, stupid choices and patterns of action) are the most acute effects of *karma* and *samsara*. They virtually define the bondage under which human beings suffer, from which yoga aims to liberate them. Ishwara dwells outside *karma* and *samsara*, free rather than enslaved. Going toward him in devotion is therefore moving toward freedom and unconditionedness and away from ignorance and constraint.

Ishwara also dwells apart from the results of action (where *karma* works itself out). This has important implications inasmuch as devotion to Ishwara would minimize or even denigrate action. Some yogas focus on purifying action, but others propose withdrawal from action. Ishwara can be untouched by action either because he is withdrawn from it or because he dwells beyond it and its fruits. Either way, devotion to him would move the disciple out of the pragmatic world of business and family life, the world where action calls most of the tunes.

The knowledge that the god possesses has no limits. What tends to be only a beginning, a seed or a germ, in others, such as the disciples laboring at yoga, is in him fully developed. He is aware as broadly and as deeply as awareness can go. For the precisely yogic focus of enlightenment, he possess a fullness that makes it completely fitting that disciples orient themselves to him, use him as a summarizing symbol, a *mandala*.

For Patanjali, devoted to Ishwara, the god has always taught yogis. From time immemorial, as long as human gurus and disciples have collaborated at the yogic task, Ishwara has been the patron, the focus and source of their success. He has been able to be this because he dwells outside of time. Each generation, each personal lifetime, can connect to him. Thus he represents the "other" that human yogic effort, which despite its best successes remains indebted to time and a mortal body, can glimpse and aspire to attain. He symbolizes how the eternal dwells in the flux of the human now and then.

If Ishwara himself is a *mandala*, a focus for human concentration, he yet has his own *mantra* (holy sound), the traditional and sacrosanct *Om*. The comparativist Geoffrey Parrinder has described *Om* as follows:

> The most sacred word of the Hindus, occurring first in the *Upanishads*. It is composed of the sounds A, U, M and a humming nasalization, and so it is said to represent the three oldest *Vedas*, and the triad of gods: Visnu, Shiva, and Brahma. Om is placed at the beginning of works, like "Hail," and at the end, like "Amen." It is written at the head of books and papers and uttered before prayers.[10]

By making *Om* Ishwara's word, Patanjali makes Ishwara the divinity in whom the whole tradition of Hindu wisdom comes to focus.

The yogic disciple ought to meditate on *Om* by chanting it softly. Taken traditionally, to heart or the depths of consciousness, *Om* can epitomize all that the disciple seeks, the basic sound on which the universe depends. Meditation is such a taking to mind and heart. Meditation is not study in the bookish sense of the term. It is not idle rumination. It is a strong effort to concentrate consciousness, to collect the disciple's awareness and attune it to reality—being and knowledge that are spiritual and not bound by *karma*. Traditionally, one repeats *Om* steadily, letting its reverberations set up a harmony in one's spirit. *Om* becomes a sound one can ride, traveling out of bondage and into greater spiritual freedom. Thus *Om* comes to represent the fire of creativity from which the universe issues.

From meditation the disciple can gain knowledge of the *Atman*. Here, Patanjali seems to understand *Atman* as the spirit or Self holding the key to or being the gist of the enlightenment that the yogi seeks. From Upanishadic times, as we have seen, the term carries both objective and subjective connotations. Usually it stands for the ultimacy in the depths of the subject, for example the disciple trying to gain enlightenment, but it has also been correlated with the more objective *Brahman*. Indeed, for a Yajnavalkya the two are closer even than two sides of one coin because there is only one reality, *Brahman/Atman*, that is the ultimate identity of everything that exists. Patanjali is less metaphysical and radical, but clearly for him reaching the *Atman* is a high success. Inasmuch as a meditational use of the *Om* overcomes obstacles to a knowledge of *Atman*, it would be highly precious, something any disciple with an ounce of common sense would employ.

There are more profound *sutras* in Patanjali than this sequence of seven that we have glossed, but our seven set up many of the sage's later inquiries and are representative of his regular style. A good summary of Patanjali's view of the venture of *yoga* overall occurs in his description of its eight limbs:

> The eight limbs of yoga are: the various forms of abstention from evil-doing (yama), the various observances (niyamas), posture (asana), control of the prana [breath] (pranayama), withdrawal of the mind from sense objects (pratyahara), concentration (dharana), meditation (dhyana), and absorption in the Atman (samadhi).[11]

The first four limbs are propaedeutic (preconditions or substantial preliminaries). Because yogis are pursuing purification of consciousness and union with *Atman*, they must have good morals. Negatively, they must abstain from doing evil. Positively, they must perform the rituals, execute the good deeds, of mainstream Hinduism. Good posture (generally, the lotus position, which anchors the body and keeps the spine erect) is a precondition for successful yogic sitting. The breath mediates between the body and the spirit, so controlling the breath, focusing the mind–body complex through its intake and outflow, is an obvious necessity.

The second four items are more substantial and more characteristic of most references to *yoga*. To withdraw the mind from sense objects is to take the first deliberate step toward exercising the spirit with less dependence on its material constraints than what a more spontaneous, nondeliberate, unintentional, undisciplined (nonyogic) use of the mind shows. If we are to advance toward anything like an experience of pure spirit, we have to control the outer senses and the imagination, become their masters rather than their slaves.

Concentration, the sixth limb, is where the block of seven *sutras* that we have already examined begins. If we consider "withdrawal of the mind from sense objects" to be the last of the preliminaries, concentration is the first of the substantial achievements. Those who gain this power become focused by fusing together their various talents and powers for maximum impact. Meditation increases the power of the concentrated spirit. Becoming one with a *mantra* such as *Om* or a *mandala* such as Ishwara, the yogi penetrates reality ever more deeply, letting it shape his or her being ever more completely.

Meditation is less thinking about ideas than engaging the deeper, submental spirit with ultimate reality as directly and comprehensively as possible. The last limb, where *yoga* reaches its goal, is *samadhi*. Here, the yogi becomes identified substantially with ultimate reality (*Atman*). The experience of *samadhi* may be likened to a trance or inner transport, where without the usual observation (and so objectification) of self by Self the yogi exists immersed in the oneness of personal self (*atman*) and cosmic Self (*Atman* or *Brahman*). *Samadhi* is a loss of self (which, paradoxically, is also a finding and exercising of truest Self) that is deeper than sleep and matches up well with the simple revelation of being that Panikkar describes as capital for the Vedic seers.

In leaving Patanjali and his classical description of yogic practice, we may profit from some cautionary words written by Eliade's Hindu teacher S. N. Dasgupta, whose own name stands high in the ranks of those who have explained Hindu mysticism well:

> It is very difficult for a Western mind of today to understand, or appreciate, the minds of the Indian seers. They felt a call from within the deep caverns of their selves—a call which must have started from a foretaste of their own true essence—which made all earthly pleasures or hopes of heavenly pleasures absolutely distasteful to them. They could feel satisfied only if they could attain this true freedom, their true self. . . . These men had no riches, and they did not seek them. Their natural needs were few, and it never occurred to them that these could be augmented or multiplied. They thought, rather, that what needs they had were in themselves too numerous and could be indefinitely curtailed.[12]

One might call *yoga* the curtailing of the needs that stand in the way of full human freedom.

The *Vedanta*

Even greater than the influence of Patanjali on *yoga* was the influence of Shankara on *Vedanta*. Both *yoga* and *Vedanta* qualify as *darsanas* (coherent systems that blend experience and doctrine, setting them in the service of liberating people from ignorance and preparing them for *mok-*

sha). *Samhyka* is also a *darsana,* differing markedly from *Vedanta,* which is monistic (inasmuch as only *Brahman* exists in the final, enlightened analysis), because it is dualistic—matter (*pakriti*) and spirit (*purusha*) co-determine reality. The teaching of Shankara, known as *Advaita* (nondual) *Vedanta,* rejects such a dualism, arguing that *atman* and *Brahman* are one. Indeed, it argues that the entire world of *samsara* is illusory and that for the enlightened there is only *Brahman.*

The traditional dates for Shankara are 788–820 C.E., but many scholars dispute them, arguing that it is unlikely that a man who died at age thirty-two could have produced the large and extremely influential corpus of works attributed to Shankara. As a recent description of his influence suggests, he has shaped all subsequent Hindu religious thought:

> Sankara (c. 700 C.E.), also known as Samkara or Sankaracarya, Hindu metaphysician, religious leader, and proponent of Advaita Vedanta. Sankara is generally acknowledged to be the most influential of all Hindu religious thinkers. The many modern interpretations and popularizations of his uncompromising intellectual metaphysics represent the dominant current of contemporary Hindu religious thought. For scholars of Sanskrit his compositions, above all his famous commentary (*bhasya*) on the Brama Sutra of Badarayana, serve as models of philosophical and literary excellence.[13]

In a popular recent study of Hindu culture, David R. Kinsley treats Shankara and the Mahatma Gandhi as "representative thinkers." A considerable Hindu lore has developed around the childhood of Shankara, from which Kinsley offers the following sketch of the sage's early years:

> Shankara was born in a small village in Kerala, in Southwest India, to a childless couple who had petitioned [the god] Shiva to grant them a son. Shankara's various biographers say that he was the embodiment of Lord Shiva himself, who came among humans to teach them the truth. Shankara is described as being a particularly precocious child, mastering the *Vedas* at a very young age. He is also described as showing little interest in worldly life and as desiring to renounce the world even as a child. His father died when Shankara was about eight years old, and as he was an only child, his mother opposed his wish to give up the world and become a sannyasi, or wandering mendicant. Shankara obtained his mother's permission, however, in a curious incident that is told in most of his biographies. While bathing in a river, he was seized by an alligator. In his predicament, he shouted to his mother for her permission to enter the stage of the sannyasi, and thinking her son near death his mother agreed. The alligator subsequently freed the boy, and with his mother's permission Shankara soon afterward left the world in search of liberation.[14]

The major works attributed to Shankara that have survived the examinations of modern textual critics are the *Brahmasutra-bhasya;* the commentaries on the *Brhadaranyaka, Taittirtya, Changogya, Aitareva, Isa, Katha, Kena* (two), *Mundaka,* and *Prasna Upanishads;* the commentary on the classical devotional text, the *Bhagavad-Gita;* the commentary on the *Mundukya Upanishad* along with the *Gaudapadtyakarika;* and a work known as the *Upadesasahasri.* Other works sometimes attributed to Shankara (for example, many devotional hymns) are of dubious origin.

One can see from this list that Shankara was most interested in the *Upanishads.* His greatest treatise, the *Brahmasutra-bhasya,* works from the Upanishadic viewpoint to develop the identity of *atman* and *Brahman.* Although the philosophical quality of his investigations is high, Shankara subordinates his analyses to the goal of liberation. He is not seeking knowledge or clarification of the Upanishadic texts for its own sake, or even to teach others the riches of the *Upanishads.* He is seeking a spiritual illumination, a direct and intuitive knowledge (*jnana*) that will show him, or any disciple following him, that realizing the nonduality of reality (*1*) dissolves almost all of the enslaving effects of *karma* and (*2*) will allow the *jnanin* at death to pass definitively from *samsara* to *moksha* and never be reborn into *samsara* again. Shankara thinks that only *jnana* can accomplish definitive liberation. The *yogas* of devotion (*bhakti*) and purified action (*karma-yoga*) can bring one to a higher state in one's next existence and so increase one's chances for gaining *moksha,* but by themselves they cannot accomplish full liberation.

In Shankara's program, liberation requires that one belong to an upper caste (only the upper castes can study the *Vedanta*). It also requires detaching oneself from worldly things (material possessions, thoughts, acquisitive desires), restraining thoughts and feelings, and pursuing tranquility. In effect, then, Shankara presupposes that those studying the *Upanishads* under his guidance are yogis in the sense of people disciplining their consciousnesses, most likely through regular meditation. They should enjoy the direction of a local guru, and the focus of their labors after *jnana* ought to be the *Upanishads.*

Boiled down, Shankara's view of ultimate reality is that only *Brahman* exists, or is, in the full sense of those words. For the enlightened person, who has passed out of karmic bondage and ignorance, *Brahman* is existence pure and simple—the ground, the inmost reality, of all that we consider to exist, both material and spiritual. *Brahman* has no qualities, no qualifications, no limitations. It is not a personal deity, and so devotion, in the ordinary sense of venerating a god, usually with considerable personal engagement and emotion, does not befit it. *Brahman* is unchanging, does not act, and has no needs because it is utterly complete. It is the ground of being and consciousness, present wherever they are; in fact their inmost reality. It is not a being because it is unlimited, and beings are finite, circumscribed. Rather, it is being pure and simple.

For Shankara there is no substantial difference between *Brahman* and

Atman. *Atman* is the eternal, unchanging self of the world, the ground of being considered to be the source of awareness and subjectivity. It, too, has no conditions or qualities. It is passive, as *Brahman* is, because it is perfect and so does not need to act to gain any enrichment. We tend to think of it as the ground of subjectivity, but it is not personal, having no emotions or ego. As identified with *Brahman*, it too is the only thing that exists fully or truly for those who have gained enlightenment and so see reality with a direct intuitive awareness of what it actually is.

One can see how *jnana*, such intuitive realization, fits the description of mysticism with which we are working. To see in a flash that there is only *Brahman/Atman* is, in our terminology, a mystical achievement. The fact that the *Vedanta* program developed by Shankara is highly intellectual does not negate this judgment. Shankara assumes, with much (perhaps the mainstream) of higher Hindu culture, that reality is ideal, spiritual, and best understood by concentrating the spirit, mind, and heart on the pure being that grounds all beings. (Correlatively, sensory knowledge keeps people in thrall to *karma*.) However, the study that he advocates, as we noted, is correlated with a typically yogic program of discipline, both moral and intellectual. As well, the end of this discipline is *moksha*, not intellectual mastery in an academic sense. So, to call Shankara the leading Hindu philosopher can be misleading. His philosophy is an invitation to mystical self-realization. It is not simply a program of study but rather a full way of life.

To conclude this section, we offer some selections from Shankara's greatest work, the *Brahmasutra-bhasya*, also known as the *Vedantasutra:*

> Now [in response to the charge] that Brahman may be well known or unknown; if it is well known, there is no need to desire to know it; if on the other hand, it is unknown, it could never be desired to be known. The answer to this objection is as follows: The Brahman exists, eternal, pure, enlightened, free by nature, omniscient, and attended by all power. When the word "Brahman" is explained etymologically, it being eternal, pure, and so on, are all understood, for these are in conformity with the meaning of the root *brh* [from which Brahman is derived]. The Brahman's existence is well known, because it is the Self of all; everyone realizes the existence of the Self, for none says, "I am not"; if the existence of the Self is not well known, the whole world of beings would have the notion "I do not exist." And the Self is the Brahman.

> Of this universe made distinct through names and forms, having many agents and enjoyers, serving as the ground of the fruits of activities attended by specific places, times, and causes, and whose nature and design cannot be conceived even in one's mind—that omniscient, omnipotent cause wherefrom the origin, maintenance, destruction of such a universe proceeds is the Brahman, such is the full meaning to be understood.

> By showing the Brahman as the cause of the universe it has been suggested that the Brahman is omniscient; now to reinforce that omniscience the author of the aphorisms [of the *Brahmansutra*] says: "Of the extensive scripture composed by the *Rig Veda,* etc., reinforced and elaborated by many branches of learning, illuminating everything even as a lamp, and like unto one omniscient, the source is the Brahman. Of scripture of this type, of the nature of the *Rig Veda* and the like, endowed with the quality of omniscience, the origin cannot be anything other than the omniscient one . . . from whom, as the source, issued forth, as if in sport and without any effort, like the breathing of a person, the scripture in diverse recensions, called *Rig Veda,* etc., which is the repository of all knowledge and is responsible for the distinctions into gods, animals, humans, classes, stages of life, etc."; this is borne out by scriptural texts like: "This that is called *Rig Veda* [and so on] is the breathing out of this Great Being."
>
> [I hold] that Brahman, omniscient, omnipotent, and cause of the birth, existence, and destruction of the universe is known from the scripture as represented by the Upanishads. How? "Because of textual harmony." In all the Upanishads the texts are in agreement in propounding as their main purport this idea. For example, "Dear one! this thing Existence alone was at the beginning." "This Brahman, devoid of anything before or after, inside or outside." "At first there was only this Brahman, the immortal one."
>
> Because the Brahman is a thing already well established, it cannot be held to be the object of perception by senses, etc.; for the truth that the Brahman is the Self, as set forth in the text, "That thou art," cannot be known without the scripture. As regards the objection that since there is nothing here to be avoided or desired, there is no use in teaching it, it is no drawback; it is from the realization that the Self is the Brahman, devoid of things to be avoided or desired, that all miseries are ended and the aspiration of man [human beings] is achieved.[15]

By way of brief commentary, we note (*1*) the "scholastic" quality of these remarks—the suggestions that Shankara is constantly debating with other philosophers; (*2*) the absolute quality of his *Brahman* and its equation with Being as such, *Ens simpliciter; (3)* the identity of this Being with the Self through which all beings know intuitively that they exist; and (*4*) the crediting of the creation of the universe to the *Brahman* (and so the crediting of some sort of action?). We also note the crediting of the *Vedas,* which are understood to be omniscient, the source of all (valuable) knowledge, to *Brahman,* imagined as the omniscient breath from which they came forth playfully. The argument from "textual harmony" that the *Upanishads* regularly represent the *Brahman* as Shankara has and that their basic refrain is the primacy of *Brahman* is important. So is the description of this teaching about the *Brahman* as something that cannot be known through the senses but must be revealed through the Vedic scriptures. Finally, we note the assertion that, far from being a drawback, the nature of the *Brahman* as offering nothing to be avoided or desired puts it clearly beyond the karmic realm, so that the realization that the Self is the *Brahman,* through jnanic enlightenment takes the disciple to the realm where desire holds no place and the samsaric miseries created by desire are no more.

Bhakti

Bhakti is devotional love. On the whole, Hindus have expressed *bhakti* principally toward their personified gods, especially Vishnu (often in the form of Krishna, his primary *avatar* or mode of appearance), Shiva, and the Mahadevi, or the Great Mother. Occasionally, the latter appears directly, but more frequently she appears in such personifications as Durga, Kali, and Parvati (consorts of Shiva), Radha (consort of Krishna), Lakshmi (consort of Vishnu), and Sita (consort of Rama, another *avatar* of Vishnu). Because it comes in for high praise in the *Bhagavad-Gita,* the most influential synthetic work in all of Hinduism, and because it is a major theme in the two great Hindu epics, the *Mahabharata* and the *Ramayana, bhakti* has risen above being considered vulgar, as are the instinctive ways in which common, unrefined people pursue union with ultimate reality. Rather, it has commanded the attention of poets, mystics, and first-rate intellectuals, such as the philosopher Ramanuja, second only to Shankara in prestige, who have thought love more central than knowledge to both the constitution of the individual and his or her attainment of liberation from *samsara.*

Beginning an article on *bhakti,* John B. Carman has written:

> The Sanskrit term *bhakti* is most often translated in English as "devotion," and the *bhaktimarga* [way of *bhakti*] is a path leading toward liberation (*moksha*) from material embodiment in our present imperfect world and the attainment of a state of abiding communion with a personally conceived ultimate reality. The word *devotion,* however, may not convey the sense of participation and even of mutual indwelling between the devotees and God so central in *bhakti.* The Sanskrit noun *bhakti* is derived from the verbal root *bhaj,* which means "to share in" or "to belong to," as well as "to worship." Devotion, moreover, may not suggest the range of intense emotional states so frequently connoted by *bhakti,* most of which are suggested by the inclusive English word *love.* God's love, however, whether answering or eliciting the devotee's love, is denoted with other words than *bhakti.* Thus *bhakti* is the divine–human relationship as experienced from the human side.[16]

The *Bhagavad-Gita* draws together the major teachings of the Hindu ascetic and mystical traditions. It is a synthesis, or a compendium, of traditional lore presented as the instruction of a disciple (Arjuna) by a guru (Krishna). The *Gita* explains the traditional pathways (*margas*) to liberation, largely by describing the mainstream yogic techniques for (*1*) gaining detachment from desire and (*2*) quieting the mind. Characteristically, it purifies a traditional pathway, such as ritual sacrifice, rather than eliminating it, and it lays great stress on purifying action (*karma-yoga*), a notion that appealed greatly to Mahatma Gandhi, who eventually worked out an idealistic politics of nonviolent resistance that liberated India from British rule. For those inclined to the way of devotional love, the *bhaktimarga,* the consummate word of the *Gita,* appears in Chapter 18, verses 64–65:

> And now again give ear to this my highest Word, of all the most mysterious: "I love you well." Therefore will I tell you your salvation. Bear Me in mind, love Me and worship Me, sacrifice, prostrate yourself to Me; so will you come to Me, I promise you truly, for you are dear to Me.[17]

The speaker is Krishna, who represents full divinity, or ultimate reality, in a personified form. The famous theophany of Chapter 11 has made clear beyond any possible doubt the fully divine status of Krishna, our need to equate him with *Brahman,* the final ultimate. The theophany begins with Krishna speaking:

> Do you today the whole universe behold centered here in One, with all that it contains of moving and unmoving things; (behold it) in my body, and whatever else you fain would see. But never will you be able to see me with this your (natural) eye. A celestial eye I'll give you, behold my power as Lord.
>
> Sanjaya [the narrator] said: "So saying Hari [Krishna], the great Lord of power-and-the-skilled-use-of-it, revealed to the son of Pritha [Arjuna] his highest sovereign form—(a form) with many a mouth and eye and countless marvelous aspects; many indeed were its divine adornments, many the celestial weapons raised on high. Garlands and robes celestial he wore, fragrance divine was his anointing. (Behold this) God whose every (mark) spells wonder, the Infinite, facing every way. If in (bright) heaven together should arise the shining brilliance of a thousand suns, then would that perhaps resemble the brilliance of that (God) so great of Self. Then did the son of Pandu [Arjuna] see the whole (wide) universe in One converged, there in the body of the God of gods, yet divided out in multiplicity.
>
> "Then filled with amazement Arjuna, his hair on end, hands joined in reverent greeting, bowing his head before the God, (these words) spoke out. Arjuna said: 'O God, the gods in your body I behold and all the hosts of every kind of being; Brahma, the lord (I see) throned on the lotus seat, celestial serpents and all the (ancients seers).' "[18]

Arjuna continues to list the realities that center in Krishna, as in the source and ground of all that exists. The salient point for our purposes is that when in Chapter 18 Krishna says that his highest word, the most mysterious of all, is "I love you well," he throws a great thunderbolt.

The most definitive word of ultimate reality, of the god who summarizes the universe in his own being, is love. True, Krishna's love is not *bhakti (isto 'si me drdham iti)*, nor does the dearness mentioned in 18:65 imply *bhakti* in the god *(priya 'si me)*, but *bhakti* is the word used in 18:65 for the love that Arjuna is to have for Krishna *(mad-bhakto)*, implying an intense emotional attachment, a whole-hearted ardor. The god is saying that Arjuna's salvation will come best through making Krishna his great passion, the love of his life, the power behind his worship, his very reason to be.

The *Gita* often praises a given conviction or technique as though that item of the spiritual life alone sufficed for progress or epitomized the way of salvation. As an overall, classical religious work, the *Gita* is eclectic, intent more on including everything useful from tradition than on ordering such materials logically according to categories of superior and lesser significance. Nonetheless, it remains provocative that *bhakti*, coming near the end of the *Gita*, receives so absolute an endorsement. In fact, as noted, those committed to the *bhaktimarga* have found in this endorsement the practical wisdom that the *Gita* considers supreme: love is the highest way. Thus *bhaktas*, those walking the pathway of love, generally revere the *Gita* greatly, reading all of it in light of the primacy of *bhakti* and considering it the great scriptural sanction for their intensely emotional commitment to Krishna (or, by accommodation, Shiva or the Mahadevi).

In the trinity of gods that developed after the Upanishadic stress on an impersonal *Brahman*, Brama has been considered the creator, Vishnu the preserver, and Shiva the destroyer. Brama is rather remote, abstract, even impersonal, and so has attracted little *bhakti*. Vishnu, frequently in such avataristic forms as Krishna and Rama, has tended to draw a highly positive, indeed frankly erotic, love, especially from women. For example, one of the great religious mythologies within the huge Hindu corpus is the dalliance of Krishna with the *gopis*, the girls who herd cows (sacred animals). The canonical interpretations of this symbolism make it a teaching about the mystical interactions between the human spirit and divinity, much as the canonical interpretations of the biblical Song of Songs make it a teaching about the intercourse of either the collective people (Israel or the Church) or the individual devotee with God.

Bhakti toward Shiva, the deity presiding over the destruction of the universe (Hindu cosmology sees the universe as continually dying and being reborn, in huge cycles of time called *kalpas*), tends to be more fearful than that shown to Krishna but not necessarily less ardent. In the case of Shaivites, *bhaktas* of Shiva, braving the destructive fierceness of the god's appearance becomes a pledge of complete devotion to him. So, for example, the following verses of Basavanna, a celebrated Shaivite poet of the twelfth century, express his longing and love in terms of a willingness to suffer for his lord and make himself abject.

> Cripple me, father, that I may not go here and there. Blind me, father, that I may not look at this and that. Deafen me, father, that I may not hear anything else. Keep me at your men's feet looking for nothing else. O lord of the meeting rivers. Don't make me hear all day, "Whose man, whose man, whose man is this?" Let me hear, "This man is mine, mine, this man is mine." O lord of the meeting rivers, make me feel I'm a son of the house. Siva, you have no mercy. Siva, you have no heart. Why why did you bring me to birth, wretch in this world, exile from the other? Tell me, lord, don't you have one more little tree or plant made just for me? As a mother runs close behind her child with his hand on a cobra or a fire, the lord of the meeting rivers stays with me every step of the way and looks after me.[19]

Here, we obviously have a lover forlorn, feeling neglected, rejected, abandoned; but he is a lover of the supreme deity, so from the outset his rights are minimal. The deity may abuse him as it wishes. Still, at the end of the day he believes that Shiva remains with him and cares for him. He believes, with the *Gita,* that he is dear to the god. So, he asks Shiva to cripple him, blind him, deafen him, if that would ensure his fidelity to such a favor. He wants to sit at the feet of and be numbered among those who have surrendered themselves to this lord and made him their sole treasure.

The poet's great comfort would be to know that the lord numbers him among his own, the people dear to him. Basavanna had found enlightenment at the confluence of two rivers in southern India, so he thinks of Shiva as connected with this spot of holy waters, its special lord. To be a son of Shiva's house would mean complete acceptance, salvation from the alienation the poet feels in the outside world. That world is dangerous, like the cobra and the fire from which the mother must protect her child. Shiva seems merciless to insist that Basavanna live there, and the poet taxes his god with cruelty. Finally, though, his faith and hope surge forth, and he asserts that the lord is always with him, though in what form or degree of union is not clear. What is clear, though, is the full emotion, the highly charged and personal love, that is working in these poems. They flow from a *bhakti* of classical proportions.

A. L. Basham, whose invaluable work *The Wonder That Was India* remains a landmark study of Hindu culture before the coming of the Muslims, has written of Ramanuja, the greatest theoretician of *bhakti:*

> This impassioned devotionalism [the *bhakti* movement] gradually affected the whole religious outlook of the Tamil [southern Indian] country. The great Sankara himself, though he maintained the rigid Upanisadic doctrine

> of salvation by knowledge, was the reputed author of some fine devotional poems in Sanskrit. It was only to be expected that the new forms of worship should receive formal shape, and be harmonized with the Upanisads. This was done in different forms by a series of Dravidian theologians who succeeded Sankara.
>
> The chief of these was Ramanuja, a brahman [member of the highest, priestly caste] who taught in the great temple of Srirangam. He is said to have lived from 1017 to 1137, but the first date is in all probability several decades too early. Like Sankara he taught in many parts of India, and claimed to base his doctrines on earlier sources, writing lengthy commentaries on the *Brama Sutras,* the *Bhagavad Gita* and the Upanisads. Ramanuja's system was founded on that of the Pancaratras [devotional lore deriving from a Vaishnavite religious observance of five nights commemorating the five day sacrifice through which Vishnu became the greatest of the gods], but his emphasis was rather different. He admitted the usefulness of ritual observances, but only in qualified measure, and he also admitted Sankara's doctrine of salvation by knowledge, but declared that those so saved would find a state of bliss inferior to the highest. The best means of salvation was devotion, and the best yoga was *bhakti-yoga,* such intense devotion to Visnu that the worshipper realized that he was but a fragment of God, and wholly dependent on him. Another means of salvation was *prapatti,* the abandonment of self, putting one's soul completely in the hands of God, trusting in his will, and waiting confidently for his grace.[20]

Basham's summary of the theological implications that Ramanuja drew from these devotional convictions is admirably lucid and brief, condensing into one paragraph more insight than what many lengthy monographs finally achieve:

> Ramanuja's God was a personal being, full of grace and love for his creation. He could even override the power of karma to draw repentant sinners to him. Unlike the impersonal World Soul [*Atman*] of Sankara, which made the illusory universe a sort of sport (*lila*), Ramanuja's God needed man, as man needed God. By forcing the sense Ramanuja interpreted the words of Krsna, "the wise man I deem my very self" [*Bhagavad-Gita,* 7:18], to imply that just as man cannot live without God, so God cannot live without man. The individual soul, made by God out of his own essence, returned to its maker and lived forever in full communion with him, but was always distinct. It shared the divine nature of omniscience and bliss, and evil could not touch it, but it was always conscious of itself as an I, for it was eternal by virtue of its being a part of godhead, and if it lost self-consciousness it would cease to exist. It was one with God, but yet separate, and for this reason the system of Ramanuja was called *visistadvaita,* or "qualified monism" [in contrast to Shankara's *advaita* or "unqualified"]. Ramanuja was not as brilliant a metaphysician as Sankara, but Indian religion perhaps owes even more to him than to his predecessor. In the centuries immediately following his death his ideas spread all

over India, and were the starting-point of most of the devotional sects of later times.[21]

Thus *bhakti* eventually gained a full theological rationale, as well as an overflowing repertoire of rituals, ceremonies, songs, and myths. It became the most popular religious pathway, the font of innumerable shrines and devotions in the home. In forms both low and high, it has marked most of subsequent Hindu culture as profoundly convinced that divinity is the key to happiness and liberation and that divinity both cares for human beings and asks for their wholehearted, fully trusting love.

Summary

We have dealt with a considerable range of Hindu texts and ideas; let us now briefly engage them with our overall inquiry into the varieties and constancies of the mysticism that we find in the world religions. In other words, let us ask both (*1*) how what we have seen of Hindu mysticism colors our working definition of mysticism as direct experience of ultimate reality and (*2*) how that working definition invites us to think about the Hindu religious complex of experiences, aspirations, and techniques.

From the general orientation to the gross history and worldview of Hinduism that we offered at the start of this chapter, we would in summary stress three things. First, Indian history is long, complicated, and rich, and any periodization is but a more or less useful fiction. Even a brief inventory of the key moments would have to include the coming of the Aryans, their interaction with the native Dravidians, and the later coming of the Muslims (their lesser interaction and tenser relations). The partially Christian influence of the British and other Westerners in recent centuries seems to have been less significant, religiously, than that of the Muslims and certainly has had nothing like the definitive, transformative impact of the Aryans. Overall, looking at the entire development of the representative Hindu religious personality, one could say that the Aryans contributed the passionate drive for *moksha;* the Dravidians contributed the firm grounding in sex and nature that eventually supported the *bhakti* movement (somewhat against Vedantic idealism); and the Muslims contributed constant challenges to Hindu anthropomorphism, iconography, ritualism, and other possible encroachments on the purity and utter transcendence of God.

Second, the concepts that we singled out as capital in the gross Hindu worldview have been at the center of our studies of Vedic, Upanishadic, yogic, Vedantic, and Bhaktic pursuits of liberation. *Dharma, karma, mok-*

sha, yoga, and *maya* have been either presuppositions or direct concerns of all the Hindus who have worked hard in those pursuits. The *Vedas* obviously are the principal source of *dharma* inasmuch as they have been considered to be the highest scriptural authority from which the teaching or tradition comes. *Karma* and *moksha* are contraries, logical companions locked into an inevitable dialectic: *karma* epitomizes the miseries of the human condition, being the moral law of cause and effect that condemns creatures to an endless cycle of death and rebirth, and *moksha* is blessed release, liberation, from this dismal, painful human condition caused by *karma* (and most simply designated as *samsara*). *Yoga,* in the broad sense of spiritual discipline, is the favored Hindu way out of karmic misery toward *moksha. Maya,* finally, is the illusion that binds one to *samsara* and stands overcome by the enlightenment that gives one *moksha.*

Against this general backdrop, we followed Raimundo Pannikar in treating the *Vedas* as revelatory of the true nature of reality. The dawn of light, the coming of wisdom, symbolized by the goddess Dawn herself, served as an example of the longstanding Hindu focus on the mutuality of being and consciousness that has shaped most of Hindu mysticism. From Vedic times, seers and seekers have concentrated on the moment of illumination, when being, the ultimate and fully real, gives itself to the receptive human spirit. In that moment, the reciprocal giving back of the human spirit, the participatory flow of the human spirit into being, imparts a sense of union, spirit to spirit, that prepares the way for the radical Upanishadic assertion "That thou art." Panikkar's Vedic seer (*rishi*) is a mystic, dealing with ultimate mysteries, which for the moment have dropped their veils.

The directness of the Vedic experience of ultimate reality no doubt varied from person to person, but by locating the core of the experience in the mutuality of the known and the knower, in the dawning light that being gives to human consciousness, the mystical tradition begun in the Vedic quests proved highly formative for the rest of Hinduism's search for liberation. Even though the majority of the texts that one finds in the *Vedas* prior to the *Upanishads* are concerned with natural phenomena like dawn more than with movements of the human spirit, the Vedists' wondering about these phenomena and lingering with the impact of dawn or darkness on the human spirit set the stage for subsequent meditative probes that went deeper. Such probes define much of what *yoga* has meant in Indian religious history and so are the source of an enormous overall influence on Hindu mysticism.

The *Upanishads* reflect the movement away from the plethora of natural phenomena in search of something more unitary and interior; this led to a powerful conviction that a single ultimacy, usually called *Brahman* or *Atman,* undergirds, is the final reality of, and may be found in everything that we experience, all that we would say exists. The texts we examined from the *Isa Upanishad* illustrate the somewhat rough and ready but still provocative forms in which we find this characteristically Upanishadic conviction becoming clear. By the time that the premier Upanishadic sage, Yajnavalkya, can in the *Brhadaranyaka Upanishad* put the crucial proposition about unity ("He is your self") with utter brevity and clarity, the border between self and other, inside and outside, has broken open. More important than how beings differ is the sameness they share as expressions or presences of being. To any of them the sage can say, "That thou art."

This conviction set up the metaphysics of Shankara and so the great influence on the quest for liberation of Vedantic views of enlightenment and the priority of intuitive knowledge (*jnana*). *Jnana* is certainly direct experience of ultimate reality, but, to use terms developed by one of our masters, Eric Voegelin, it is "noetic" rather than "pneumatic." That is to say, it addresses the mind, or knowing in a highly intellectual sense, more than the full human spirit, which is addressed by love. *Bhakti,* therefore, is the great counterbalance to *jnana,* and the interplay of these two Hindu versions of noetic and pneumatic direct experiences of ultimate reality is perhaps the backbone of at least the articulate body of Hindu mysticism.

Shankara and Ramanuja, noeticist and pneumaticist, are analogous to the Christians, Thomas Aquinas, on the one hand, and Augustine or Bonaventure on the other. Together, they invite us to see how either knowing or loving and how both knowing and loving take the human spirit toward its ultimacy, which turns out to be the ultimacy of all that is. The mysticism in this journey is the experiential core, where the dominant force is not ideas but encounters, participatory movements, heart to heart, being to being.

The description of mysticism with which we are working casts the Hindu mystics as more intellectual and monistic than many others, but it also casts them as deeper, more metaphysical or ontological in a praiseworthy sense, than most others. Along with some classical Buddhists, they remind us that Indo-European thought (Greek and Celtic as well as Aryan) has turned over the primordiality of being, of is-ness, with an especially, perhaps distinctively, acute touch, correlating this primordiality with human spirituality to make knowing and loving divine activities.[22]

NOTES

1. See R. C. Zaehner, *Hinduism* (New York: Oxford University Press, 1966), 1–13; also Mircea Eliade, *Yoga* (Princeton, N.J.: Princeton University Press/Bollingen, 1969), 3–7.

2. Romila Thapar, *A History of India 1* (New York: Viking Penguin, 1966); Percival Spear, *A History of India 2* (New York: Viking Penguin, 1970).

3. R. N. Dandekar, "Vedas," in *The Encyclopedia of Religion,* ed. Mircea Eliade (New York: Macmillan, 1987), 15:214. See also James A. Santucci, *An Outline of Vedic Literature.* (Missoula, Mont.: Scholars Press, 1976); Wendy Doniger O'Flaherty, *The Rig Veda: An Anthology* (London: Penguin, 1981); Satsvarupa dasa Gosvami, *Readings in the Vedic Literature* (New York: Bhaktivedanta Book Trust, 1977); and J. C. Heeszerman, *The Broken World of Sacrifice* (Chicago: University of Chicago Press, 1992).

4. Raimundo Panikkar, *The Vedic Experience: Mantramanjari* (Berkeley: University of California Press, 1977); for the ritualistic use of mantras, see Harold Coward and David Gor, *Mantra* (Chambersburg, Pa.: Anima, 1991).

5. William K. Mahony, "Upanisads," in *Encyclopedia of Religion,* 15:147. See also two classical introductions: Paul Deussen, *The Philosophy of the Upanishads* (1905; New York: Dover, 1966), and S. Radhakrishnan, *The Philosophy of the Upanisads* (London: George Allen & Unwin, 1924).

6. Robert Ernest Hume, ed., *The Thirteen Principal Upanishads* (New York: Oxford University Press, 1971), 9 (v. 9).

7. Sarvepalli Radhakrishnan and Charles A. Moore, eds., *A Sourcebook in Indian Philosophy* (Princeton, N.J.: Princeton University Press, 1970), 83; also Karl Polter, *Presuppositions of India's Philosophies* (New Delhi: Motilal, 1991).

8. Eliade, *Yoga,* p. 7. See also Mircea Eliade, *Patanjali and Yoga* (New York: Schocken Books, 1975), and Ernest Wood, *Yoga* (Baltimore: Penguin/Pelican Books, 1962). On yoga and mysticism, see Karel Werner, ed., *The Yogi and the Mystic* (London: Curzon Press, 1989).

9. Patanjali, *How to Know God: The Yoga Aprhorisms of Patanjali,* trans. Swami Prahhavananda and Christopher Isherwood (New York: Mentor Books, 1953), 36–39.

10. Geoffrey Parrinder, "Om," in *A Dictionary of Non-Christian Religions* (Philadelphia: Westminster Press, 1971), 208.

11. Patanjali, *How to Know God,* 97.

12. S. N. Dasgupta, *Hindu Mysticism* (1927; New York: Ungar, 1977), 66–67.

13. David Lorenzen, "Sankara," in *Encyclopedia of Religion,* 13:64. A good example of Vedantic influence on modern Hindu philosophy might be S. Radhakrishnan's influential work, *An Idealistic View of Life,* 3rd ed. (1932; New York: Barnes & Noble, 1964). For descriptions of Vedantic texts, see Eliot Deutsch and J. A. B. van Buitenene, *A Source Book of Advaita Vedanta* (Honolulu: University of Hawaii Press, 1971). On the general philosophical context, see Polter, *Presuppositions of India's Philosophies.*

14. David R. Kinsley, *Hinduism: A Cultural Perspective,* 2nd ed. (Englewood Cliffs, N.J.: Prentice-Hall, 1993), 96. One of the classical studies of Shankara (which compares him to the medieval Christian mystic Meister Eckhart) is Rudolf Otto's *Mysticism East and West* (1932; New York: Macmillan, 1960). For recent efforts at Hindu–Christian dialogue, see Howard Coward, ed., *Hindu–Christian Dialogue* (Maryknoll, N.Y.: Orbis Books, 1989), and Kana Mitra, *Catholicism–Hinduism* (New York: University Press of America, 1987).

15. Ainslie T. Embree, ed., *The Hindu Tradition* (New York: Vintage Books, 1972), 202–205.

16. John B. Carman, "Bhakti," in *Encyclopedia of Religion,* 2:130. See also Klaus K. Klostermaier, *Mythologies and Philosophies of Salvation in the Theistic Traditions of India* (Waterloo, Ont.: Council on Religion 1984). Some of the

psychic impact of *bhakti* on ordinary devotees may be gleaned from Sudhir Kakar, *Shamans, Mystics, and Doctors* (New York: Knopf, 1982).

17. R. C. Zaehner, trans., *The Bhagavad-Gita* (New York: Oxford University Press, 1973), 400.

18. Ibid., 305–306. On bhakti to Kali and Krishna, see David R. Kinsley, *The Sword and the Flute* (Berkeley: University of California Press, 1977).

19. A. J. Ramanujan, *Speaking of Siva* (Baltimore: Penguin Books, 1973), 70–71. The verses come from vacanas 59, 62, 64, and 70 in the Basavanal edition of 1962.

20. A. L. Basham, *The Wonder That Was India* (New York: Grove Press, 1959), 332.

21. Ibid., 332–333.

22. Good overviews of the Hindu spiritual tradition, with considerable attention to mysticism, include Krishna Sivaraman, ed., *Hindu Spirituality: Vedas Through Vedanta* (New York: Crossroad, 1989); S. Cromwell Crawford, *The Evolution of Hindu Ethical Ideals,* 2nd ed. (Honolulu: University of Hawaii Press, 1982); Charles A. Moore, ed., *The India Mind* (Honolulu: University of Hawaii Press, 1967); and Nikunja Vihari Banerjee, *The Spirit of Indian Philosophy* (New Delhi: Arnold-Heinemann, 1974).

3

Buddhism

General Orientation

We have described Buddhism as a challenge to Vedic Hinduism. Along with the Mahavira, who founded Jainism a generation before him, Gautama, the Buddha (perhaps 563–476 B.C.E., perhaps as much as a century later), asked for a reform of the sacrifices and outwardness that characterized the religion of his day. We deal with the enlightenment of Gautama himself in the next section, treating it as the paradigm of Buddhist searches for liberation. Here, the key historical point is that Gautama considered that he had found a better way than that provided by the popular Vedism of his time.

The history of Buddhism is the story of Gautama's way—how it fared, first in India and then after moving east; how it articulated itself through such pan-Buddhist forms as "the three jewels" (the *Buddha*, the *dharma*, and the *sangha* [the community]) and the three principal religious concerns, or "pillars" (meditation, morality, and wisdom); and how from these beginnings and developments it has come down the historical road to present times as "The Middle Way," a graceful humanism almost always poised to become an arresting mysticism.

The Vedic religion to which Gautama objected tended to stress sacrifice to the gods for material or spiritual benefits. The metaphysical foundations that Gautama wanted to overhaul included *karma*, *samsara*, and *moksha*. The (mythical) stories about Gautama's passage to Buddhahood describe him as a prince caught in an existential crisis. Having, through forays outside the protected palace, encountered old age, disease, and death, he wondered deeply about the meaning of life—whether anything could survive destruction, where a wise person would go to escape suffering. He found neither meaning nor escape in the Vedic system of sacrifices. The Vedic gods gave him no *moksha,* for they were as caught in the web of *karma,* as entrapped in *samsara,* as he and the rest of his culture were. So he left the luxury of the palace and his wife and child in search of a way to defeat suffering, a better way than that offered by the priests of Vedism.

The beginning of Buddhism as a social force, a somewhat organized religious way of life, was the moment when Gautama decided to teach others the nontheistic solution that he had found to the bedrock human problem of suffering. By choosing to accept disciples, Gautama (now thinking of himself as the Buddha, the "Enlightened One") turned his personal experience into a movement. Eventually, his followers developed many of the trappings that we find in other religious movements: canonical teachings, social institutions, stipulated or favored devotional practices. First, however, came his decision to share his enlightenment, a decision that apparently drew an enthusiastic response from many of his fellow Indians of the early fifth century B.C.E. Apparently, many of them were also dissatisfied with the Vedic sacrifices, the priesthood, and the welter of gods and goddesses. Apparently, many of them, too, sought something deeper, simpler, more luminous. Some Indians of Gautama's era turned to the *Upanishads,* perhaps even participated in creating the *Upanishads.* Others joined the community forming around the Enlightened One.

By the time of the Buddha's death, a full generation after his enlightenment, he had prepared a phalanx of disciples, more than enough to keep his movement going. The core of this phalanx, the powerhouse of his community, was monastic and male. Those who ruled the early Buddhist community imitated Gautama in seeking enlightenment as a full-time occupation. They quit the world of business and family life, practicing celibacy. They lived under a monastic rule that held them to material poverty, obedience to their religious superiors, and much attention to meditation. Ideally, they begged their food from supportive laity and led a wandering life, settled nowhere but in their quest for enlightenment, the liberating insight that would free them from *karma* and carry them to *nirvana,* the Buddhist equivalent of *moksha.*

Buddhist tradition says that shortly after the death of the Buddha his followers met in congress to consolidate his achievements and plan for the continuance of his community. The main lines of both doctrine and practice were clear from Gautama's own teaching and example, but, perhaps inevitably, disputes arose over how to interpret certain points, and different factions developed. Some traditions say that the Buddha only reluctantly admitted women to the *sangha,* largely on the urging of his venerable aunt, but eventually nuns joined the *sangha* and developed a Buddhist monasticism for women.

From the outset Gautama and the first disciples had relied on the material support of laypeople. In return for food, shelter, and other necessities, the Buddha and his leading disciples would instruct their lay supporters in spiritual matters. Whereas Gautama thought that gaining enlightenment was the crux of a successful life, he accepted the realities of *samsara* and granted that not all of his hearers would embrace the monastic life. Thus the full import of the word *sangha* became *all* followers, monks and nuns first but devout laity as well.

As his followers pondered both the Buddha's practice and his teaching, they placed diverse emphases on the rights and capacities of laypeople. Along with differences in how the followers conceived of saintliness, these diverse emphases led to the divergence of the Theravada and Mahayana schools. By the end of the last centuries B.C.E., the Mahayana schools had adapted Buddhist piety considerably to accommodate the needs of the laity. The Theravadists remained rooted in the old monastic traditions, which certainly had an orientation to help laypeople, while the Mahayanists often promoted devotional figures (*bodhisattvas*) and a this-worldly conviction that *nirvana* both lies outside *samsara* and works in its midst.

Politically, the years 273 to 236 B.C.E. were the high watermark for Indian Buddhists. This was the period when the Buddhist emperor Ashoka tried to run a regime based on the teachings of the Enlightened One. Gautama's way was nonviolent, stressing an absence of desire. Ashoka sought a gentle rule that would bring peace to all his subjects. Buddhist political dominance, in fact, proved short-lived, and the development of Hindu reforms (Upanishadic, yogic, Vedic) and theistic devotions (*bhakti*) eventually won India for Hinduism rather than Buddhism. Still, Ashoka remains a fascinating study in the possibilities for Buddhist political rule.

After Ashoka Theravadist traditions continued to exert a strong influence in the far south (Sri Lanka), but by the middle of the third century C.E., Buddhism had begun to move eastward. By the sixth century C.E. missionaries had carried Buddhism, the *dharma,* and the *sangha* to Burma, Cambodia, Laos, Vietnam, Korea, China, and Japan. Eventually, Buddhists became stronger in Eastern Asia than in their native India. (The general pattern in this missionary emigration has been that Theravadist traditions have held most sway in the lands closest to India, while Mahayanist traditions have held most sway in more Eastern Asian lands.)

By the end of the sixth century C.E., Buddhism had gained sufficient influence in Japan to be proclaimed the state religion. By the middle of the eighth century there were Buddhist monasteries in Tibet, where a distinctive set of schools, neither Theravadist nor Mahayanist but Vajrayanist, developed. By the middle of the ninth century Buddhists were being persecuted in China because they were perceived as threatening native traditions.

Regularly, Buddhist ideas and practices both called forth the opposition of native religious groups and amalgamated with them. In Japan the most dramatic interaction was with Shinto groups. In China both Confucians and Taoists adapted Buddhist ideas and were in turn adapted themselves. In Tibet native shamanistic practices (*Bon*) were the main focus of such mutual adaptation. Regularly, Buddhism bore both the intrigue and the stigma of being a foreign import. It offered new ideas and religious techniques, which numerous natives considered deeper philosophically and more useful ascetically than their traditional practices; but it also threatened traditional cultural patterns, which had shaped family life, burial rituals, political organization, work, and the rest of a local, holistic way of life from time out of mind. In times of emotional security Buddhism seemed a great enrichment. In times of emotional turmoil, when the Japanese or Chinese or Koreans lacked confidence, Buddhism could either seem a threat or become a scapegoat.

For more than the past thousand years, Buddhism has been an important feature of virtually all East Asian cultures. Generally, ordinary people have not felt obliged to choose between their Shinto (or Confucian or Taoist or native shamanistic) ways and those stemming from Gautama. They could be both Buddhists and Confucians. They could bury their dead in Buddhist funeral rites and structure their extended families according to Confucian principles. A few specialists—monks, nuns, priests—might try to live by purely Buddhist principles, but the local culture as a whole was usually a rich mixture of many different traditions, so interwoven after centuries of interaction that it was hard to discern a purely Buddhist or Taoist strand. Nowadays there are perhaps 325 million Buddhists worldwide, making Buddhism the fifth largest bloc (after Christians, Muslims, nonreligious people, and Hindus).[1]

We have mentioned some of the key ideas that have shaped the outlook of representative Buddhists, or the Buddhist worldview. From contemporary Indian culture Gautama accepted *karma, moksha* (*nirvana*), *yoga,* and other staples. His enlightenment can be viewed as a creative personal appropriation and reworking of the longstanding Indian ascetic tradition. The three jewels (*Buddha, dharma, sangha*) reset the Buddhist sense of this tradition.

One becomes a Buddhist, formally, by "taking refuge" in the Buddha, the *dharma,* and the *sangha.* The Enlightened One becomes one's privileged teacher, the unique source and exemplar of the truth, the light, that can free one from *samsaric* suffering. His *dharma* (teaching) is the truth of reality itself, as primordial as the Vedic revelation, the ancient disclosures of being to the Aryan seers; and his community keeps alive both his memory and his teaching. Generation after generation, it hands on the living spirit of enlightenment, detachment, and monasticism, as well as practical instruction in meditation, morality, and wisdom.

The Buddha, smiling serenely, is the personal, iconographic, somewhat historical focus of the Middle Way. Disciples hope that by following this Way they can become like the Buddha—enlightened, free, fulfilled, indeed Buddhas in their own right.

The *dharma* of the Buddha is summarized in his Four Noble Truths: (*1*) All life is suffering. (*2*) The cause of suffering is desire. (*3*) The way to make suffering cease is to stop desiring. (*4*) The way to stop desiring is to follow the Noble Eightfold Path of right views, right intention, right speech, right action, right livelihood, right effort, right mindfulness, and right concentration. The first three aspects of the Noble Eightfold Path suggest wisdom. Aspects four through six suggest morality. Aspects seven and eight suggest meditation.

The *sangha* of the Buddha has remained both monastic and lay. It has cared little for international uniformity or political cohesion, but virtually everywhere that Buddhists have emigrated (for example, the West Coast of North America) monks have followed and Buddhist "churches" have arisen. At the center of Buddhist orthodoxy and survival has been the personal relationship between gurus (usually monks or nuns) and lay disciples, as this has focused on the disciples' efforts to meditate. In teaching serious followers of Gautama to meditate, the *sangha* has assured that its views of wisdom and morality remained tied to the formative experience of the Buddha himself, his enlightenment.

The core of Buddhist morality (*sila*) stands forth in the five precepts that bind all Buddhists. No follower of Gautama's Middle Way is to kill, steal, lie, be unchaste, or take intoxicants. Naturally, Buddhist moralists have refined this core, elaborating it for monks and laity, men and women, people of one cultural area or another. Still, the precepts themselves suggest a pan-Buddhist ideal of nonviolence, freedom from desire, utter honesty, sexual control, and psychosomatic purity (simplicity, naturalism).

Buddhist wisdom, finally, appears in the worldview of the typical, representative disciple as an organization of life, a configuring of reality, as truly nirvanic and accidentally karmic. *Prajna,* wisdom, is symbolized as a goddess who stands beyond the karmic realm, free of desire, in the full though ineffable light of *nirvana,* the unconditioned state where the flame of desire has ceased to burn and to distort human existence. Wise Buddhists orient their course by this primacy of *nirvana.* Through meditation they strive to gain the enlightenment that makes *nirvana* plain so that it possesses them entirely. Through moral discipline they strive for action and a social life both conducive to *nirvana* and expressive of it. Inasmuch as they believe that *nirvana* is the gist of all reality, present and at work in the pseudoreality of *samsara,* astute Buddhists often manifest great peace and joy. The world itself and their personal realities are intrinsically good, full of light. To become a complete success, they need only let this goodness, this primary reality, this "suchness," emerge for them as as basic and primary as it is in itself.[2]

The Enlightenment of Gautama

Gautama, the historical Buddha, is the best candidate for such titles as the "founder," "model," or, to use a Western term, "incarnation" of Buddhist wisdom. Many lay Buddhists have prayed regularly to him, often as Shakyamuni (sage of the Shakyas), a title reflecting his clan origins, though they have also prayed to other Buddhas, such as Amitabha, the Buddha of Light, or Maitreya, the Buddha to Come. As Buddhism developed its sense of how the Enlightened One, or how any enlightened one, fits into the cosmic scheme of things, Gautama faded somewhat in significance. Buddhahood and Buddha nature (the intrinsic enlightenment in all beings) came to the fore, while all historical and personal embodiments of Buddha nature receded to the background.

Only the most sophisticated Buddhists, deeply immersed in dialectical studies of wisdom or intense meditation, pursued this line of development to the point where Gautama was dispensable. The development itself, however, reminds us that such comparable religious founders as Buddha, Confucius, Jesus, and Muhammad have come to occupy at least slightly different places in the regulative worldviews that their followers have developed.[3] Overall, Gautama probably has been to his followers more ontological (the incarnation of Buddha nature) than a sage like Confucius, yet less ontological than a Son of God like Jesus (not so unique or exclusive), and more sapiential than a prophet like Muhammad.

Gautama has also been perceived differently from each of his fellow religious founders because the niche he finally assumed in the "system" he himself had generated presented him as a nonunique, repeatable enfleshment of the enlightenment at the core of all beings. In addition to generating other Buddhas, his system retained samsaric circularity or repetition. The systems of the other founders were more linear, more "historical" in our modern sense of the term, where each individual is unrepeatably unique.

Historically, Gautama was the only founder of The Middle Way (in our era or *kalpa*). Metaphysically, he is not the only founder or exemplar or linchpin, though Buddhist thinkers can legitimately configure the cosmos, indeed the full sweep of reality, around him. What we study in this section is therefore both an experience traceable to Gautama and a paradigm of the mystical attainment available in principle to all human beings, indeed to all sentient beings inasmuch as *karma* may bring any of them to human form. It is both something carrying traces of his unique imprint and something that all enlightened beings have experienced.

We have referred to Prince Gautama's sheltered life in the palace and the encounters with old age, sickness, and death that led him to leave his wife and child and to pursue release from the human vulnerability to suffering. Literary sources do not exist for a modern biography of Gautama, but those that do exist report at least plausibly that he studied and experimented for some years before striking out on his own and finally winning victory. The (Hindu) teachers he consulted were masters of meditation, but they never led him to the *jnana* of *moksha*. The asceticism that he practiced, supposedly so severe that when he touched his navel he felt his backbone, never broke him completely free from karmic bondage. So finally he sat himself under a ficus tree and resolved not to leave until he had conquered ignorance and suffering, *samsara* itself, the painful human condition. He would either gain freedom or die trying. He would stake everything in a last definitive battle.

The traditional accounts of this battle show Gautama passing through the upper ranges of yogic attainment (gaining complete control of his body and mind, coming to know all of his previous lives, coming to realize the karmic condition or position of all living things) and then being tempted by Mara, the Indian Satan. The temptation was to refuse to share his rapidly developing enlightenment with other beings and instead to narrow in on his own private liberation. Gautama was bludgeoning the foundations of Mara's desire-driven kingdom. The last resistances were crumbling and the first scents of a new order were perfuming the air.

George Marshall, who has been so bold as to attempt a biography of the Buddha, has imagined the decisive leap that made Gautama the Enlightened One. It came when Gautama had rejected the temptation of Mara and vowed to remain in his fight until he gained the knowledge, the wherewithal, to liberate all suffering beings. This determination carried him to the final insight, which Marshall describes as follows:

> Suffering is the root of all evil. People suffer because they hold on to this existence too tightly, and believe that all life is inscribed in this one temporal existence. They see themselves as bounded by the finite limitations of their bodies, and of the brief years of growth, maturity and aging, end-

ing at death. Out of the illusion that this comprises all of life comes the sense of suffering. People cannot let themselves ride freely in the great whirling wheel of existence, which gives them not only this life but many others. Hence they are striving for success or subsistence most of the time, and pressing for fulfillment now. The result is the unhappiness seen on every side.[4]

In the *Buddhacarita*, a devotional summary of the key moments in the life of Gautama, traditionally attributed to the first-century C.E. poet Ashvaghosa, the crux of enlightenment is not precisely the insight that suffering comes from desire. It is that, for the wise, there is no self to generate desire, no subsistent "I" in which desire can repose. Quite succinctly, Ashvaghosa presents the achievement of Gautama as both reaching the peaks of sagehood and causing the entire cosmos to celebrate:

> When the great seer had comprehended that where there is no ignorance whatever, there also the karmic formations are stopped—then he had achieved a correct knowledge of all there is to be known, and he stood out in the world as a Buddha. He passed through the eight stages of Transic insight, and quickly reached their highest point. From the summit of the world downwards he could detect no self anywhere. Like the fire, when its fuel is burned up, he became tranquil. He had reached perfection, and he thought to himself: "This is the authentic Way on which in the past so many great seers, who also knew all higher and all lower things, have traveled on to ultimate and real truth. And now I have obtained it!"
>
> At that moment, in the *fourth watch* of the night, when dawn broke and all the ghosts that move and those that move not went to rest, the great seer took up the position which knows no more alteration, and the leader of all reached the state of all knowledge. When, through his Buddhahood, he had cognized the fact, the earth swayed like a woman drunken with wine, the sky shone bright with the Siddhas who appeared in crowds in all the directions, and the mighty drums of thunder resounded through the air. Pleasant breezes blew softly, rain fell from a cloudless sky, flowers and fruits dropped from the trees out of season—in an effort, as it were, to show reverence for him.[5]

A third, quite creative interpretation of the enlightenment of Gautama places it as the tenth within a drama of twelve acts. The tension builds from act eight, when Gautama chooses the bodhi tree as his mandala, his sacred circle within which he can labor safe from the assaults of evil:

> Siddhartha [another name for the historical Buddha] set out to find the Bodhi Tree, that particular specimen of *Ficus Religiosus* under which, it is said, all Buddhas have realized enlightenment (bodhi). Finding it, he went around it to determine the Bodhi-mandala, "the Enlightenment Circle," where he must sit. The ground sank and swayed out on all sides except the east: here was the center of the earth for him. He made a sitting-pad

> of sacred grass, and composed himself on it with the firm resolve not to move until enlightenment had been attained. This was the source of power, the pivot which does not shift, "the still point of the turning world." As long as he remained on it, he would be unconquerable.[6]

Act nine is the attack of Mara and Gautama's resistance. Mara is Lord of the karmic realm, *samsara*. He does not give up his sway lightly. So he assaulted Siddhartha with fear and hate. Then he tried greed and lust, sending his ravishing daughters to tempt Gautama to desire, but to no avail:

> Siddhartha looked at the luscious ladies, realized that they were in fact skin bags of bone and excrement, only temporarily delightful, and saw them decay into repulsive old hags. Mara was vanquished, but, bolstered by his huge army, he called upon the solitary Siddhartha to surrender, for he seemed to have no one to witness for him in the joust. "The Earth," said Siddhartha, "is my witness." He touched it with his right hand (a gesture often reproduced in Buddhist art) and it quivered in agreement. Mara left Siddhartha in peace.[7]

Act ten is the enlightenment that follows, now that Siddhartha has faced Mara down and gained his freedom from *samsara*. Notice how this summary of Gautama's enlightenment both repeats what we have seen in the first two accounts and goes beyond them:

> In the middle of the night, Siddhartha began to observe his own former lives, the lives of others, and the entire space–time continuum concentrated in an extensionless, eternal point. He saw the universality of suffering (*dukkha*), the pain of cyclic existence, in which beings trap themselves in ignorance and desire, like an animal walking around in a circle in a cage. Cutting the circle at the right point would bring liberation: he relinquished desire (attachment), desirelessness (aversion), and indifference (mixed attachment/aversion), and, as dawn broke upon him, cried, "Now is birth-and-death finished! The ridge-pole of that house built over many lives is broken!"[8]

Modern interpretations of the enlightenment of the Buddha clearly vary, depending on how the given interpreter understands both Buddhism and human psychology. Traditional Buddhist interpretation is controlled by Gautama's own exegesis, which reposes in three central forms of the teaching that he embarked upon after enlightenment. We have seen two of these forms: The Four Noble Truths and The Eightfold Path (the Fourth Noble Truth, which becomes an epitome of Buddhist practice). The third form in which Gautama expressed the gist, or the intellectual content, of his enlightenment was the doctrine of *Dependent Coarising*. In our textbook on the world religions we have described this doctrine as follows:

> Often Buddhists picture the doctrine of Dependent Coarising, which provides their basic picture of reality, as a wheel with twelve sections or a chain with twelve links (the first and last are joined to make a circuit). These twelve links explain the round of samsaric existence. They are not an abstract teaching for the edification of the philosophical mind, but an extension of the essentially therapeutic analysis that the Buddha thought could cure people of their basic illness.
>
> The wheel of Dependent Coarising turns in this way: (1) aging and dying depend on rebirth; (2) rebirth depends on becoming; (3) becoming depends on the appropriation of certain necessary materials; (4) appropriation depends on desire for such materials; (5) desire depends on feeling; (6) feeling depends on contact with material reality; (7) contact depends on the senses; (8) the senses depend on "name" (the mind) and "form" (the body); (9) name and form depend on consciousness (the spark of sentient life); (10) consciousness shapes itself by samsara; (11) the samsara causing rebirth depends on ignorance of the Four Noble Truths; and (12) therefore the basic cause of samsara is ignorance.[9]

From ignorance to aging and dying, samsaric existence is a chain that most human beings never break. From aging and dying to ignorance, it is a wheel running them over, pushing them down into the ruts of karmic suffering. The plausibility, perhaps even the intelligibility, of the relations among the twelve links of dependent coarising depends on the Indian philosophy of Gautama's time. What is clear, though, is that the Enlightened One found ignorance to be as basic as desire. What desire is to the will or core personality, ignorance is to the mind. The Four Noble Truths address the ignorance that keeps people in *samsara.* If people will embrace them, dedicating themselves to the program implied in the Fourth Noble Truth (agreeing to walk along The Noble Eightfold Path), they will gradually see how these truths dispel ignorance, just as they will feel how enacting the program slowly dispels slavery to desire and offers liberation. Gradually, it will become credible that *nirvana* is the human vocation. Slowly, steadily, it will make sense that reality in itself and human seekers in themselves are essentially light, being, and goodness and are only accidentally ignorance, mortality, and evil.

The Buddhist mystics regularly recapitulate the enlightenment of Gautama. With expectable variations, they replay the realization of The Four Noble Truths or The Noble Eightfold Path or Dependent Coarising that is at the center of Gautama's victory under the bodhi tree. Much as the death–resurrection of Christ has been a structural element in Christian mysticism, so the enlightenment of Gautama, his clarification of the role of desire and ignorance through conquering them both, has been a structural element, a cornerstone or central arch, in Buddhist mysticism.

Theravada

The three principal Buddhist schools, Theravada, Mahayana, and Vajrayana, have sponsored vigorous traditions of meditation. By implication, all have nurtured Buddhist mysticism, which needs healthy roots in morality and wisdom but flowers most directly through meditation. The direct experience of ultimate reality that Buddhists seek is an enlightenment, a realization, a *satori*, that can come while one is strolling through the park or watching a cherry blossom fall or taking to heart the suffering of a sick friend. It need not flow from years of intense meditational practice, but most Buddhist masters assume and expect that such practice is the best, even the normative, way to prepare for it. Certainly, some masters urge people to strive for enlightenment with might and main, while others counsel an approach in which meditation is "just sitting," a privileged exercise of the basically enlightened being that one is always. In other words, there is considerable variety of opinion on which the student of Buddhist mystical experience can draw. Let us, then, consider an authoritative sage from the Theravadist tradition to begin the task of coloring mysticism in Buddhist hues.

One of the most important texts on meditation in the Theravadist tradition is the *Visuddhimagga* of the fifth-century C.E. sage Buddhaghosa, written while he was laboring in Ceylon (now called Sri Lanka). Edward Conze's popular introductory work *Buddhist Meditation* draws heavily upon the *Visuddhimagga* and notes that at one point Buddhaghosa lists forty topics for meditation that had become standard by his day.[10] Perhaps listing these forty topics will provide us with an inventory of what Buddhists of the Theravadist tradition focus upon when they settle themselves in the lotus posture, regulate their breathing, clear their minds, and begin to deal with topics that they suspect can lead them out of samsaric illusion toward enlightenment and the personal appropriation of The Four Noble Truths.

Conze's digest of the forty topics begins with ten "devices": earth, water, fire, air, blue, yellow, red, white, light, and enclosed space. These devices appear to be circumstantial matters, representing the physical world in which meditators find themselves. By penetrating to the essence of such circumstantial matters and finding how they can be objects of desire and so enslaving, meditators (people pursuing what Buddhagosa called *samadhi*) can begin to put such devices in perspective and estimate the significance they should and should not receive.

A second standard set of ten subjects for meditation collects "repulsive things," all closely connected with death: swollen corpses, bluish corpses, festering corpses, fissured corpses, gnawed corpses, scattered corpses, hacked and scattered corpses, bloody corpses, worm-eaten corpses, and skeletons. Now, in many religious traditions death is the great teacher. "Look to the sure end of bodily human existence," many religions have said, "and you will intuit in germ all that you need to make your way through life soberly, with proper detachment."

Coming to our present case, look at death with a strong Buddhist desire to eliminate desire, and you will find the great lever: Who can take any passing pleasure or achievement fully seriously so that it binds him or her karmically to the samsaric world, while realizing that nothing this-worldly is stable or completely fulfilling or able to deliver the joy and peace that the human heart craves? By making death, in the image of the corpse, repulsive—by emphasizing its rotting, decaying nature—meditation masters such as Buddhaghosa have tried to detach their disciples from lust, gluttony, vanity, and a host of other vices, large and small, that depend on taking the body seriously, as though it were important, lasting, worthy of homage. In other words, they have tried to make the body—something so intimate that we have to include it when we think about our ordinary identity, something on which we are bound to lavish much care and worry—a constant reminder that nothing physical can bear a life's weight or is worthy of passionate desire.

A third group of ten subjects suitable for Theravadist meditation is composed of "recollections," topics that faithful Buddhists will bring to mind on a regular basis: the Buddha himself, the *dharma,* the *sangha,* morality, liberation, the *devas* (gods), death, the body, breathing, and peace. The first three, as we have noted, are the three "jewels" at the core of Buddhism. In remembering Gautama, disciples recall their historical beginnings and the prime exemplar of The Middle Way. The teaching that has flowed from Gautama, whether one takes it in such narrow gauge as The Four Noble Truths or looks at the full range of the Pali Canon, the huge collection of Theravadist scriptures, elaborates the implications of an enlightened view of reality. The *sangha,* both monastic and lay, is the social context for all Buddhist thought and activity, whether it bears on wisdom, morality, or meditation.

Morality (*sila*) condenses itself into the five negative precepts that we have noted: not to kill, lie, steal, be unchaste, or take intoxicants. However, one can also meditate on any of the many extensions of *sila* that Buddhist commentators have developed, for example, the rules peculiar to monks. Liberation is *nirvana,* the great goal of all Buddhist striving. The *devas* are the gods, who can help one's efforts but are ultimately irrelevant in the great battle, which involves the conquest of self. As we have noted, death is the standard teacher in religious matters, as prominent in Buddhist ascetic tradition as in other traditions concerned passionately with human destiny and fulfillment.

As noted also, the body both is and is not the concrete definer of what human beings can be. We must serve the body through eating, resting, working, and like activities that maintain it. For mainstream Buddhism, we must do this in a balanced way, keeping the body both healthy and disciplined so that it serves our spiritual ambitions rather than thwarting them. However, the body is less than the spirit. The mind is more significant than the belly or loins. Respiration is the link between body and spirit, flesh and mind. When we breathe regularly in meditation, we harmonize the body with the spirit, producing momentarily what we would constantly maintain were we living as the Enlightened One did. Lastly, peace is perhaps the most characteristic feature of the Enlightened One, adorning his face and emanating from his entire being. He is at rest in the light that his victory under the bodhi tree has brought him. He has no desire driving him, no need compelling him, no force from either the front or the back marring his serenity. He has no special stake in the work that he does, the acceptance or rejection of his teaching. He is where he wants to be and has been made to be, where all beings can find that they are intrinsically perfect.

Four quite positive subjects for meditation have gone by the traditional Buddhist name "stations of Brahma." They are friendliness, compassion, sympathetic joy, and evenmindedness. The metaphor is that Brahma, the first god in the traditional Indian (Hindu) trinity, resides in these virtues. All four are attitudes that one can bring to social life. All four are also basic dispositions that one can bring to the entirety of existence, which includes non-human beings. Buddhist compassion (*mahakaruna*) is the signal virtue of the *bodhisattva*, the Mahayanist saint who postpones *nirvana* to labor for the salvation of all living things. It is present in the Theravadist saint, the *arhat*, as well, usually not with so cosmic an accent as what we find in descriptions of the *bodhisattva* but under the conviction that anyone who realizes the gist of Gautama's enlightenment and consequent teaching will manifest the compassion for which Gautama was so notable.

Theravadist meditation masters have also traditionally offered their disciples four topics known as "formless states": endless space, unlimited consciousness, nothing whatsoever, and neither perception nor nonperception. With these four states, or "stations," we enter more fully the potential of the human spirit to transcend imagination and operate by a purer reason, for while any thought about these states is sure to generate images, usually pale or weak, the center of such thought is an idea, an outreach of the unlimited portion of the human makeup toward the "all" that provides us with a horizon of infinity.

Endless space abridges the rights of matter by breaking through its limitation, its borderedness, which at first we may naively assume is intrinsic to matter but later may intuit need not be. Unlimited consciousness abridges similarly the rights of the imagination and also of the ordinary ratiocinative mind by suggesting that there is an awareness more primitive and comprehensive than what we can picture fancifully or what we can gain by plodding from premises to conclusions. Intuitive knowledge, the *jnana* so dear to Hindu philosophers such as Shankara, is a prime specimen of this unlimited consciousness, as is the *samadhi* that Patanjali places at the depths of yogic attainment.

Buddhist enlightenment, and the wisdom (*prajna*) that it always entails, is an awareness similarly unlimited because resident on the far side of *samsara,* in the unconditioned state (*nirvana*) that one enters when the flame of desire has blown out. This state may be described as nothing (no-thing) whatsoever. It is indefinite, indeterminate, more like an all than a something, but too primal, pervasive, and simple to be denominated clearly or set in its place in the world's pile of playthings. It is neither something that we can perceive (grasp exactly, conquer, and pigeonhole) through our senses or upper minds nor something that we cannot perceive in this way. We can never master it with our upper, perceptual faculties but only engage it well, spirit to spirit, from our depths. Still, it is so real, so much the source and foundation and even stuff of the ordinary world, *samsara* (which, as the enlightened see, has whatever reality it does have only from *nirvana*), that it is in our midst, ingredient in our perceptions, closer to us and more operative on us than we suspect, until we have gained enlightenment.

Topic thirty-nine in Buddhaghosa's list of forty standard foci for meditation is a perception of the disgusting aspects of food. For monks, this meditation would have been an enticement to eat with detachment, keeping to the light diet, sufficient for nourishment but no kindling agent for gluttony, that the *sangha* has usually sought. The traditional precept "no belly more than two-thirds full" and the traditional practice of not eating after mid-day suggest what long-standing monastic asceticism in the matter of food has been. The Buddha had found that starving himself did more harm than good, but he was also convinced that eating, like most other aspects of bodily life, was best dealt with functionally, lightly, by making it so nondescript, so matter-of-course, that it never distracted monks from their prime business, which was spiritual liberation, most directly through meditation.

The fortieth topic is analysis. Specifically, the disciple takes the human body as a compound of the four primary elements prominent in then-contemporary cosmology: earth, air, fire, and water. By breaking this compound down analytically, disciples could prepare themselves for the insights that the body is not the essence of the human being and that nothing bodily justifies the notion of a self, an *atman.*

In the Buddhist perspective the body, indeed the body–mind composite, is simply a temporary bundling of elements (*skandhas*) that give us both a place in the physical world and human consciousness. These elements unbundle at death but reassemble in new ways if we have not broken fully the karmic affinities among them. We suffer the overwhelming illusion that they constitute a unique identity, make up a human self both substantial at the core and capable of accidental change through the life cycle, but in fact when they come unglued (through our analytic anticipation of their dissolution at death) we realize that our core was always empty, a participation in nirvanic nothingness. By analyzing the body into frangible parts, the fortieth meditation prepares the way for a liberating insight into our no-selfness.

The *Visuddhimagga* is a complete treatise on consciousness, a classic of the *Abhidharma* (Pali *Abhidhamma*) literature, the portion of the Buddhist scriptures dealing with psychological and philosophical analysis, especially of the experiences that one meets in meditation and the realities (*dharmas,* in the sense of entities, either mental or physical) that meditation spotlights. As such a complete treatise, it deals with much more than forty traditional topics on which to focus one's meditational awareness. It behooves us to sketch briefly the entire contents of the *Visuddhimagga* to suggest the full context of, for example, a meditation on friendliness or death.

Buddhaghosa begins with a study of virtue—strength of mind and will, vigor, and health in the fight for enlightenment. First, he deals with definitions and kinds of virtue, then with what defiles virtue, and finally with the ascetic practices that develop virtue. Part Two of the *Visuddhimagga* treats concentration, focusing the mind, developing a Buddhist yoga. Here, the main subtopics are, first, a detailed description of concentration, developed from a careful study of what goes on in typical meditations (on topics such as the forty that we have noted); second, an illustration of concentration through examining the *kasina* (universal idea) of the earth—the import of the element earth for meditation; third, a consideration of the other *kasinas*—water, fire, air, blue, yellow, red, white, light, and limited (enclosed) space—most suitable for meditation (recall that these ten *kasinas* comprised our first grouping of traditional topics for meditation); fourth, an illustration of concentration through meditation on foulness; fifth, a description of six recollections; sixth, a description of four more recollections (these six and four recollections comprise the third group of traditional topics for meditation that we described earlier); seventh through ninth, further illustrations of aspects of concentration.

Part Two of the *Visuddhimagga,* then, is a nearly exhaustive survey of what the meditator can do practically to get down to the business of transforming consciousness from ignorance to enlightenment. Buddhaghosa assumes that those beginning meditation will be disciples of masters, but he offers far more content than Patanjali did in similar circumstances. Whereas Patanjali merely set down *sutras,* knots on a string summarizing traditional teachings, Buddhaghosa gives full descriptions of each of the topics that meditation ought to handle. Indeed, he does not leave the matter of concentration, does not complete the second major part of his treatise, until he has described the benefits of concentration (including the supernormal powers that it can bring), the direct knowledge that the meditator can gain (including the recollection of past lives), and "the divine eye" (knowledge of the karmic state of all living things). If we recall the quasi-canonical traditions about the powers that Gautama gained during his ascent to enlightenment, we see that Buddhaghosa has fitted his descriptions of the results of an ideal concentration to the paradigm of Gautama's victory under the bodhi tree.

Part Three of the *Visuddhimagga* moves from the capital experiences of meditation to the understanding that grows with progress toward enlightenment. After defining "understanding" and surveying its different levels, Buddhaghosa explains the soil from which understanding grows, the faculties of the senses and mind. Then he explains The Four Noble Truths and describes dependent origination as the conclusion or acme of enlightened understanding, how the person whose meditational concentration has succeeded fully views the world. Finally, Buddhaghosa explains how to purify (keep developing) the understanding generated through successful meditational practice by deepening one's appreciation of the relations between mentality and materiality, by overcoming doubt, by clarifying regularly what lies on the proper Buddhist path and what is peripheral or a detour, by working on further meditational topics (to increase one's knowledge and vision of the Way), by dealing with more advanced aspects of both knowledge and vision, and, finally, by reminding oneself of the great benefits of meditational understanding (ultimately, *nirvana*).

The major concerns, then, of this masterwork of Theravadist religious philosophy are virtue, concentration, and understanding. We might describe them, respectively, as preparation (of character), meditational practice, and meditational fruits. They take a highly mental, ideal view of the individual seeker, of the activity crucial to liberation, and of the knowledge that liberation brings. The first chapter of the *Dhammapada*, a much beloved book of simple moral advice found in the Theravadist (Pali) canon, puts all this mentality succinctly: We are how we think. If we think good thoughts, we shall be good and happy. If we think evil thoughts, we shall be evil and unhappy.

Buddhaghosa gives a full, analytical elaboration of what is essentially the same conviction. To gain liberation, the being–awareness–bliss of *nirvana*, he would have us grasp as firmly and profoundly as we can how to orient our beings toward the mental light of suchness, the world as it really is; how to concentrate our awareness to best facilitate the dawning of that mental light; and how to estimate the knowledge that we gain, the yield that our successful orientation and concentration brings. In the course of his treatise, he collects an immense amount of traditional Buddhist lore, the fruits of centuries of meditational experience. Underneath the dry exterior of his many lists of topics live moist, vital clues to a maximal human awareness, Buddhahood itself, or mystical fulfillment.

Students interested in testing the psychology that Buddhaghosa lays out might do best by attempting either of the two meditational complexes that Buddhists have considered unfailingly advantageous: that on friendliness and that on death. Noting how practitioners have traditionally striven to appropriate a profound Buddhist appreciation of either of these topics, students can easily sense how deeply Buddhaghosa invited his readers to go. The foundation of friendliness, great compassion, is nothing less than the positive nature of reality itself, its nirvanic, intrinsically enlightened character. The same for the foundation that death touches. When we have realized the implications of human decay, we see that our most important aspects are not our bodies but our minds or spirits, the light that is our own suchness, the *nirvana* in us that both concentrates and is concentrated upon.

We shall grapple with the significance of Buddhist mysticism more fully later, but here we note that the yield that Buddhaghosa hopes, indeed expects, to reap from concentration, in understanding, comes close to our working definition of direct experience of ultimate reality. Meditation aims to purify the mind so that it becomes utterly realistic. Utter realism entails both an awareness of all the aspects of samsaric reality, its diversity of mental and material forms (including, especially, the impediments it raises to enlightenment), and an awareness of the undistorted suchness that lies under, beyond, and in the midst of *samsara*—of the reality that the wise, rather than the ignorant, appreciate.

Buddhaghosa's mysticism, if he would ever use the term to describe the attainments he promotes, is not ecstatic or dramatic. It has a place for supernormal powers, but that place is certainly peripheral to the main thrust of concentration, perhaps even accidental to the main achievement of understanding. Buddhist mysticism in his model is more "enstatic" than ecstatic, as it is for most other Indian, yogic ventures in liberation. What flowers in Buddhist realization or enlightenment is the potential self-possession of the enspirited being, though what is "possessed" is not a self in the sense of a solid "own being" but rather an "emptied" or extinguished former candle of desire that now exists without conditions, fully free.

Mahayana

We have mentioned that shortly after the death of Gautama Buddhists divided somewhat in their interpretations of relatively minor points of doctrine and discipline, leading to the split between the Theravada and Mahayana schools. Although all Buddhists continued to reverence the Buddha, the *dharma*, and the *sangha*, as well as to base their lives on

The Four Noble Truths and to imagine the world in terms of dependent coarising, Mahayanists thought that their way was broader than that of Theravadists and better suited to the needs of both the laity and nonhuman creatures. In their view, the *bodhisattva,* who postponed entering *nirvana* to labor for the liberation of all creatures, more fully exemplified the compassion of the Enlightened One than did the Theravadist saint (the *arhat*), whose social concern was less manifest. The Mahayanists developed fuller devotions for the laity and also a more profound metaphysical analysis of the relations between *samsara* and *nirvana,* at least some of it motivated by a desire to show lay people how ultimate reality was not beyond them but rather was the stuff of their daily lives.

The monastic lives of the two branches differed little, as did their formal meditational practices. All Buddhists remained held to the five essential ethical precepts of *sila.* So, to catch the distinctive overtones of Mahayanist Buddhist mysticism, one does well to move into wisdom, the understanding of ultimate reality and the mind that the foremost Mahayanist teachers developed. Two notions worth stressing in this regard are "emptiness" and "mind-only."

From early times, Buddhist teachers trying to detach their disciples from the experiences of the samsaric, unenlightened world described the elements of that world as characterized by three tell-tale marks. All were (*1*) painful, (*2*) fleeting, and (*3*) selfless. When they generalized about these three marks, they reflected on emptiness. The following might be representative meditations on the three marks and emptiness:

1. Nothing that we find in space and time, nothing that we pursue with desire, is a pure pleasure or profit or joy. It is always flawed, finite, unable to give our bodies or minds or hearts complete or permanent fulfillment. Therefore, it always brings us at least the minimal pain of frustration and alienation. Once again, we have failed to find our destiny, to gain our true home of being, awareness, and bliss.

2. Equally, all the things of ordinary experience are fleeting. None lasts or can be leaned upon; all fall apart, change, decay, or die. Samsaric life is quintessentially change. It is flux rather than substance, becoming rather than being. One can try to move with it and change alongside it, but that merely brings out the phenomenal, unanchored character of one's own human existence. Samsaric life is a running rather than a resting. Living it, one is bound to grow weary, sick at heart or spirit. Do all the grasses wither, all the flowers fade? Is there no field beyond the turn of the seasons, beyond death and rebirth?

3. The selflessness (*anatman*) of all samsaric existence becomes clear when we investigate the substance of the things that we experience as painful and fleeting, as well as when we turn our gaze inward and inquire about our own substance, the core of our own spirit, or self. Even there, we find, nothing is permanent, unstamped by variation.

Even there, nothing is without pain or purely joyous. We are not the beings we would like to be, and the change in us that we cannot avoid or control more than superficially can be for the worse as easily as for the better. "I" does not name the same being today that it did twenty-five years ago. "I" may be richer, fatter, less kind, or the reverse. In fact, there is no "I," as Buddhist meditation masters see things. There are only the *skandhas,* the "heaps" that stick together precariously throughout one turn of the samsaric wheel but come unglued at death.

Generalizing about these marks, the leading Mahayanist sages summarized in a single word their view of the reality that the unwise, or unenlightened, deal with desirously: "empty" (*sunya*). Thus the *Heart Sutra,* a brief but representative specimen of the vast assortment of devotional speculation known as the *Prajnaparamita,* the "Wisdom That Has Gone Beyond," begins with a foundational perception of emptiness:

> Homage to the Perfection of Wisdom, the Lovely, the Holy! Avalokita [Avalokitesvara], the Holy Lord and Bodhisattva, was moving in the deep course of the Wisdom that has gone beyond. He looked down from on high. He beheld but five heaps, and he saw that in their own-being they were empty. "Here, O Sariputra, form is emptiness and the very emptiness is form; emptiness does not differ from form, form does not differ from emptiness; whatever is form, that is emptiness; whatever is emptiness, that is form; the same is true of feelings, perceptions, impulses, and consciousness." [11]

As a prayer, the *sutra* begins with homage. It addresses the perfection of wisdom as a goddess. We may interpret her as the lovely face of the knowledge, the enlightenment, that leads us to ultimate reality. She is holy—pure, fully real, no resident of this lower, samsaric world in which everything is tainted. By orienting our minds and hearts toward her, we prepare our meditation for maximal benefit.

Avalokita is a famous *bodhisattva,* one of the foremost enlightened beings who has postponed entry into *nirvana* to labor for the liberation of all other beings. He lives in the fathomless sea of the *Prajnaparamita.* What he has to say comes from the fullness of reality, the beyond where ignorance and desire no longer mottle either perception or being. Looking down upon our lower world, he beholds no people, no substances, nothing but what is painful, fleeting, and selfless. He beholds only the *skandhas,* the five provisional, temporary heaps. Generalizing, he sees that in their "own-being," what we might call the identity at their core, they are empty. The implication for meditation is that one should try to appreciate the no-thing-ness of all experience. The implication for morality is that one should cling to nothing and detach one's desire from everything because one has been convinced by the wise that all things are empty. Both mentally and morally, a grasp of emptiness should be a catapult to freedom.

Thus far, Mahayanist speculation has not advanced greatly beyond what one might find with the Theravadists, for example, with Buddhaghosa. The generalizing of the three marks by focusing on emptiness may be a characteristic, even a distinctive Mahayanist move, but it is not especially dramatic or radical. More distinctive is the dialectical cast of mind that we find in the description of the relations between the *skandhas* and emptiness. Lecturing Sariputra, one of Gautama's favorite disciples, Avalokita makes it clear that one cannot separate the *skandhas* and emptiness as though they belonged to two separate realms. Form, feelings, perceptions, impulses, and consciousness are shot through with emptiness, while emptiness, in turn, manifests itself through these *skandhas.*

Dialectically, the mind being carried by the *Prajnaparamita* toward enlightenment comes to think in terms of both/and, neither/nor, and much silence. At term, it can say that *nirvana* and *samsara* are one. In saying this it does not mean that desire and *karma* and suffering no longer exert great influence. It means that for the wise, those seeing mystically, those who hunger to know the real and live by it, there is only Suchness, the way that things are. The way that things of our experience are is empty. However, if we deal with their emptiness actively, meditatively, in effect emptying them out by letting go of our illusions, we can move into their emptiness and find *nirvana,* the unconditioned, the fully emptied that we may equally call the completely replete, perfect, without lack.

Mahayanist dialectics want to defeat the ordinary, ignorant mind so as to heal and save it. The ordinary, ignorant mind does not appreciate emptiness or think holistically, with the depth and simplicity necessary to engage with nirvanic ultimacy and so find liberation. Meditating, not simply playing in a logic class, the devout Mahayana Buddhist might be carried to a point of arrest, where the mind stops and the spirit simply is, abiding with quiet, unworking attention. There, it might be possible to intuit the ineffable mystery of what makes anything or everything be. There, the disciple might see that when grasping stops ultimacy reveals itself.

Nagarjuna (ca. 150 C.E.), perhaps the foremost Mahayana religious philosopher, is famous for his efforts to defeat the mind dialectically. Associated with the Madhyamikas, a centrist school focused on the *Prajnaparamita,* Nagarjuna underwrote much of what later became Zen and Tibetan points of view. The following analysis of time is a good example of his nearly ruthless assaults on ordinary thinking. Remember that the point to such analyses is, in our terms, mystical. This analysis is not a scholastic exercise as much as it is an effort to clear the way to direct experience of ultimate reality:

> 1. If "the present" and "future" exist presupposing "the past," "the present" and "future" will exist in "the past."
> 2. If "the present" and "future" did not exist there [in "the past"], how could "the present" and "future" exist presupposing that "past"?
> 3. Without presupposing "the past," the two things ["the present" and "future"] cannot be proved to exist. Therefore neither present nor future time exist.
> 4. In this way the remaining two [times] can be inverted. Thus one would regard "highest," "lowest," and "middle," etc., and oneness and difference.
> 5. A non-stationary "time" cannot be "grasped," and a stationary "time" which can be grasped does not exist. How, then, can one perceive time if it is not "grasped"?
> 6. Since time is dependent on a thing (*bhava*), how can time [exist] without a thing? There is not any thing which exists; how, then, will time become [something]?[12]

One would need a specialist in Madhyamika dialectics to lay bare the full subtlety of this analysis, but the gist is clear. Every attribute of a partial, samsaric reality is questionable precisely because such a reality is both partial and related to other partial realities, which together compose a system or network. Thus nothing that we experience is independent, utterly clear or fixed. Not even such basic entities as time and space can support or meet our craving for stability, certainty, definition.

Moreover, the relations between past, present, and future are tricky. Time is not time unless it flows, yet if time flows, how can we grasp it? Similarly, we tend to attach time to things, yet meditational analysis shows that there are no things. How then does time in fact work in our world? Can we ever know? What then happens to this supposedly familiar reality, time?

What Nagarjuna wants is to have time explode. As disciples learn about the arbitrariness of all mental categories, which reflects the related and empty character of all samsaric realities, they may stop creating such categories. They may back away, leave their primitive either/or mind frame, and begin to sense from their innermost spirits something better.

Time is only a wave on the ocean of suchness and the wisdom that has gone beyond. The ocean exists through the waves, but no wave is more than a passing, provisional focus. The point to Buddhist discipline is to take one beyond the stream of suffering. The point is to find and join the ocean of suchness, *prajnaparamita, nirvana*. The ignorance of the unenlightened mind is the greatest obstacle to gaining this point and realizing this goal. If we can overcome such ignorance by blowing ordinary logic apart, we should do it. Certainly, other masters have proposed other ways to rout ignorance, most of them less intellectual than that of Nagarjuna. However, this master has helped a great many disciples. As well, he has stimulated some of the most profound Buddhist metaphysical reflections (characteristically, on emptiness).

If "emptiness" is the watchword of Nagarjuna, the Madhyamikas, and the *Prajnaparamita* literature, "mind-only" is the watchword of Vasubandhu, the Yogacarins, and the *Lankavatara Sutra*—a second major movement within Mahayanist metaphysics and mysticism. Vasubandhu, an Indian monk of the fourth century C.E., began in the Theravadist tradition but developed ideas that flourished more vigorously in Mahayanist circles. In the following excerpt from one of his tracts, we may glimpse how the passionate insistence of the Yogacarins that all reality is mental (that only *idea* or *light* is ultimate) was intent on freeing disciples from dualistic illusions (for example, that one ought to distinguish between a subject that knows and an object that is known):

> But when cognition no longer apprehends an object, then it stands firmly in consciousness [mind]-only, because, where there is nothing to grasp there is no more grasping.[13]

The cognition that interests Vasubandhu, the knowing or exercise of intelligence, is not concerned with objects, either material or conceptual. It is pure, active, intransitive—light without an object, target, or tether. It is an awareness that simply is, or flows, grasping nothing, finding nothing to grasp. This cognition, reminiscent of the *jnana* of the Hindu mystics and the flood of light described in Buddhist accounts of Gautama's awakening, is for the Yogacarins the best indication of what reality, or ultimacy, is like, of what suchness itself consists. Mind-only is the Yogacarins' best index or specimen of *bodhi*, the wisdom that came to Gautama under the bodhi tree and made him the Enlightened One.

Practically, the Yogacarist proposition is that if we focus on pure mind or intellectual light, and on that only, we shall not get lost in dualisms, material or mental. We shall hold to the course, stay in the stream, of enlightening wisdom and so be carried across suffering into ultimate wisdom and *nirvana*. Yogacarist mysticism, then, is distinctively idealistic. By experiencing directly that ultimate reality is a simple mental light, the Yogacarins became convinced that idealistic (mind-ordered) meditation on texts such as the *Lankavatara Sutra* was the best way to bring Buddhist seekers to perfection.

Vajrayana

The third Buddhist tradition or school is one that has roots in the Indian tantric tradition but that flourished also in Tibet. The Indian tantric tradition employed "magic" in the search for enlightenment. Sacred sounds (*mantras*) and sacred spaces (*mandalas*) have intrigued Tantrists, as has a general sense that enlightenment requires liberating the most powerful psychosomatic energies. For example, Tantrists have sometimes deliberately violated prohibitions against eating certain foods, such as meat, drinking intoxicants, and engaging in sexual intercourse. By taking control of the aversions and repressions, as well as the positive libidinal energies involved in these matters, they have tried to set explosive charges deep in the personality of the one seeking enlightenment to blast away ignorance.

In Tibet Indian traditions have been more influential than Chinese traditions and Mahayanist thought more powerful than Theravadist thought. Consider, for example, the following excerpt from a song of Mila Repa, one the most fabled Tibetan Buddhist saints:

> It is in such a lonely place as this, that I, the yogin Mila Repa, am joyous in the Clear Light of realization of the Void, joyous exceedingly at its many ways of appearance, joyous at its greatness of variety, joyous with a body free of harmful karma, joyous in confusion of diversity, joyous midst fearful appearances, joyous in my freedom from that state where distractions rise and pass away, joyous exceedingly where hardship is great, joyous in freedom from sickness, joyous that suffering has turned to be joy, joyous exceedingly in the mandala of spiritual power, joyous in the dance of bringing offerings, joyous in the treasure of triumphant songs now uttered, joyous exceedingly at the sounds and signs of multitudinous syllables, joyous at their turning into groups of words, joyous in that sphere where mind is confident and firm, joyous exceedingly at its spontaneous arising, joyous at its manifestation in diversity.[14]

The prototypical Tibetan saints were eccentrics. Stereotypically, they apprenticed themselves to gurus who abused them, and they wandered in the snowy mountain wastes, striking their ordinary contemporaries as crazed. Indeed, one of their basic tenets was that wisdom seems crazy to ordinary people and that bizarre behavior is a fitting sign of enlightenment. To the fore in this song of Mila Repa is the great joy that enlightenment has brought him. Compared to it, any disparaging judgments of ordinary people are but trifles.

Generalizing about the wandering saints who have been the great heroes in Tibetan Buddhism, Stephan Beyer has said:

> During and after the Gupta Dynasty, a new type of contemplative appeared in Buddhism: long-haired, mad wanderers who sought an enlightenment they called the Whore, for she opened herself to all who came to her. These wandering saints mocked the established traditions of the monastic institutions, and sang their mystic songs and riddles in the language of the marketplace. Their enthusiasm could infect even the academics themselves. The story is told, for example, of how the sober scholar *Naropa* gave up his academic career to wander the road seeking the half-clad and crazy *Tilopa* as his guru.[15]

However, the Tibetan saints honored the orthodox Buddhist convictions about the three marks and emptiness. Notice the presence in Mila Repa's song of staple Mahayanist doctrinal elements: realization of the void (emptiness), freedom from *karma,* and a dialectical view of sickness and suffering (what seems negative may become positive if it brings wisdom). Notice also the indications of a ritualistic Buddhist practice: the mandala of spiritual power, the dance of bringing offerings, the songs of triumph. Mila Repa's song is a good example of how mainstream Mahayanist convictions have undergirded distinctively Tibetan interests in magic and ritual.

The native, pre-Buddhist Tibetan religious traditions, with which immigrant Buddhism was bound to interact, were shamanistic. The native *Bon* ritualists were curers and protectors against death. From its interactions with such native realities Tibetan Buddhism developed a distinctive preoccupation with death. Indeed, perhaps the most famous Tibetan Buddhist text is *The Book of the Dead,* which describes what people experience at death and how they may use their experiences to gain liberation from the wheel of death and rebirth. Early in this text one reads:

> The factors which made up the person known as E. C. [the person being addressed, but in reality any human being on the verge of death] are about to disperse. Your mental activities are separating themselves from your body, and they are about to enter the intermediary state. Rouse your energy, so that you may enter this state self-possessed and in full consciousness!
>
> First of all there will appear to you, swifter than lightning, the luminous splendour of the colorless light of Emptiness, and that will surround you on all sides. Terrified, you will want to flee from this radiance, and you may well lose consciousness. Try to submerge yourself in that light, giving up all belief in a separate self, all attachment to your illusory ego. Recognize that the boundless Light of this true Reality is your own true self, and you shall be saved![16]

The moment of death, which turns out to be prolonged, does not introduce new teachings about the makeup of the human personality or the illusions it must overcome if it is to escape from karmic bondage. Rather, the moment of death becomes a privileged time for realizing the staple Buddhist truths about no-self and emptiness. If a person can endure the light of emptiness and surrender (illusory) selfhood, death can be the moment when final enlightenment dawns, craving ceases, and *nirvana* begins.

Tibetan Buddhist mysticism probably strikes the average comparativist as more imaginative than what one finds in other traditions. As gurus developed the "craziness" bequeathed them by the famous mountain saints, along with the tantric Indian traditions and the *Bon* shamanistic traditions, they tended to stress meditations in which disciples moved imaginatively into circles of sacred space and songs of sacred sound which might restructure their consciousness to Buddhist standards. Tibetan gurus were not loathe to employ *bodhisattvas* as regal focuses for such meditations, inviting disciples to identify themselves with these figures of wisdom and so realize their own *bodhi* nature.

In a recent work endorsed by the current (fourteenth) Dalai Lama, the Tibetan master Tsong-kha-pa puts this imaginative identification with a divinity (characteristic of Vajrayana, here called the Mantra Vehicle) into rather esoteric language:

> At the time of the fruit, the base—a body adorned with the major and minor marks—and the mind of non-apprehension (of inherent existence) which depends on it abide at one time as an undifferentiable entity. In the same way, at the time of the path, the method is that the yogi's body appears to his own mind in the aspect of a Tathagata's [Buddha's or *Bodhisattva's*] body, and at the same time his mind becomes the wisdom apprehending suchness—the non-inherent existence of all phenomena. These two are a simultaneous composite, undifferentiable in the entity of one consciousness. This should be understood as (the meaning of) undifferentiable method and wisdom (in the Mantra Vehicle). Through cultivating the yoga of joining these two at the same time one attains the state in which non-dualistic wisdom itself appears as Form Bodies to trainees.[17]

Let us pause to explain these knotted sentences. First, they make it plain that Tibetan Buddhist mysticism takes seriously the role of wisdom as an explanation of the reality that enlightenment discloses. The metaphysics assumed and described in this paragraph is the objective correlative of the subjective experience of enlightenment. The assumption of the teacher is that the more clearly the student understands how the world appears to the enlightened, the more likely the student is to meditate successfully. Therefore, what may strike the non-Buddhist reader as dry and foreign can well be music to the soul of the faithful Buddhist disciple, a cadence carrying overtones of the blissful achievement of Buddhahood.

Second, the first distinction of note is between two times, that of fruit and that of path. Fruit is success, enlightenment, access to *nirvana*. Path is the struggle prior to fruit, progress toward enlightenment (which many teachers optimistically describe as the steady dawning of light, the irresistible invasion of *nirvana*.) The base or foundation of fruit is the body of a Buddha; that is what the marks, major and minor, indicate. (Those who are in a fortunate existence [turn of the karmic wheel], from which they are going to exit into *nirvana*, carry heavenly marks.)

From early times, when mythologists embellished the biography of Gautama, such marks became standard, a guarantee or valorization of the fittingness that this holy person was about to become a Buddha. The mind of such a candidate or *bodhisattva* would display nonapprehension, that is, no dualism. Equally, the mind of the *bodhisattva* is never lost in the illusion of "inherent existence," the error of thinking that things have selves or are anything other than empty. These two, a properly "marked" body and a mind free of illusion, abide together. There is no dualism within the *bodhisattva,* any more than there is dualism (or nondualism) in the world that an enlightened one perceives. There is only Suchness, which is also emptiness.

Third, even while striving to make progress, the meditator's body should appear to his or her mind as an aspect of a Buddha's body. (The Tathagata is a Buddha under the aspect of Suchness, as part of the objective metaphysical landscape, as a fact of light and realization entering into the structures of the world that all beings compose.) With such a body, the mind of the meditator becomes "wisdom apprehending Suchness," light making him or her aware of such an objective metaphysical landscape, as well as of his or her own place in it (as a *bodhisattva*).

The gist of this wisdom is emptiness, the noninherent (selfless, substanceless) existence (being) of all phenomena (everything that the meditator can perceive). This body, appearing as an aspect of the Tathagata's body, and this empty, wise mind are not two separate aspects of consciousness but one integrity. They make a composite reality and cannot be differentiated or separated adequately. In Vajrayana, neither the meditational stage (method, path) nor wisdom (enlightenment, fruit) can be distinguished fully from its partner. Whatever distinctions the mind makes between them, for the sake of teaching students the elements of the integral reality with which meditation deals, fall away when one has either the experience of enlightenment or the wisdom to explain the reality into which enlightenment takes a Buddha. At the moment when one gains nondualistic wisdom, the method of developing this *yoga* (discipline) of joining the path and the fruit brings the meditator to the stage of Form Bodies, images in which the ultimacy or divinity of the meditator's own mind or spirit or final reality can take shape.

> If the supreme method—the appearance of a deity—is devoid of wisdom and bereft of the unmistaken cognition of the nature of one's own mind, one cannot progress to Buddhahood. Therefore, it is necessary to have a composite of the two. Jnanapada's *Self-Achievement* says, "Since a Subduer having immeasurable effulgence of light serves as a source of limitless marvels for oneself and others, even if this which has the character of being the supreme right method were manifestly cultivated but bereft of wisdom, it would not be a means of achieving all marvels; therefore, the nature [of the divine body] should be known without mistake." [18]

Here, we have a pause in the exposition to exhort students to honor wisdom—actual understanding, light from the far side of *nirvana*—and not get lost in the techniques. The essence of the wisdom that brings enlightenment is a sure understanding of the (empty) nature of one's own mind. One's own mind, as the Yogacarins especially had stressed, is the crux of one's reality, just as only mind is the crux of reality at large. Thus the proper method is to join one's meditational efforts with a dawning realization of the nature of one's own mind. A "subduer" (*bodhisattva*) is a source of brilliant light. This light comes not from technique but only from actual achievement. Actual achievement depends on wisdom, knowing how one's body and mind relate to the divine body, the Tathagata, with which one joins in liberational meditations, and how it is empty.

> The wisdom cognising non-inherent existence and appearing in the aspect of a deity is itself one entity with the mind of deity yoga, the vast. However, method and wisdom are presented as different by force of the convention of different opposites of negatives in dependence on the fact that their opposites are different. Wisdom is established from the viewpoint of being the opposite of a mind that mistakenly conceives the meaning of suchness. For knowledge of the ultimate (emptiness)—the final object of knowledge—is the supreme knowledge. Method is established from the viewpoint of being the opposite of that which does not have the capacity of attaining its fruit, Buddhahood. For the methods of Buddhahood are capable of achieving that state. . . . [19]

From the integrity of the mind and body of the meditator, as he or she identifies imaginatively with the suchness or Tathagatahood that rules both the subject and the object in any cognitional relationship, we move to the oneness of the meditator and the (vast, unlimited) mind, light, of the deity at the center of mantric discipline (*yoga*). In fact, the deity being visualized and the meditator doing the visualizing are not two but one. The deity is the substance of the meditator's mind, sprung from the meditator's imagination to provide a Form Body through which the oneness of the meditator with it may come home with full force.

At such a homecoming, or climax, any distinctions between method and wisdom that conventional awareness of their differences may present (to keep us in thrall to dualism) have once again to be brushed aside. Certainly, to the ordinary, ignorant mind wisdom is the opposite of the mind that botches the business of Suchness. The basic botching is always a matter of not grasping (that is, of ungrasping, or simply letting be) the emptiness or noninherent character of all things. Knowledge of this emptiness is the supreme knowledge, the gist of enlightenment.

The final paragraph of Tsong-kha-pa's chapter on deity *yoga* offers a summary of sorts:

> Thus, a Form Body is achieved through the appearance of the wisdom apprehending (emptiness) as a divine mandala circle, and a Truth Body is achieved through the cognition of its nature—emptiness. One should know that joining such method and wisdom non-dualistically is the chief meaning of the method and wisdom and of the yogas set forth in the Mantra Vehicle.[20]

The Form Body that focuses the Vajrayana disciple's meditation on the divinity of his or her own mind comes from the wisdom that persistence on the path has generated, wisdom about the emptiness of all entities. It takes shape as a sacred space, a privileged circle (*chakra*), where the meditator may picture objectified his or her own subjective buddhahood or divinity. A Truth Body (related more to the fruit, or the climactic condition of enlightenment, than to the path) can appear as a similar projection, though in this case a projection from a definitive understanding of emptiness, one so thorough as to make emptiness more prominent than any form, personal or impersonal, that its divinity or sacredness or ultimacy might assume to help the meditator focus properly. For Tsong-kha-pa, the union of method and wisdom, path and fruit, and now Form Body and Truth Body is the special genius of the disciplines of the mantric Vehicle—of the Vajrayanist techniques that have dominated Tibetan Buddhism.

It bears noting that the highly intellectual character of the treatments of meditation and metaphysical reality that one finds with many Tibetan masters stems from a strong tradition of "university" study and disputation. Tibetan Buddhism has traditionally sponsored a full curriculum of studies, on the whole modeled on higher Buddhist education in India, that encouraged students to engage in "*dharma* combat." Somewhat like the scholastic disputations sponsored in the late medieval European faculties of theology, these intellectual clashes honed the mind by sharpening both its capacity and its inclination to make distinctions. Logic became very important, sometimes perhaps inordinately so. In reaction, some masters moved meditation away from mental dialectics toward a simpler, more holistic striving. The Ch'an and Zen techniques that we discuss in the next chapter illustrate this movement, though historically they were not reactions to Tibetan developments as much as to Indian and Chinese.

What is intriguing about the Tibetan dialectics is not simply their reworking of such staple notions as emptiness and the lack of any adequate distinctions between body and mind, path and fruit, and the like, but also their use of mandalas and mantras. Although what we find in Tsong-kha-pa is relatively abstract, in some of the one-to-one regimes tailored by (possibly eccentric or "crazy") Tibetan gurus for the specialized needs of particular (gifted) disciples, the imaginative meditations could become powerfully tantric. Within the special mandalic circle, meditators might not only break taboos about food, liquor, or dealing with corpses, they might also imagine sexual union with goddesses or gods (Form Bodies), representing and focusing the divinity of their own minds.

In what were probably rare practices, sometimes known as Left Hand Tantra, disciples might even have actual intercourse with a partner sharing the mandalic circle and imagined to be the goddess or god whose expression of the empty wisdom of their own minds they were trying to appropriate. At all times when the tantric tradition was functioning in health, these exercises were fully disciplined and not at all licentious. The point was not pleasure but breaking radically with the conventions, the moral as well as noetic dualisms, that kept disciples from a full realization of emptiness and Suchness.

Devotional Buddhism

By now it may be clear that Buddhist mysticism, like that of mainstream higher (educated as opposed to folk) Hinduism, has been enchanted by spirit. The main difference between the Buddhist and the Hindu versions of this characteristically Indian enchantment has been the Buddhist insistence on *anatman,* with its correlative hymning of emptiness. However, just as *bhakti* has complicated Hindu mysticism, adding personal gods and an erotic love seeking to experience ultimate reality in passionate union, so devotional Buddhism has complicated any completely cool, detached reading of the direct experience of ultimate reality warranted by the enlightenment of Gautama. In this section we reflect on some aspects of this complication, moving from Vajrayanist to Mahayanist to Theravadist species of devotion. Although laypeople in all three Buddhist families have both stimulated the devotional practices that we describe and used them extensively, the practices have also served monks and nuns. In other words, they have been more than mere accommodations to popular needs and forms by the high traditions.

In Tibetan tantric devotions, perhaps the most popular deity or *bodhisattva* is Tara, a goddess of several colors. Here, we witness a contemporary guru, Tara rinpoche, in the first stages of a ceremony called "initiation into life." To begin this ceremony, he identifies himself with the white Tara. Notice that this identification amounts to the projection of the divinity, or the holiness or wisdom, of his own (empty) mind or being. Characteristically Tibetan is his tantric move to focus both meditation and ritual on imaginary deities such as Tara. The first two of the four initial prayers that Tara rinpoche utters are standard and might appear in any Buddhist school, though the emphasis on the gurus is also typically Tibetan. By the time that he gets to "visualization," however, in his third prayer, the tantric devotional slant has taken over. The fourth prayer, and the description of the white Tara that follows, are unmistakably tantric:

> I and all beings—all sentient beings as infinite as space—from this time until we attain the terrace of enlightenment, go for refuge to the glorious and holy gurus; we go for refuge to the perfect blessed Buddhas; we go for refuge to the holy teachings; we go for refuge to the noble community of monks. . . .
>
> I prostrate myself and go for refuge to the guru and the precious three jewels. I pray you: empower my stream. . . .
>
> It is for the sake of all beings that I should attain the rank of perfect Buddhahood, and it is for that reason that I shall experience this visualization and recitation of the life-bestowing Cintacakra (White Tara). . . .
>
> I myself instantaneously become the Holy Lady, and on my heart, above the circle of a moon, is a white syllable TAM. Light radiates forth therefrom and, in the sky before me, the holy Cintacakra, surrounded by the hosts of gurus in the lineage of the "life teaching," Buddhas and Bodhisattvas. . . .
>
> All becomes Emptiness. From the realm of Emptiness PAM and from that a lotus; from A is the circle of a moon, above which my own innate mind is a white syllable TAM. Light radiates forth therefrom, makes offerings to the Noble Ones, serves the aim of beings, and is gathered back in, whereupon in my mind the syllable is transformed, and I myself become the holy Cintacakra: her body is colored white as an autumn moon, clear as a stainless crystal gem, radiating light; she has one face, two hands, three eyes; she has the youth of sixteen years; her right hand makes the gift-bestowing gesture, and with the thumb and ring finger of her left hand she holds over her heart the stalk of a lotus flower, its petals on the level of her ear, her gesture symbolizing the Buddhas of the three times, a division into three from a single root, taking the form of an open flower in the middle, a fruit on the right, and a new shoot on the left; her hair is dark blue, bound up at the back of her neck with long tresses hanging down; her breasts are full; she is adorned with divers precious ornaments; her blouse is of varicolored silk, and her lower robes are of red silk; the palms of her hands and the soles of her feet each have an eye, making up the seven eyes of knowledge; she sits straight and firm upon the circle of the moon, her legs crossed in the diamond posture.[21]

The full description largely speaks for itself. Clearly, the guru is working with very traditional language, imagery, and symbolism. The key words function as mantras, echoing with special resonance. The key images, such as the goddess's eyes, suggest such capital notions as wisdom. Because ultimate reality and the wisdom through which it resides in the guru are both empty, the visualization can take place without troublesome questions about the status, or the ontological stature, we should accord the goddess.

However, the visualization is not undisciplined but controlled by a firm line of interpretation handed down from guru to guru. Tara rinpoche draws on this line to make present for those attending the initiation into life a dramatic form through which holy emptiness may seize their imaginations, minds, and hearts. What transpires in his imagination as we watch him preparing for the ceremony takes objective form for the crowd on the altar, carefully prepared, where he will soon preside. Depending on their degree of familiarity with the symbols of the White Tara, they will enter more or less fully into the priestly words and gestures through which he tries to mediate to them her power; but the devotion of the lay attendants, whether superficial or profound, will, like his own devotion, focus on this imaginative deity and be filled with love and hope directed toward her.

In Mahayanist circles, including such advanced ones as those composed of disciples meditating on the *Prajnaparamita,* a mantric element could also be important. So, for example, at the conclusion of the *Heart Sutra,* after the demanding dialectical analysis of the equality of emptiness and form, meditators raised their sights to the Lady Prajnaparamita and spoke her name as a spell:

> Therefore one should know the prajnaparamita as the great spell, the spell of great knowledge, the utmost spell, the unqualified spell, allayer of all suffering, in truth—for what could go wrong? By the prajnaparamita has this spell been delivered. It runs like this: Gone, gone, gone beyond, gone altogether beyond, O what an awakening, all-hail!—This completes the Heart of perfect wisdom.[22]

The language is geared to seduce the ear and encourage repetition so that "gone" works its way deeper and deeper into the disciple's consciousness. The image of the lovely goddess is geared to seduce the eye so that emptiness and wisdom become beautiful, all that the heart desires. Buddhist meditation became holistic and humanized in these ways. Time and space, which it sought to escape, became allies, what the West sometimes calls "sacraments," or in this case containers and gestures of emptiness. Long experience with actual meditational, mystical practice showed the assets and liabilities of tantric, magical methods. The mainstream of gurus in all schools avoided low interpretations of how to manipulate imagery and libidinal forces, but this mainstream was well aware of the mysteriousness of ultimate reality, its more-than-rational richness, and thus the utility of evoking it through special images and sounds.

We find both an interest in spells and a concern for merit in one of the most influential Mahayanist devotional texts, the *Saddharma-pundarika,* or *Lotus Sutra.* The conceit behind this text is that it epitomizes the Buddhist *dharma* so fully that by reciting it a disciple may fulfill all the necessities of good practice. Such recitation has shaped the piety of many lay Mahayana Buddhists, whose religious ambition has been less to gain enlightenment than to improve their karmic condition (merit) so as, in a more favorable future lifetime, to achieve definitive liberation. Notice in the following quotation, first, the assurance of such merit and, second, the provision of protective (magical) words:

> Thereupon the Bodhisattva Mahasattva Bhaishagyaraga rose from his seat, and having put his upper robe upon one shoulder and fixed the right knee upon the ground lifted his joined hands up to the lord [Buddha] and said: How great, O Lord, is the pious merit which will be produced by a young man of good family or a young lady who keeps this Dharmaparyaya of the Lotus of the True Law, either in memory or in a book? Whereupon the Lord said to the Bodhisattva Mahasattva Bhaishagyaraga: Suppose, Bhaishagyaraga, that some man of good family or a young lady honors, respects, reveres, worships hundred thousands of kotis of Tathagatas equal to the sands of eighty Ganges rivers; dost thou think, Bhaishagyaraga, that such a young man or young lady of good family will on that account produce much pious merit? Yes, Lord; yes, Sugata. The Lord said: I announce to thee, Bhaishagyaraga, I declare to thee: any young man or young lady of good family, Bhaishagyaraga, who shall keep, read, comprehend, and in practice follow, were it but a single stanza from this Dharmaparyaya of the Lotus of the True Law, that young man or young lady of good family, Bhaishagyaraga, will on that account produce far more pious merit.
>
> Then the Bodhisattva Mahasattva Bhaishagyaraga immediately said to the Lord: To those young men or young ladies of good family, O Lord, who keep this Dharmaparyaya of the Lotus of the True Law in their memory or in a book, we will give talismanic words for guard, defence, and protection; such as, anye, manye mane mamane itte karite same, samitavi, sante, mukte. . . . [23]

The first paragraph illustrates the opulent oral style of much Indian Mahayanist devotional literature. Repetition and magnification (building image upon image—hundreds of thousands of myriads, the sands of eighty Ganges) lure lay hearers to consider the court of the Buddha, the heavenly splendor of the Lord, the most desirable of all the images they might contemplate. Further, this paragraph invites them to place all their confidence in the *Lotus Sutra,* as though the Buddha had made it the scripture par excellence, complete and sufficient, needing none other. Finally, what they can gain from reciting verses of the *Lotus,* as well as contemplating its teachings, is maximal merit, progress along the Buddhist way. No other devotional exercise will profit them more.

The second paragraph raises the matter of magical protection through talismanic words. Notice again the sounds of the words: chanted for any length of time, they would break free from ordinary tonation and denotation and move on their own, becoming loose and reminding the chanter, however obliquely, of the frailty of the ties between sound and meaning, breath and being.

The realities that we try to control by our words are not so inert or fixed as, in our panic at the potential chaos of reality, we hope to make them. We cannot pin them to the board like butterflies and assure that they will fly no more. However, if a Buddha controls them, they may cluster around us protectively, becoming a magical circle of sound and meaning within which we can dwell securely. This very Indian text is old enough to conjure up the ancient Vedic association of words and sacrifice. In it, words still possess sacred, numinous power because it is still credible that they are the stuff of the universe, the atoms of being.

In virtually every premodern culture, ordinary people have prayed and even spoken in the marketplace with considerable awareness of the forces of light and darkness that their words might summon. They have employed stock phrases ("God willing," "knock on wood"), as they have carried lucky charms and hidden their children from the "evil eye."

Devotional religion has always flowed in these pathways or ruts of superstition, concern for luck, need for healing, and general anxiety about the danger of the world, as well as along superior highways. The words with which people have prayed to the *bodhisattvas,* the Christian or Muslim saints, or the Hindu gods and goddesses have usually been impure, complicated, directed to both high and low aims at one and the same time. Usually they have carried desires to praise God or emptiness simply for themselves, as well as desires to bend God or emptiness to the will, the needs, the psychic survival of the people praying through them. Religious words, magical or plain, have been extraordinarily human, a rich epitome of our complicated, very mixed status as beings both material and spiritual, both nervous and rational.

The last specimen of devotional Buddhism that we probe for its mystical implications comes from Theravadist sources. In Melford Spiro's fascinating anthropological study of recent Burmese religious practices, the focus of the piety of most lay Buddhists emerges not as the gain of definitive liberation, *nirvana,* but as the gain only of the merit that might bring a more blessed next life. So, contrary to orthodoxy, much Burmese Buddhism has become not a radical attack on suffering but a vehicle for future material pleasure and prosperity. Laypeople multiply the devotional and charitable acts from which they may expect to receive merit, seeing such acts as the coin with which they may buy a happy future life, nor are they necessarily bashful or restrained in how they imagine such a life:

> Meanwhile, before I reach Nirvana by virtue of this great work of merit I have done, may I prosper as a man, and be more royally happy than other men. Or as a spirit, may I be full of colour, dazzling brightness and victorious beauty, more than any other spirit. More especially I would have a long life, freedom from disease, a lovely complexion, a pleasant voice, and a beautiful figure. I would be the loved and honoured darling of every man and spirit. Gold, silver, rubies, corals, pearls and other lifeless treasure, elephants, horses and other living treasure—may I have lots of them. By virtue of my power and glory I would be triumphant with pomp and retinue, with fame and splendour. Wherever I am born, may I be fulfilled with noble graces, charity, faith, piety, wisdom, etc., and not know one speck of misery; and after I have tasted and enjoyed the happiness of spirits, when the noble law of deliverance called the fruit of Sanctity blossoms, may I at last attain the peaceful bliss of Nirvana.[24]

One may speak of a "mystique" of merit, a fascination with piling up credits for a good afterlife, but "mysticism" seems inappropriate here. Since the goal is not direct experience of ultimate reality, this Burmese Buddhist devotionalism does not qualify for our inventory of mysticism. Inasmuch as it abdicates the struggle to overcome, root and branch, the painfulness of all existence, it appears to be a compromise or halving. Certainly, it offers laypeople the hope of overcoming their sufferings in the next life. Equally, it marshals in their imaginations reasons for maintaining the Buddhist rituals and performing acts of charity. Indeed, *nirvana* remains a potent concept. The laypeople's concern for merit cannot forget that *nirvana* is the ultimate goal, but a certain settledness, even complacency, has entered in, diluting the original fire and passion. A certain magical mediocrity is knocking at the door.

The most interesting questions here for students of mysticism may be, "How much fire and passion for radical, comprehensive solutions to the problem of suffering (imperfection, sin) are necessary for a truly mystical quest?" and "where does the border lie between humanizing mystical fire so that people can find ultimacy in the world (*nirvana* in *samsara*) and bastardizing it—losing its legitimacy?" All religions face the challenge of incarnating in everyday forms the passionate search of their founders for ultimate reality, finding sacred meaning in eating, drinking, praying, sleeping, making love, laughing around the campfire. For example, in Buddhist statuary the peaceful, compassionate Gautama has become attractive humanistically as a form of enlightenment that his people can love. Thus devotional religion, with its almost naked patches of crudeness, offers touching lessons in both the difficulty and the inevitability of the work of the religions to incarnate mystical holiness in everyday forms.

Summary

Our general orientation sketched the history of the movement founded by Gautama, probably in the fifth century B.C.E., as pulsating in two main phases, Indian and East Asian. The origins of historical Buddhism lie in the decision of the Indian Prince Gautama to promulgate the solution to the problem of suffering that he had gained in enlightenment. Similarly, the basic social structure of the Buddhist community that has prevailed down the centuries derives from the decision of Gautama to found a community of monks, people who would abandon ordinary family life to concentrate full-time on gaining enlightenment. The Buddha also provided for laypeople, but throughout Buddhist history monks and nuns have constituted the inner core of the *sangha.*

While Buddhists have enjoyed periods of cultural rule in India and have survived as the predominant religion in Sri Lanka, by the beginning of the second millennium C.E. Hinduism had organized a resistance that had won the day against Buddhism among the common Indian people. Buddhist missionaries had already made strong inroads to the east, becoming dominant in such countries as Burma, Thailand, and (eventually) Tibet. The farther east they went, the more deeply into the cultural zones dominated by China, the more they had to share influence with Confucianism and Taoism. As well, in Japan they had to share influence with Shintoism. Nevertheless, everywhere in East Asia, Buddhists adapted with great creativity, ensuring that The Middle Way took solid root and often gained great cultural power.

In sketching the core of the Buddhist worldview that has prevailed thoughout this two-pulsed history, we stressed the three jewels in which Buddhists take refuge when they make their formal profession of faith: the Buddha, the *dharma,* and the *sangha.* These jewels have remained the backbone of The Middle Way both sociologically and doctrinally. Second, we described the three pillars on which traditional Buddhist religion has rested—meditation, morality, and wisdom—and offered a quick epitome of each. For example, we noted the five precepts of *sila* that form the minimal Buddhist ethics: not to kill, not to steal, not to lie, not to be unchaste, not to take intoxicants.

Our first topic, the enlightenment of Gautama, began our effort to locate the mystical dimension that has thrived in The Middle Way within the general Buddhist history and worldview. Gautama became the Buddha through his experience of awakening and realization under the bodhi tree. That is the experience that empowered him to found The Middle Way and become its first jewel. As well, that is the paradigm for the experience that Buddhist meditation masters, moralists, and philosophers have tried to inculcate in their disciples.

Inasmuch as we can glimpse it from the canonical Buddhist sources, which are as much mythology as history, the experience of Gautama under the bodhi tree brought him to a direct, comprehensive, but intuitive understanding of ultimate reality. Articulated, his experience took the form of The Four Noble Truths, the first two of which are the linchpins: (*1*) all life (reality that human beings experience) is suffering and (*2*) the cause of suffering is desire.

The traditional accounts of how the Buddha rose to this surpassing moment of illumination and how nature responded to his great victory fit him into the profile of a master of meditation come to the peak of consciousness. At the peak, his powers drew all of nature into obeisance and praise. Along the way of his struggle toward the peak, Gautama both defeated the forces of evil, epitomized in Mara, and gained the lesser powers of shamans and middle-range meditation masters, such as the power to know his own previous lives and the power to know the karmic condition of all other living things. These secondary powers have meant less to later meditation masters than has the peak experience of the Buddha. The peak experience, as articulated in The Four Noble Truths, has structured how all later Buddhists have thought about gaining *nirvana* (the psychology) and representing nirvanic ultimate reality (the ontology). As such, the peak experience of Gautama has been the archetypal mystical referent.

Second, moving to the Buddhist school that thinks of itself as the eldest, or the best, representative of the original traditions, we studied the mysticism of a classic of Theravadist meditation, the *Visuddhimagga* of Buddhaghosa. This text illustrates the major concerns of traditional meditation masters, the subjects they offered their disciples as the materials most apt to quench desire and so liberate them from karmic suffering. Two meditations, on friendliness and death, stood out as being always efficacious.

In addition to proposing traditional topics for meditation, the *Visuddhimagga* discourses on virtue, the strengths of character that undergird progress toward enlightenment. The zealous disciple will work to develop such virtue, showing the ties between meditation and morality. Practically, enlightenment seldom comes to people who are not good, or well developed morally and free of major vices. For Buddhists, enlightenment seldom comes to people who have not mastered their bodies and their appetites, both material and spiritual. Until a person has routed desire, *nirvana* is but a dream.

Buddhaghosa's third interest is the understanding that constitutes enlightenment and leads into *nirvana*. This understanding affirms and realizes experimentally the orthodox characterization of proximate, samsaric reality as painful, fleeting, and selfless. As well, it affirms the orthodox Indian characterization of ultimate, nirvanic reality as an ineffable being, bliss, and awareness. Buddhahood is a dwelling in the light, a consciousness of how reality itself, properly grasped, is light, intelligibility, lack of imperfection and suffering, simple Suchness (being oneself, without falsification).

Our third topic was Mahayana Buddhist mysticism. As illustrated by such classical writings as the *Heart Sutra,* the dialectical analyses of the Madhyamika sage Nagarjuna, and the Yogacarist elaboration of mind-only, this family of schools builds on the three marks (painful, fleeting, selfless) that mainstream meditation masters and philosophers had come to consider staple descriptions of samsaric realities. Characteristically, however, the Mahayanists generalized these marks into the powerful, comprehensive symbol of emptiness.

Negatively, emptiness showed the illusion, the danger, the unreality of what most people pursued, craved, dedicated themselves to most of the time. Positively, emptiness characterized *nirvana* itself, ultimate reality, what the wisdom that had escaped from *karma* and gone beyond samsaric illusion, found to be Suchness, or plain actuality.

The *Prajnaparamita* literature exalted emptiness and equated it with the five *skandhas* comprising the samsaric self. Nagarjuna showed that *nirvana* and *samsara* could be one, but that the logical permutations of this identification stopped the mind and called the integral spirit, the sole potential source of enlightenment, to move wordlessly, imagelessly, to something far beyond logic. The Yogacarist predilection for mind-only, for equating the ultimately real with the ideal or the intelligible, located emptiness and Suchness in the light of *bodhi,* the core of Gautama's realization.

At term, then, Mahayanist mysticism became characteristically paradoxical and free, capable of adapting itself well to whatever needs its laypeople, Indian or East Asian, presented.

Our fourth topic, Vajrayanist mysticism, focused on the development of Tantrism in Tibet. The Tibetan mountain gurus tended to take the Mahayanist convictions about emptiness and the equation of *nirvana* and *samsara* to eccentric extremes. As well, they utilized mantras, mandalas, and breeches of taboo to draw the entire personality of the disciple—mind, emotion, body, imagination, and subconscious libidinal energies—into the pursuit of enlightenment. Under the influence of native shamanic interests in death, Tibetan masters produced a classic, *The Tibetan Book of the Dead,* that counseled the dying person about the stages involved in the dissolution of the mind–body complex on the way to the next round of the karmic cycle and pointed out the opportunities for definitive liberation that these stages offered. Finally, study of a present-day Tibetan master's understanding of "deity yoga" suggested a contemporary version of the long-standing tantric inclination to express the divinity of the mind (its Buddha nature) by having disciples identify themselves with vivid deities such as the goddess Tara.

Our analysis of a fifth and final topic, devotional Buddhism, began with the ritualistic practices of another present-day Tibetan guru. Observing his personal meditational preparations and anticipating the ceremony that he would perform for lay Buddhists, we saw just how vivid Tara could become and how extensive could be the mandalic imagery associated with her. Although she continued to be the projection of the divinity of the master's (any intense Buddhist's) own mind, she sat enthroned on a richly decorated altar as a focus of veneration little different from a goddess of any other religious tradition. The guru understands this imagery in terms of emptiness, and no doubt instructs his disciples to this effect, but the ritual activities themselves celebrate external sights and sounds, images and smells, more than interior appreciations of emptiness.

The *mandalas* and *mantras* so prominent in Vajrayana occur less dramatically but still importantly in Mahayanist texts, such as the *Heart Sutra,* where the lady Prajnaparamita is understood to be not only a great goddess but also a powerful spell. The words celebrating the "going beyond" of wisdom have a mantric cadence, while contemplating the lady as the embodiment of ultimate wisdom has the effect of making her a magical sign. Similarly, the mantric words that we meet in one of the most popular Mahayanist devotional texts, the *Lotus Sutra,* show that magical, talismanic sounds have played an important role in lay meditations, while the heavenly scenes depicted in this text have often functioned as *mandalas,* sacred imaginative spaces concentrating the gist of liberation from karmic bondage.

Our last specimen of devotional Buddhism, drawn from the concern of present-day Burmese laity to gain merit rather than *nirvana,* raised interesting questions about the compromises and declines that the mystical intensity of religious founders can suffer. Because many lay Burmese concentrate on gaining merit and samsaric happiness rather than solving the radical problem of suffering, they give up the quest to experience ultimate reality directly. This reminds us that even the highest mysticism has tended to exist alongside middling or low alternatives and that religious ritual as a whole has been a very mixed human enterprise.

When it comes to characterizing the core or most revealing mysticism in the considerable range of Buddhist religious striving that we have witnessed, emptiness comes to the fore. Implied in Gautama's realization that desire, grasping after samsaric things as though they were full or valuable, is the source of the pervasive pain that human beings experience, emptiness gradually developed into the paramount mark of all reality.

The Theravadists stressed the painful, fleeting, and selfless character of all *dharmas*, all items of existence. Mahayanist philosophers riveted this interest onto emptiness with an almost ruthless intensity. Vajrayanist masters assumed it in their tantric meditations and rituals. Significantly, emptiness fades when we come to mainstream devotional Buddhism, as though it were too daunting for ordinary layfolk, whose need for colorful deities and rituals to incarnate general Buddhist doctrine into daily life pushed emptiness to the background. Indeed, by the time of the Burmese development of "karmic Buddhism" (concentration on merit), emptiness lay beyond the practical pale.

Inasmuch as Buddhist mysticism loves emptiness, it stands apart somewhat from Hindu mysticism. *Anatman*, selflessness, moves a step away from the Vedantic equation of *atman* and *Brahman*. How different the experiences of ultimate reality need be is hard to determine, but Buddhist mystics, especially those shaped by East Asian cultural influences, strike us as more fluid, spontaneous, in tune with the flow of nature than their Indian counterparts. This characterization is more descriptive than evaluative. It says nothing immediate about the comparative depth or richness of the two mystical complexes involved. However, it suggests some of the key implications of the romance of the Buddhist mystics with emptiness.

The last summary question that we raise is how Buddhist mysticism in general and emptiness in particular ought to incline us to think about our working description of mysticism as direct experience of ultimate reality. Minimally, it ought to incline us to think that the descriptions that mystics themselves work with can vary considerably. "Emptiness" is not a term prominent in the usage of Western mystics (though a little probing can turn up cognate, analogous terms: "nothingness," "contingency").

The Buddhist use of emptiness does not oppose it to light or mind. The figure is therefore flexible, as well as provisional (*nirvana* is ineffable, all figures are inadequate, more or less convenient). Similarly flexible is the movement of the human spirit of many Buddhist meditators. Still, whole, alert, poised, it is more like a feather than a rock, a bird than a hippopotamus. It is a dancer, a pilgrim, a wanderer that just may find the world fully congenial because it places no conditions on the world, strives to let the world be, move, just as the world chooses to do.

NOTES

1. See, for example, *1993 Britannica Book of the Year* (Chicago: Encyclopaedia Britannica, 1993), 270. Figures on the supposed adherents of the religions vary widely; one can only hope that a scholarly authority uses its criteria consistently and evenhandedly.

2. For a good geographical sketch of Buddhist history, see the articles in Mircea Eliade, ed., *The Encyclopedia of Religion* (New York: Macmillan, 1987), 2: 334–439. Supportive information is available in Charles S. Prebish, ed., *Buddhism: A Modern Perspective* (University Park: Pennsylvania State University Press, 1975). In addition to articles in these sources dealing with key Buddhist teachings and texts, see Walpola Rahula, *What the Buddha Taught,* rev. ed. (New York: Grove Press, 1974), and Edward J. Thomas, *The History of Buddhist Thought* (New York: Barnes & Noble, 1951).

3. For a comparative study of the spiritualities of the Buddha, Confucius, Jesus, and Muhammad, see Denise Lardner Carmody and John Tully Carmody, *In the Path of the Masters* (New York: Paragon House, 1994).

4. George N. Marshall, *Buddha: The Quest for Serenity* (Boston: Beacon Press, 1978), 61–62. See also Michael Carrithers, *The Buddha* (New York: Oxford University Press, 1983), 53–78.

5. Ashvaghosa, "The Legend of the Buddha Shakyamuni," in *Buddhist Scriptures,* ed. Edward Conze (Baltimore: Penguin, 1959), 50–51.

6. Roger J. Corless, *The Vision of Buddhism* (New York: Paragon House, 1989), 10.

7. Ibid., p. 11.

8. Ibid.

9. Denise L. Carmody and John T. Carmody, *Ways to the Center: An Introduction to World Religions,* 4th ed. (Belmont, Calif.: Wadsworth, 1993), 128.

10. Edward Conze, *Buddhist Meditation* (New York: Harper Torchbooks, 1969), 14. See also Bhadantacariya Buddhaghosa, *The Path of Purification,* 2 vols. (Berkeley, Calif.: Shambhala, 1976). Helpful studies of Theravadist meditation include Nyanaponika Thera, *The Heart of Buddhist Meditation* (London: Rider, 1969), and Winston L. King, *Theravada Meditation: The Buddhist Transformation of Yoga* (University Park: Pennsylvania State University Press, 1980).

11. "The Heart Sutra," nos. 1–9, in *Buddhist Wisdom Books,* ed. Edward Conze (New York: Harper Torchbooks, 1972), 77–81.

12. Nagarjuna, *Mulamadhyamakakarikas,* no. 19 (*kala*), in *Emptiness: A Study in Religious Meaning,* ed. Frederick J. Streng (New York: Abingdon Press, 1967), 205. See also T. R. V. Murti, *The Central Philosophy of Buddhism* (London: Allen & Unwin, 1955).

13. Vasubandhu, *Trimsika,* 28, in *Buddhist Texts Through the Ages,* ed. Edward Conze (New York: Harper Torchbooks, 1964), 210.

14. Mila Repa, *mGur-hBum,* in *Buddhist Texts Through the Ages,* ed. Conze, 258.

15. Stephan V. Beyer, "Buddhism in Tibet," in *Buddhism,* ed. Prebish, 240.

16. *Bar-do thos-grol* (Summary), in *Buddhist Scriptures,* ed. Conze, 227.

17. Tsong-kha-pa, *Tantra in Tibet* (London: George Allen & Unwin, 1977), 1:126.2.

18. Ibid., 126–127.

19. Ibid., 127.

20. Ibid., 127–128.

21. Stephan Beyer, *The Cult of Tara: Magic and Ritual in Tibet* (Berkeley: University of California Press, 1978), 378–379. For an overview of Buddhist goddesses and women in Buddhism, see Diana Y. Paul, *Women in Buddhism* (Berkeley, Calif.: Asian Humanities Press, 1979).

22. "The Heart Sutra," no. 50, in *Buddhist Wisdom Books*, ed. Conze, 101–102.

23. H. Kern, trans., *Saddharmapundarika or The Lotus of the True Law* (New York: Dover, 1963), chap. 21, 370–371.

24. Melford E. Spiro, *Buddhism and Society*, 2nd ed. (Berkeley: University of California Press, 1982), 79.

4

Chinese and Japanese Traditions

General Orientation

We begin with a brief sketch of the histories and worldviews of both China and Japan. Today nearly a quarter of the world's population comes from these two countries. China is the vaster geographical realm, extending nearly 3.7 million square miles. Estimates of the population of China run as high as 1.3 billion.

Archaeological remains suggest a presence of *Homo erectus* in China more than 300,000 years ago (perhaps even 500,000).

The Neolithic era developed in China during the sixth millennium B.C.E. Recorded history, based on written records, begins around 3000 B.C.E. with the shadowy period of the Three Sovereigns, who recede into prehistory. More definite is the period about 2700 to 2200 B.C.E., ruled by five semilegendary emperors. The Hsia Dynasty, running from perhaps 2200 to 1800 B.C.E., yields few extensive records, but with the Shang Dynasty (also known as the Yin), which covers the period 1800 to 1200 B.C.E., history in a modern, critical sense begins.

The Chou Dynasty (1111–255 B.C.E.) is the truly formative era, when higher Chinese culture laid down its foundations (for example, through Confucian thought). The Han Dynasty (206 B.C.E.–220 C.E.) represents the successor leadership that consolidated much of the social solidarity implied in the Chou. Later larger historical periods include the Six Dynasties Period (220–589 C.E.), when numerous different rulers prevailed in the eastern and western portions of the country); the Sui Dynasty (581–618 C.E.), during which considerable consolidation of the empire occurred; the Tang Dynasty (618–907 C.E.); the Sung Dynasty (960–1279 C.E.), when the arts, especially landscape painting, flourished; the Mongol (Yuan) Dynasty (1206–1368 C.E.); the Ming Dynasty (1368–1644 C.E.); and the Manchu (Ch'ing) Dynasty (1644–1912 C.E.). China was riven by various factions throughout much of the first half of the twentieth century C.E., until the Communist victory of 1949, which has supplied the recent rulers.

Japan has drawn much of its higher culture from China, but indigenous traditions, roughly known as Shinto, have kept Japanese culture distinctive.

Human beings at the Paleolithic stage of culture may have appeared in Japan as early as 30,000 years ago. From 10,000 years ago to the second or third century B.C.E., a distinctive pottery, *Jomon*, prevailed as a major artifact, and people developed agriculture. Recorded history begins about 2,700 years ago, with the first (legendary) emperor Jimmu. The period from about 250 B.C.E. to 250 C.E., known as the Yayoi, saw a new cultural influence move from the south northward and eventually supersede the Jimmon. Chinese culture, which stabilized during the corresponding (Han) era, appears to have reached Japan through Korea. In addition to the system of social arrangements that we associate with Confucianism, this Chinese influence mediated such practical innovations as the cultivation of rice. During the later Han era Japan appears for the first time in Chinese chronicles.

Some historians call the period 250 to 710 C.E. "ancient." The Yamato court became the center of a relatively unified country. In the sixth century C.E., Buddhism reached Japan, quickly becoming influential. Moved by Buddhist ideals, in 604 C.E. Prince Shotoku published a constitution in seventeen articles that set out both the goals of the state and a code of conduct for all citizens. Other significant historical eras, named for the clan that prevailed, include the Nara (710–784 C.E.), the Heian (794–1185 C.E.), the Kamakura (1192–1333 C.E.), the Muromachi (1338–1573 C.E.), the Tokugawa (1600–1867 C.E.), and the three modern dynasties—Meiji, Taisha, and Showa. The last governed Japan until the end of World War II. Since 1945 Japan has operated under Western political principles, though increasingly it has interpreted them according to its own cultural and sociological instincts.

When it comes to the Chinese and Japanese worldviews, several capital notions stand out: the Way, ritual, goodness, filial piety, not-doing (inaction), and (for Japan) the *kami*. The Way (*Tao*) is the path that nature lays out for all creatures, which the sages of yore described for human beings. (The sages, in fact, knew how to harmonize human affairs with those of the cosmos.) Although Taoists such as Lao-tzu and Chuang-tzu make more of the *Tao* of nature than Confucians do, Confucius and Mencius are aware that human affairs, as they come to an especially acute focus in ritual, ought to move in tandem with the natural and bodily affairs that heaven oversees. So *Tao* is a central Chinese concept. Nature provides a way, reveals a path to wholeness, that human beings neglect at their peril.

Ritual (*Li*) is more beloved of Confucians than Taoists. It reflects the instinct of Confucians that human beings are socialized, stay human, and manage their affairs best when they observe the social amenities. Ritual, protocol, and etiquette form attitudes, states of soul, working from the outside in. (Confucians insisted, however, that ritual was innate, or instinctive.) The hierarchical cast of the Confucian mind fits this ritual mentality well. Master/pupil, ruler/subject, parent/child, elder/younger, male/female—these are hierarchies that Confucius thought came from nature and were part of humanity's social if not genetic code. So, the bows and ceremonies, both formal and informal, by which people made their way through the different stations on the social track and the varying needs of the calendar year were quintessentially human. To not know how to behave or to not strive to behave well was to stamp oneself as a boor, a renegade, a problem to the commonweal.

Goodness (*Jen*) is the Confucian master virtue. It ensures the rare but achievable goals of benevolence, love, and a human personality that rings true. The Master (Confucius) thought that long study and self-discipline could inculcate *Jen*. Mencius, the premier disciple of Confucius, was more sanguine: human nature is good intrinsically. Either way, the path to fully realized human nature is largely ethical and practical. The good person does what is right. He or she obeys the appropriate social etiquette, the rituals proper to his or her station. Gradually these rituals mold a person to benevolence. Sensing how human affairs ought to be a microcosmic reflection of the macrocosmic system of nature, the mature human being is kind, helpful, and gladly benevolent as a reflection of a nature that generally is generous.

Filial piety (*Hsiao*) is the Confucian virtue that dominates domestic life. Children can do nothing finer than honor their parents by obedience and by maintaining a good reputation. When their parents die, the first responsibility of children is to mourn fittingly (ideally, for three years) and make well the sacrifices that ensure the parents' enjoyment of a happy afterlife. Parents give children life and sustenance. As though driven by a deep appreciation of this bedrock bequest, Confucian thinkers vied to outdo one another in praising filial piety. The Master himself was no slouch; many texts in the *Analects* make it plain that he considered reverence for one's parents the foundation of all other virtues.

Not-Doing (*Wu-wei*) is associated with the Taoists. When they contemplated how the *Tao* of nature operates, they discovered an active passivity. As water wears away rock, an infant dominates a household, a woman influences a man, a valley succeeds better than a mountain, so does nature operate: subtly, indirectly, patiently, by moving with the grain or the current rather than against it. Taoists see the dangers of overintervention, to say nothing of violent force. They think that most problems will settle themselves if one proceeds patiently, working behind the scenes. The race is long and goes to those with the best stamina. The better people submit themselves to the rhythms, the disciplines, of nature, the better the dance of their lives will be. The less they intrude themselves, the more self-effacing and seconding they become, the longer and better they will survive. It is the gnarled, ugly tree that survives. It is the unobtrusive civil servant who keeps his limbs intact (loss of a limb was a common punishment for falling into disfavor). It is the space that the walls contain that makes a house useful. So, learn to be empty, wait, discern the opportune moment, move to the music of the spheres and you may enter upon a graceful old age. Remain headstrong, obstreperous, a kicker against the goad and you will neither succeed nor be happy.

All of these notions, Taoist and Confucian alike, made their mark on Japan, as did such Buddhist notions as *karma* and *enlightenment.* Native to Japan, however, was the notion of the *kami.* The *kami* are the gods, spiritual powers, "geniuses" of the land, which in Japan has always been sacred. Any significant, outstanding hill, cave, spit at the seashore, tree, hero, or heroine might be a *kami.* Tradition said there were 800,000 of them, so the roster was generous, nearly open-ended. After Buddhism had made a formidable impact in Japan, the *kami* tended to fuse with the *bodhisattvas.* Either way, people remained greatly interested in localizations of the power of nature, intimations of sacred, ultimate significance.

Folk religion in both Japan and China has always stayed close to nature and been fascinated by displays of power, especially those that facilitated healing. A shamanistic heritage remained powerful in the "little tradition," the peasant stratum of culture, in both realms. So the *kami,* like the Chinese gods that accredited astrology and divination, did a brisk business. Inextricable from the long-standing Japanese love of the beauty of their islands and their veneration of the *kami,* Shinto became the native Japanese way, in contrast to foreign imports, drawing the people's pride in the direction of chauvinism and joining with their ethnic solidarity to produce a powerful sense of dignity, even superiority.

Confucianism

Most scholars do not stress mysticism when they describe Confucianism. They note the restraint of Master Kung himself (531–479 B.C.E.) regarding the things of heaven. The usual slant of his teaching in the *Analects* is worldly. He wants to clarify human relations and make plain the virtues necessary for a good, harmonious social life. In this work he draws on earlier traditions, attributed to legendary leaders standing at the dawn of recorded history. These leaders carried the reputation of being able to move people by their example. The goodness and moral strength in them exerted a quasi-magical attraction. Confucius appreciated the difficulty of developing this degree of moral strength, but he thought that the Way handed down by the ancient sages could still guide people to a robust humanity.

Because he thought this way, Confucius arranged his life, when he thought back on it, in terms of his increasing identification with the Way:

> The Master said, At fifteen I set my heart upon learning. At thirty, I had planted my feet firm upon the ground. At forty, I no longer suffered from perplexities. At fifty, I knew what were the biddings of Heaven. At sixty, I could hear them with a docile ear. At seventy, I could follow the dictates of my own heart, for what I desired no longer overstepped the boundaries of right.[1]

Here is an autobiographical digest that suggests a spiritual journey. Indeed, it suggests a steady clarification of the will of heaven in Confucius's regard and a corresponding willingness on his part to abide by this will. In fact, the will of heaven, the Way, took on a romantic aura, as though becoming an object of love. To hear its voice became the Master's great delight, as the following famous verse suggests: "The Master said, In the morning, hear the Way. In the evening, die content!" [2]

Should we consider these and a few other texts that seem to take the Master into the mysteries of heaven signs that he was a mystic communing with heaven regularly? Probably not, or if so only with caution. As noted, the general bent of his conversation with his disciples bore on human affairs. For example, when a disciple asked about spirits and the dead, the Master counseled him to place first things first:

> Tzu-lu asked how one should serve ghosts and spirits. The Master said, Till you have learnt to serve men, how can you serve ghosts? Tzu-lu then ventured upon a question about the dead. The Master said, Till you know about the living, how are you to know about the dead?[3]

It is possible that Confucius was in awe of the spirits and so cautioned the disciple not to approach them until prepared well, but overall the

Analects, the text in which we seem to come closest to the Master's own voice, is more interested in human affairs than in theology.

Moreover, the great ambition of Confucius was to participate in public life. He aspired to counsel political rulers so that all the people might benefit from his learning and enjoy prosperity and peace. Teaching promising young men had its consolations, but Confucius thought his life unfortunate because, like Plato, he never found a philosopher-king to implement his wisdom.

The ritual context in which the Master lived subordinated human affairs to the cycles and powers of nature. The rites at which the emperor presided during the imperial festivals of the calendar year presented him as the link between heaven and earth. The things happening on earth, below, ought to take their pattern from the things of heaven, above. The rites sought to please the forces of heaven, for the well-being of all creatures on earth; but to admit that Confucius lived in a culture open to heaven, at least somewhat aware of transcendence and genuine mystery, is not to say that he sought or achieved direct experience of such an ultimate reality. We have scant evidence of that. On the whole, the Confucian corpus remembers its founder as a sage, a person of great practical insight and virtue, not a mystic.

To anticipate our later analysis of Chinese, heavily Confucian approaches to ultimate reality, let us suggest here that the Chinese sage generally appears in the garb of a practical analyst of reasoning nurturing a humane mind. Reaching back to Eric Voegelin's distinction between the noetic and the pneumatic, we might call such an analyst a practical noetician. Inasmuch as Confucius set the great model for the "Teacher" that the Chinese were brought up to revere (eventually becoming a deity), the upbringing of the mainstream targeted self-control, an appreciation of manners, shrewdness about human motivation, and an ability to manage practical affairs. The devoted Confucian was idealistic in the sense of being committed to the persuasive powers of virtue, but the Confucian gaze seldom wandered off into mystical tracks, if such tracks required other-worldliness or any serious disparagement of the body, space and time, or secular responsibilities or enthusiasms. In contrast to classical Greece, where a solid line of philosophers, from the pre-Socratics to Aristotle, worked on the speculative problems of how the mind knows and how its knowing correlates it with being, Confucianism shows only a grudging interest in speculative noetics, the implications of foundational thought and meditation, some centuries after the advent of Buddhism, when its poverty on this score had become a considerable embarrassment. So, whereas Confucian China is a great place to go if one wants epigrammatic advice about human interactions, it is not a great place to go for metaphysical or mystical depth. There is a reverence toward the ancestral spirits and a corresponding concern for close attention to ritual occasions, but there is little labor to experience ultimate reality as such, apart from the particulars in which it might manifest itself.

The second most influential Confucian thinker, sometimes described as Confucius's St. Paul, is Mencius (372–289 B.C.E.), who interpreted the Master's thought (without making it markedly more mystical or metaphysical) more than a century after his death. Like Confucius, Mencius tutored promising young men in what today we would call political science and moral philosophy. Like Confucius, he believed that the key to good rule is the virtue, or the moral force, of the rulers. So, like Confucius, he was much interested in character. A ruler of good character was bound to rule well. A ruler of bad character would always run into trouble.

Mencius went so far as to link this conviction with the mandate of heaven. Traditionally, Chinese rulers had claimed that they received their warrant to rule from heaven. Indeed, they claimed that heaven had commanded them to gain rule and sanctioned their decrees. So when Mencius advanced the notion that heaven could withdraw its mandate, most likely because of a ruler's vice, he opened a door to radical criticism, perhaps even to revolution.

Consider the following text in which Mencius makes clear the freedom of heaven to uproot old rulers and plant new ones:

> Mencius said, "When the Way prevails in the Empire men of small virtue serve men of great virtue, men of small ability serve men of great ability. But when the Way is in disuse, the big serve the small, the strong serve the weak. Both are due to heaven. Those who are obedient to Heaven are preserved, those who go against heaven are annihilated. . . . The *Book of Odes* says, 'The descendants of Shang exceed a hundred thousand in number, but because God so decreed, they submit to Chou. They submit to Chou because the Mandate of Heaven is not immutable. The warriors of Yin are handsome and alert. They assist at the libations in the Chou capital.' "[4]

The Duke of Chou was a great hero to Confucius. The *Book of Odes* was an ancient text that nurtured the Master's reflections. Mencius is following the example of Confucius. When he wants to make the point that virtue is the crux of maintaining the mandate of heaven, he reaches for an illustration that would have pleased his teacher. The greatness of Chou, focused in the duke, was not its population or army. The greatness was the favor that heaven showered upon it because, to Mencius's mind, of its uprightness. It was a realm run honestly and with compassion and decency. Shang and Yin had more secular power, larger numbers, and more prepossessing warriors, but they did not have the mandate of heaven because they were not worthy of it as Chou was.

There is a sense of wonder running through this text. Mencius is intrigued that heaven should work as it seems to. Still, this intrigue does not lead to a regular meditation on heaven itself—what the sacred realm is in its own being, how human beings happen to be able to know it, what full openness to it and transformation by it would entail. These are the questions of mystics, at least those with a speculatively noetic bent. The questions of mystics with a pneumatic bent would veer toward love of heaven, eros for its beauty, hope to become united with it, depth calling to depth.

The questions that we find in *Mencius* fall into the same genus as the questions that we find in the *Analects*. Direct experience of ultimate reality lies outside the basic horizon. At most, it nibbles at the fringe, where the sun rises, as a further possibility, a land yet to pursue.[5] The mystic is always only one step from being a radical social critic, asking for a reworking of the foundations of human polities. There is little radical about either Confucius or Mencius. Both are critical, but both are also traditional and pragmatic, little inclined to waste time on why there is something rather than nothing or how one can solve the problem of suffering in a bedrock way so as to transform human nature to the core.

Hsun-tzu (ca. 312–238 B.C.E.) followed Mencius and promulgated both his ideas and Confucius's. However, whereas Confucius took a median view of human nature, thinking that education could make it viable, and Mencius took an optimistic view, thinking that people were born good, Hsun-tzu thought that human nature was naturally bad, so people needed a strong government to control them. Consider, for example, the following famous paragraphs:

> The nature of man is evil; his goodness is only acquired through training. The original nature of man today is to seek for gain. If this desire is followed, strife and rapacity results, and courtesy dies. Man originally is envious and hates others. If these tendencies are followed, injury and destruction follow; loyalty and faithfulness are destroyed. Man originally possesses desires of the ear and eye; he likes praise and is lustful. If these are followed, impurity and disorder result, and the rules of proper conduct and justice and etiquette are destroyed. Therefore to give rein to man's original nature, to follow man's feelings, inevitably results in strife and rapacity, together with violations of etiquette and confusion in the proper way of doing things, and reverts to a state of violence. Therefore the civilizing influence of teachers and laws, the guidance of rules of proper conduct and justice are absolutely necessary. Thereupon courtesy results; public and private etiquette is observed; and good government is the consequence. By this line of reasoning, it is evident that the nature of man is evil and his goodness is acquired.
>
> Crooked wood needs to undergo steaming and bending to conform to the carpenter's rule; only then is it straight. Blunt metal needs to undergo grinding and whetting; only then is it sharp. The original nature of man

> is evil, so he needs to undergo the instruction of teachers and laws, then only will he be upright. He needs the rules of proper conduct and justice, then only will there be good government. . . . Mencius said, "The fact that men are teachable shows that their original nature is good." I reply: This is not so. This is not understanding the nature of man, not examining the original nature of man, nor the part played by acquired elements.[6]

This debate about original human nature, and also about the functions of nurture or education, might have taken Hsun-tzu into such deeper waters as the relationship between human nature and the *Tao* of heaven. It might have stimulated him to raise questions about the communion with the *Tao*, noetic or pneumatic, that reveals the light by which philosophers such as himself proceed. Either of these issues could have pushed Hsun-tzu toward a mystical experience of the ultimate reality on which Confucian wisdom depended, but neither did. There is little indication that Hsun-tzu became ecstatic for the *Tao*, enstatic (yogic) in an effort to reorient his human nature toward a greater goodness (perhaps through a greater docility to social formation; see his writings on ritual). The blanker, more contemplative mind on which mysticism depends is here submerged in the judicious mind of practical reason. The range of images, fancy, playfulness, and prayer is limited. Here is a good, clear, honest thinker, but one rather pedestrian, at times a snorer. Here is a pundit rather than a seer; here is not a man who has been rapt to the seventh heaven or pushed down to the depths of his spirit, where the ocean of being laps.

Our last specimen of Confucian thought comes from the leading Neo-Confucian thinker Chu Hsi (1130–1200 C.E.), who has exerted enormous influence from his own day to the present. Like the medieval Christian scholastic thinkers, he attempted a synthesis, or systematization, of the prior (Confucian) tradition. In part, his motive was to rebut the impressive systems that Buddhism had developed. Perhaps the closest that Chu Hsi comes to mysticism in this effort is when he reflects on the Great Ultimate:

> The Great Ultimate is merely the principle of heaven and earth and the myriad things. With respect to heaven and earth, there is the Great Ultimate in them. With respect to the myriad things, there is the Great Ultimate in each and every one of them. Before heaven and earth existed, there was assuredly this principle. It is the principle that "through movement generates the yang." It is also the principle that "through tranquillity generates the yin." . . . The Great Ultimate has neither spatial restriction nor physical form or body. There is no spot where it may be placed. . . . The Great Ultimate is similar to the top of a house or the zenith of the sky, beyond which point there is no more. It is the ultimate of principle. Yang is active and yin is tranquil. In these it is not the Great Ultimate that acts or remains tranquil. It is simply that there are the principles of activity and tranquillity. Principle is not visible; it becomes visible through yin

> and yang. Principle attaches itself to yin and yang as a man sits astride a horse. . . . The nature is comparable to the Great Ultimate, and the mind to yin and yang. The Great Ultimate exists only in the yin and yang, and cannot be separated from them. In the final analysis, however, the Great Ultimate is the Great Ultimate and yin and yang are yin and yang. So it is with nature and mind. They are one and yet two, two and yet one, so to speak. . . . The mind means master. It is master whether in the state of activity or in the state of tranquillity. It is not true that in the state of tranquillity there is no need of a master and there is a master only when the state becomes one of activity. By master is meant an all-pervading control and command existing in the mind by itself. The mind unites and apprehends nature and feelings, but it is not united with them as a vague entity without any distinction.[7]

In stressing principle, Chu Hsi was backing away from the materialism that had ruled Chinese philosophy in the recent past. Principle was more ideal, a source of intelligibility, oriented to the mind. The Great Ultimate was the principle of principles, the ultimate idea or source of intelligibility. It pervaded all things, though yin and yang were its most important localizations. These bipolar forces were imagined as female/male, dark/light, moist/dry, emotional/rational. They ran throughout the universe, imparting the need for balance and complementarity. If harmonious, they ensured that things, personalities, ran well. If out of kilter, they caused upset and pain.

The Great Ultimate occurred in both the tranquillity of yin and the activity of yang. They gave it its primary presences, its mount for riding through reality. Just as the intimacy, the reciprocal relationship, between the Great Ultimate and yin and yang does not mean that they are identical, so the intimacy and reciprocity between mind and what it apprehends, nature or feelings, does not mean that we must not distinguish the two. Mind is more basic. It ought to be the master, for it possesses within itself an intrinsic source of control.

Were Chu Hsi to have been interested in correlating such an intrinsic source of mental control with the Great Ultimate's control of reality as a whole, he might have entered upon properly mystical horizons. As it is, he seems to have worked only rationally, even rationalistically, with little overt dependence on meditation or yogic interiority. His probings of the human spirit when reflecting on *Jen* took him into human subjectivity, but although he wanted the Great Ultimate to be a nonmaterial principle, in this text he approaches it fully objectively, as though it were simply the greatest of things. How it is the thing, the force, the prepresent and prepotent reality grounding his own being and how his own being, as able to know, becomes flooded with light and participates uniquely in the Great Ultimate, making it something subjective as well as objective, is unclear and not, apparently, of passionate interest.

As we have seen, it is precisely questions such as these that were of passionate, compelling interest to the Indian philospher–mystics, as they were to the classical Greeks. So although Chu Hsi advances beyond the classical Confucians in his probings of the correlated structures of nature and mind, he remains less than a fully liberated, adventurous noetic speculator. He is not at home with the dynamics of his own spirit, let alone with the idealistic proposition that this spirit creates the world from within as much as it receives it from without.

Classical Taoism

"Taoism" is a problematic term. Confucians loved the *Tao* of the ancients and tried to promulgate it. Indeed, Confucius went out of his way to assert his traditionalism: he was no innovator, rather a disciple of the great sages of old. Significantly, therefore, Confucians were Taoists in that they were partisans of the ancient Way.

The Taoism of such seminal thinkers as Lao-tzu and Chuang-tzu also claims to be ancient, but its roots lie less in centuries-old ritual or political traditions than in contemplations of the Way of nature. The *Tao te Ching* of Lao-tzu and the *Chuang-tzu* collect sayings and stories about masters who rode their spirits out of everyday, pragmatic reality, delighting in what a Westerner might call natural transcendence and mystery. Seldom do their rides take them to a personal divinity. Usually they remain within what Eric Voegelin has called "the cosmological myth," a master story that makes natural creation the entirety of reality (and that has no divinity independent of the natural world creating it from nothingness). However, their travels in imagination and spiritual transport allow the classical Taoist poets to overturn the workaday world. Indeed, these poets frequently mock the Confucians as dullards.

A third meaning of "Taoism," with which we deal best under the rubric of folk religion, is the strand of native shamanic practices and vaguely meditative–hygienic regimes for gaining immortality that were influential among the common people. When the dictum "In office a Confucian, in retirement a Taoist" arose, it was the poetic, refined Taoism of Lao-tzu and Chuang-tzu that came to mind. The dictum applied to the educated and highly cultured. For the common people, amulets, tapers at the shrine of the ancestors, diets to balance yin and yang, even exorcisms meant more. Astrology, geomancy, and divination controlled when to build a house and how to position it, when to marry, and how to take the next steps into the future. Buddhist notions of *karma* interacted with the Taoist fatalism about death and the turning of the Great Clod (the whole clump of natural creation). Confucian social values and hierarchies shaped family life, kowtowed before the lord of the manor, and guided dealings with the local bureaucrat.

In this folk context, Taoism carried along some of the most ancient, preclassical (pre-Confucian) animism and stupor before the prodigies of both nature and the human spirit. After the advent of Buddhism, which featured a well-formed cadre of monks and nuns, Taoism also developed a priesthood and a church. At times Taoist millennialists, no doubt influenced by Buddhist ideas about the coming of Maitreya, the future Buddha, struck the government as dangerous radicals or political revolutionaries. So, overall, popular Taoism was a wild, mixed, wooly bag, fascinating to the cultural anthropologist but seldom impressively mystical in our stated sense of the direct experience of ultimate reality.

Just as the first chapter of the Buddhist *Dhammapada* sets the tone of what follows and so is deservedly famous, so the first chapter of the *Tao Te Ching (The Way and Its Power)*, attributed to the legendary Lao-tzu (perhaps a contemporary of Confucius in fifth-century B.C.E. China, though the compilation of the *Tao Te Ching* seems to have occurred one or two centuries later), sets the tone of what follows and is also deservedly famous:

> The Way that can be told of is not an Unvarying Way; the names that can be named are not unvarying names. It was from the Nameless that Heaven and Earth sprang; the named is but the mother that rears the ten thousand creatures, each after its kind. Truly, "Only he that rids himself forever of desire can see the Secret Essences"; He that has never rid himself of desire can see only the Outcomes. These two things issued from the same mould, but nevertheless are different in name. This "same mould" we can but call the Mystery, or rather the "Darker than any Mystery," the Doorway whence issued all Secret Essences.[8]

The Way can be told, the *Tao* can be named, but not completely, not even adequately. The author has perceived more than what he can say. Already, then, we are hearing a report or a warning from the threshold of mysticism. Already the author is leaning over the abyss, mustering the courage to look down into what may be ultimate. Why do the names that we can utter vary? Because they come from the workings of the abyss on our minds and spirits and we cannot control the abyss. Always ultimate reality is more than what we can handle, so always the permutations it displays across our imaginations vary, like the patterns of glass in a kaleidoscope.

Heaven and earth, whether we take them as prosaic names for above and below or sacramental symbols for the divine and the human, vary and are not ultimate. Something more primal, primitive, powerful gives them their rise. This something escapes our ability to name it. (In many ancient cultures, to possess the name of something was to control it, to imprison it even in a ritual guaranteed, magically, to make it one's servant.) Lao-tzu intuits that the whole realm of naming, of what the Buddhist thinks of as name and form, is not the ultimate reality; it is, in Buddhist terminology, samsaric rather than nirvanic.

The ten thousand things are the entirety of creation—all of the rocks and trees, animals and people, stars and waves, in their endless varieties. These come from the named: "nature," "heaven and earth," "the Great Clod." It is their mother, bringing them forth year after year, from the fertility of earth below to the sky above, from the creativity of the human mind–body complex. However, they change, pass, submit somewhat to our control, either physical or mental. More permanent, and to Lao-tzu more interesting, is what lies behind them, at their core, as their source.

How does one get to this source? A first step is to rid oneself of desire. Apparently without Buddhist influence, the classical Taoists realized that the world of the ten thousand things was a world of psychic investment, biological urgency, and passionate attachment. Living in it, one was hard pressed to sense or find the Way that is unvarying, the source that is nameless, darker than any mystery. So, in retirement if not before, sages tried to drop desire, to purify their spirits to deal with deeper things, "secret essences." They tried to step back from "outcomes," results, successes and failures, probably in the spirit of Indian *karma-yogins,* people trying to purify their action or work and deal more simply with what is, what exists more primordially than the psychic and the political, the surges in our emotions and the fortunes in our towns.

Both the secret essences and the outcomes issue from the same mold, the mysterious source. We name them differently, with good reason, but they remain equally penultimate. The mold, or the source, is the more fascinating reality. We can call it only the mystery because as soon as we sense it we realize that it defeats our understanding. We are primed to deal with finite entities, things of space and time. Infinite or purely spiritual entities we have to deal with negatively, through denials of space and time, boundaries and limits. The light of our minds seems to go out, and we have to grope through unknowing, at the depths of our spirits, like miners deep in the shafts.

We never get to the end of the shaft, through the doorway, unless the darker-than-mystery takes us by the hand and drags us through. Classical Christian mystics have described such a transition as "infused contemplation" and considered it to be mysticism proper, the prime analogue. It is passive, rather than active, a free work of divine grace, rather than anything that human effort can achieve. Ultimate reality takes us to itself, and we cannot resist. On our own (though guided by the mystery constantly), we may get to the doorway, if fortune is favorable. To pass to the far side and get to the wisdom beyond, to dwell in the interior castle onto which the doorway opens, is beyond us. The end of the journey, the consummation of the love, depends on the slow yet strong hands of an Other.

Already in this book we have gravitated to the exchange, the spirit-to-spirit flow or intercourse, the *metaxy* (to draw on Plato) wherein ultimate reality and human intelligence form one another mutually. Perhaps the text in which Lao-tzu best shows us the impact of the darkness of the final mystery on his own spirit, becoming poignantly personal, is Chapter 20 of the *Tao Te Ching*. After a call for banishing learning and getting back to simplicity, the sage lets his bruised soul cry out:

> All men, indeed, are wreathed in smiles, as though feasting after the Great Sacrifice, as though going up to the Spring Carnival. I alone am inert, like a child that has not yet given sign [become fully alert, able to discern its world]; like an infant that has not yet smiled. I droop and drift, as though I belonged nowhere. All men have enough and to spare; I alone seem to have lost everything. Mine is indeed the mind of a very idiot, so dull I am. The world is full of people that shine; I alone am dark. They look lively and self-assured; I alone depressed. I seem unsettled as the ocean; blown adrift, never brought to a stop. All men can be put to some use; I alone am intractable and boorish. But wherein I am most different from men is that I prize no sustenance that does not come from the Mother's breast.[9]

The mother's breast is the unnameable *Tao*, the mystery darker than any wonder we can conceive. A true mystic, Lao-tzu wants only the milk of this mother. He knows in his bones that all realism depends on granting the *Tao* full priority. Until the ten thousand things are perceived as taking their patterns from the ultimate, the source without further source, we do not sense them aright and so use them badly. Moreover, only communion with the ultimate gives Lao-tzu full satisfaction. None of the ten thousand things, being partial, can fill the hunger that the mother's breast stimulates. Once a mystical spirit has tasted the All, the partial seems like near beer or watered wine.

Until they have completely acclimated themselves to loving the *Tao* wholeheartedly, mystics usually feel out of sorts. They are lonely because the lemming crowd rushes off in other directions. They feel awkward, clumsy, dull because the quick, smooth, bright things of ordinary life, daily business, hold no appeal. Often they are not good at these things and have little facility for small talk. When you combine a sense of social dislocation with an even deeper sense of being lost in ultimate mysteries, you can easily create a personality come into crisis, suffering profound alienation. Lao-tzu feels alien to the entire realm of the ten thousand things. He is a creature uncomfortable with creation, especially its human realms. Nature may appeal to him because often nature seems more transparent to the *Tao* than social relations do, but even the things of nature, the rocks and birds and winds, are so much less than their source, the ultimate, the one mother, that they can oppress the mystic's spirit. Thus darkness becomes blessed, much more so than superficially clear light, and tastelessness, dryness, lack of feeling become preferable to anything vivid because anything vivid, sensible, emotionally charged becomes cloying or threatens to obscure the dark primordial mystery which increasingly seems much richer, much realer, than anything else the mystic has known.

In a full study of the *Tao Te Ching* we would pay considerable attention to the leading symbols that Lao-tzu developed for the *Tao*, such as the uncarved block, the infant, the female, the valley, and more. We would also stress the not-doing (*wu-wei*) characteristic of the Taoist sage, the active passivity or seconding.[10] Here, spatial limitations suggest moving on to a second much beloved Taoist classic, the *Chuang-tzu*, to fill out our sense of the personal orientation, the cast of heart and mind, that most pleased those who retired to draw their best nourishment from Taoist sources.

Chuang-tzu was another legendary master, who probably lived around 300 B.C.E., and whom some commentators consider more creative than Lao-tzu, though the book carrying his name has never gained the scriptural status that Taoists have accorded the *Tao Te Ching*. The following text illustrates the cast of mind characteristic of Chuang-tzu:

> Chien Wu said to Lien Shu, "I was listening to Chieh Yu's talk—big and nothing to back it up, going on and on without turning around. I was completely dumbfounded at his words—no more end than the Milky Way, wild and wide of the mark, never coming near human affairs!"
>
> "What were his words like?" asked Lien Shu.
>
> "He said that there is a Holy Man living on faraway Ku-she Mountain, with skin like ice or snow, and gentle and shy like a young girl. He doesn't eat the five grains, but sucks the wind, drinks the dew, climbs up on the clouds and mist, rides a flying dragon, and wanders beyond the four seas. By concentrating his spirit, he can protect creatures from sickness and plague and make the harvest plentiful. I thought this was all insane and refused to believe it."
>
> "You would!" said Lien Shu. "We can't expect a blind man to appreciate beautiful patterns or a deaf man to listen to bells and drums. And blindness and deafness are not confined to the body alone—the understanding has them too, as your words just now have shown. This man, with this virtue of his, is about to embrace the ten thousand things and roll them into one. Though the age calls for reform, why should he wear himself out over the affairs of the world? There is nothing that can harm this man. Though flood waters pile up to the sky, he will not drown. Though a great drought melts metal and stone and scorches the earth and hills, he will not be burned. From his dust and leaving alone you could mould a Yao or a Shun [ancient sages]! Why should he consent to bother about mere things?"[11]

Several central themes run through this dialogue. First, Chuang-tzu is keenly aware that what we can see, what we are willing and able to accredit as real, depends on our perspective and stature. To the blind, colorful things might as well not be. To the small-minded, the vistas of the large-souled make no sense. Everything is quite relative. Only the sage who reaches toward the unity behind the ten thousand things, whose spirit has penetrated toward what Lao-tzu calls the unvarying Way and darker than any mystery, verges upon something not relative, something absolute, ultimate.

Second, the holy man depicted as living on the far-away mountain is both alienated from ordinary existence and full of spiritual power. He lives removed from worldly affairs, but his removal has given him the ability to heal and see. The many stories of magical power that we find in Chuang-tzu, along with the marvelous poetic images, seem best understood as flowing from perceptions of spiritual movement, darts and journeys of the alert, contemplative person's soul. If these darts and journeys take the contemplative to union with the *Tao*, the source of all order and life, the contemplative may become a channel of power, a focus for healing, both physical and spiritual. The way to gain union with the *Tao* is to clear one's mind, empty one's spirit, ride the inspirations that the *Tao* gives, be willing to descend to the depths and brave the darkness that ultimacy entails.

Thomas Merton, no scholar of Chinese texts but an American monk attuned to the Taoist temper, once rendered a famous section of the *Chuang-tzu* as follows. Prince Wen Hui has admired the work of his cook, who is a master-carver and a great exemplar of the *wu-wei* characteristic of the *Tao*. He asks the cook what his method is, but the cook doesn't think in such prosaic terms:

> What I follow is Tao, beyond all methods! When I first began to cut up oxen, I would see before me the whole ox, all in one mass. After three years I no longer saw this mass, I saw the distinctions. But now I see nothing with the eye. My whole being apprehends. My senses are idle. The spirit, free to work without plan, follows its own instinct. Guided by natural line, by the secret opening, the hidden space, my cleaver finds its own way. I cut through no joint, chop no bone.
>
> A good cook needs a new chopper once a year—he cuts. A poor cook needs one every month—he hacks! I have used this same cleaver nineteen years. It has cut up a thousand oxen. Its edge is as keen as if newly sharpened. There are spaces in the joints; the blade is thin and keen, as if newly sharpened. When this thinness finds that space, there is all the room you need! It goes like a breeze![12]

"This thinness" stands for our human spirit, honed and ready. "That space" stands for the Way through reality, the ten thousand things, the field that human beings have to negotiate. The task is to find the space, the passage, the opening that takes us through. The best way to do this is to concentrate, to make ourselves thin and small, to notice the low trail that most previous passersby have overlooked. Once under way, on our own, apart from the madding crowd, we may discover that the low trail rises up, giving us plenty of room. We may start to swing along briskly, a wind at our back, a breeze encouraging us. The way that we can walk or sail may not be the ultimate, comprehensive way, but our little way draws upon such an ultimate. Indeed, the farther along we go, the better pilgrims or sailors or flyers we show ourselves to be, then the closer we are likely to come to the Great Way, the path that becomes the pattern of all natural things.

When we come to that Great Way, we have, in Taoist terms, become mystics. What is ultimate has become ultimate for us, the formative Other showing us how things actually stand, what the wise really see. The classical Taoists were poets more than philosophers. They did little with the technical side of noetic analysis, though a fair amount with the imaginative side. However, they give enough hints of meditative regimes, the love of solitude, and the contemplation of nature as a whole to make us confident that they knew their business. They sought the mother's breast, the ultimate thule, the *ne plus ultra,* which for them was the Way coincident with the real, the holy, the source of fulfilling and healing power, the last and first and best that they could imagine.

Chinese Buddhism

Buddhism was operating in China by the middle of the second century C.E.[13] The gross characterization that the manuals offer is that the Chinese adapted Indian abstraction to their native concreteness. They preferred poetic images to analytical precisions, wholes to distinctive parts. Chinese family traditions made Indian monasticism suspect, and though the *sangha* took hold well, a movement to rehabilitate lay life ended by glorifying married *bodhisattvas* such as Vimalakirti.[14] The Chinese reverence for ancestors and lineage helped to establish a strong tradition of gurus handing on wisdom person to person, generation after generation. Finally, the more precisely meditative Chinese regimes, such as Ch'an (or in Japan Zen) drew on Taoist terminology for their translations from Sanskrit, moving Buddhist meditation close to the spiritual world of Lao-tzu and Chuang-tzu.

The account of the enlightenment of the sixth Ch'an patriarch Huineng (eighth century C.E.) certainly serves polemical and political ends, but enough of the experiential basis for the history or legends about this influential master remains for us to appreciate what Buddhist mystical aspiration had become in China after perhaps 500 years of acculturation.

In the story, the fifth patriarch has set his disciples a test. The one who writes best of enlightenment will become his successor. The leading candidate writes, "The body is the Bodhi [wisdom] tree, the mind is like a clear mirror. At all times we must strive to polish it, and must not let the dust collect." [15] This is a good effort, but it does not please the patriarch completely.

Hui-neng is an illiterate boy working in the kitchen of the monastery. Eventually, he hears the verse of the leading candidate and offers his own, different verse in response. His verse runs, "Bodhi originally has no tree, the mirror has no stand. Buddha nature is always clean and pure; where is there room for dust?" Alternatively, he writes, "The mind is the Bodhi tree, the body is the mirror stand. The mirror is originally clean and pure; where can it be stained by dust?" [16]

The whole story, including its climactic establishment of Hui-neng as the sixth patriarch, glorifies both simplicity and purity. Enlightenment, full mystical experience of ultimate reality, is not a matter of learning so much as of letting one's original nature flourish. The school that Hui-neng came to champion believed in sudden enlightenment. Certainly, people benefited from leading good moral lives and meditating, but wisdom could not be contrived, manipulated, or expected as the fruit of mere technique. The problem with the first candidate's verse was that it implied wiping away dust—ascetic effort based on a misperception. The mind itself, the integral human spirit, is free of all dust, or imperfection. It is light, wisdom, ultimacy in us, pure and simple. How we manage to darken it and trap ourselves in karmic concerns that are only proximate is unclear. Ignorance is as great a problem for Buddhists as is sin for Christians.

Nonetheless, the practical import of the sixth patriarch's view of the mind and ultimate reality is a great vote of confidence in peace. Let disciples simply do what they ought, letting all ambitions and agitations fall away, and gradually their pure, original natures will exfoliate. One day, they will bud into enlightenment. This may involve some effort, but it will not require strain. In Taoist terms, it will proceed through *wu-wei,* an active attention, letting go, moving with the light, the being, that always holds the primacy.

The Mahayana Buddhist stress on emptiness plays into both this style and this rationale. If there is no dust, there is only purity, which can present a far face of emptiness. The Yogacarist and Madhyamikan developments that emphasized the intrinsic lightness of the mind (the ultimacy of mentality) and the lack in all things of an own-being buttressed the Ch'an predilection for simplicity and wholeness. Disciples could sit in meditation simply, with or without *koans* to anchor their minds or *mantras* to focus their breadths. The mind itself could be the empty focus.

In Japan Ch'an (*Zen*) came to animate such diverse cultural forms as the tea ceremony, swordsmanship, other martial arts (for example, archery), and flower arrangement. It gave native Shinto sensibilities depth and focus, letting the beauty of moss or green bamboo signal the wonderful light, being, ultimacy which reveals itself to an alert spirit, a mind confident and poised peacefully because aware of its intrinsic perfection. In these ways, mind and emptiness became staples of a religious aesthetic, a way of angling into the world, especially the natural world, that hoped to find purity and light.

Another important school of Chinese Buddhism, Hua-yen ("garland"), accepts the intrinsic perfection of the mind but qualifies the suddenness of enlightenment. The act of enlightenment itself has to come in a flash, carrying the disciple across what had been a block or divide into a new zone of light, but the causal factors at work in this act and the necessity for disciplined training before the act and further work after it may be more important than Hui-neng suggests.

Hua-yen loves the interconnectedness of all realities, seeing them as endless facets of a primordial mirror. The light shines from one to another, and often it is hard to know what is the primacy instance and what the reflection. Thus the part of enlightenment that comes prior to the flash of light may be more intrinsic to the crucial moment, more a presence of the essential bodhi nature, than what common sense or average psychology suspects. Cause and effect may be entwined, both psychologically and metaphysically, like strands of a web, rather than related simply as stages.

In Francis Cook's study of Hua-yen, this metaphysical suspicion supports a more methodological approach to enlightenment than what we find in Hui-neng's southern Ch'an school:

> Now, if the effect or fruit of the Bodhisattva's causal practices is Buddhahood, and if cause and effect are identical, Fa-tsang's [Hua-yen's] comment is tantamount to saying that the causal state must be Buddha-result itself. This is a thorough mixing of cause and effect, because what is conventionally considered to be the fruit of a long sequence of acts is now said to be that which is present right from the beginning and, indeed, that which is performing the causal acts. . . . All schools of Buddhism agreed that the enlightenment event was of necessity a sudden one. Rather, the question centered around the issue of whether enlightenment had to be preceded by a progressive course of moral improvement. The "suddenists" such as Hui-neng denied the necessity of prior training, and their basis for this was the conviction that beings are already enlightened, albeit ignorant of their enlightenment. . . . The [cognate Hua-yen] conclusion was that the Buddha seed is present right from the beginning, requiring only the nourishment of practice in order to sprout, flourish, and bloom. It is really the intrinsic Buddha-nature which acts to realize itself, not the individual. That is, Buddhahood is its own cause. Such an idea, so pregnant with possibilities for the career of the Bodhisattva, eventually culminated, in fact, in the perceptive and profoundly religious observation of [the Zen master] Dogen, who said that "no ordinary beings ever became Buddhas; only Buddhas become Buddhas." This can only be a world in which the Buddha realizes himself in and through beings in an act of cosmic, eternal self-limitation.[17]

We see here the push of the Buddhist metaphysical mind to find how Buddhahood comes to awareness, dawns, in the human spirit. Moved by a conviction that Buddhahood itself must be the foundation and prime agent of enlightenment, the Ch'an/*Zen* masters debated how best to prepare for the moment of dawn. In the process, the more speculative among them, such as Fa-Tsang, realized that cause and effect, training and result, overlap and flow together. They do not relate simply as successive pages, obvious prior and posterior. Inasmuch as the whole process depends on Buddhahood, at every stage the key element is the same.

It is a matter of convenience how we describe the workings of Buddhahood. It is a matter of experience (Hua-yen would say) that practice both before enlightenment and after helps Buddha nature to work. Buddha nature can work as it wishes, and any good result comes from Buddha nature, rather than the individual worker. However, the fact seems to be that Buddha nature regularly works progressively, processively, and this fact invites the speculatively mystical mind to diversify "enlightened being" so that this synonym for Buddha nature can ripple or shine at many points, in many reflective mirrors, at the same time. The reality, the ultimacy, into which enlightenment, Chinese Buddhist mystical consummation, takes us can be rich as well as simple, layered and staged as well as all of a piece.

All figures finally fall away and are shown to be inadequate, but it is significant that a school such as Hua-yen felt the need, the desire, to challenge the less intellectual, at times the anti-intellectual, simplicity of some Ch'an/*Zen* masters. Buddha nature need not appear only in the mode of undifferentiation or the void. It can also dazzle us, like the angles of a diamond cut splendidly or the glistening net of a god (Indra) who dips into the sea of being, collects this and that, and ties everything into the comprehensive, ecological web that the process of enlightenment generates, the movement of the Buddhas in and out of time and limited mind.

Chinese Folk Religion

The peasant strand of Chinese religion, which has always been the most populous, has taken into itself all the "higher" developments. Confucian, classical Taoist, and Buddhist ideas, practices, and sentiments have mingled with prehistorical loves and fears of local gods, demonic spirits, and agents of fortune. Folk religion has been naturalistic, in the sense of being tied intimately to the cycles of summer and winter, planting and harvest. It has sought help and orientation from the stars and planets. Diviners, healers, and shamans have played large roles, as have priests and exorcists. Reverence for the ancestors, the sacred line down which life had come, has dominated most households. From the time of Buddhist influence, concern for good *karma* and a better future life have colored peasant morality. Children would still burn paper money or set aside special gifts to help their parents in the afterlife, but a brighter notion of paradise could blaze, inspired by Buddhist imagery.

Little in this rich swirl of peasant religion was mystical in the formal sense of experiencing ultimate reality directly, but most of it could bear a mystical interpretation if a Taoist or Buddhist sage chose to relocate it in terms of the ultimate Way or the Buddha nature. On its own, apart from such relocation, it strikes the comparativist as what some relevant texts call "magic" (the power to move spiritual forces to one's side). In this section we suggest how Chinese folk religion might appear in such a relocation, even as we fill out the picture of daily life, the average Chinese effort to harmonize human existence with the Way of heaven, that we have been assembling.

Marcel Granet, a leading early-twentieth-century scholar of Chinese religion, has provided the following image of how traditional Chinese peasants experienced their natural world at festival time:

> They experienced the presence of a tutelary power whose sanctity sprang from every corner of the landscape, blessed forces which they strove to capture in every way. Holy was the place, sacred the slopes of the valley they climbed and descended, the stream they crossed with their skirts tucked up, the blooming flowers they picked, the ferns, the bushes, the white elms, the great oaks and the wood they took from them; the lit bonfires, the scent of the nosegays, the spring water in which they dipped themselves, and the wind that dried them as they came from bathing, all had virtues, unlimited virtues; all was a promise given to all hopes. And the animals which teemed and also held their seasonal assemblies, grass-hoppers gathering under the grass, the arrested flights of birds of passage, ospreys gathered together on sandy islets, wild geese calling to one another in the woods, all were part of the festival and shared in the holiness of the place and moment.[18]

Granet does not deny that this romantic interpretation has to square with the grinding facts, the obdurate poverty, of most peasants' lives throughout most of the year; but he reaches for an inkling of what might have kept the common people going, what peak moments, what mystical intuition, might have made their hard lives worthwhile. A sense that now and then the earth itself revealed a holy beauty, offered a winning fertility, might have been enough. A hope that a few times a year the sacred forces would manifest their reality and display their reliability could have done the job. Human beings do not require a great deal. It is enough that they can continue to hope that their lives will make sense, their pains not prove completely vain. The horror threatening any life, tempting any human being to despair, is that existence, time, being might turn out to be worthless, even sadistic. When peasant rituals tamed this horror, pushing despair away, they accomplished the prime evolutionary function of religion, which has been to keep the venture going, make winning the game still seem possible.

Max Weber, a more famous pioneer than Granet in the sociological analysis of religion, has described the divinity of the Chinese people as follows:

> On the one hand, Chinese antiquity knew a dual god of the peasantry (*Shi-chi*); for every local association, it represented a fusion of the spirit of fertile soil (*She*) and the spirit of harvest (*Chi*). This God has already assumed the character of a deity meting out ethical sanctions. On the other hand, the temples of ancestral spirits (*Chung-miao*) were objects of worship. These spirits together (*She-chi-chung-miao*) were the main object of the local rural cults. As the tutelary spirit of the local community it was probably first conceived naturalistically as a semi-material magical force or substance. Its position roughly corresponded to that of the local deity in Western Asia but at any early time the latter was essentially more personalized. With the increase of princely power the spirit of the ploughland became the spirit of the princely territory.[19]

This rather prosaic view of traditional Chinese religious life reminds us that agriculture and trade, family life and feudal service, tended to run together. What went on at the family shrine related to how the crops were growing, what the local prince required, whether one's children were healthy or sick. People needed enough prosperity to survive, ideally to fall asleep some nights contentedly. They sought this prosperity wherever they might find it: their own hard work, the help of the local prince, the favor of their gods. If making gifts or sacrifices to local gods, spirits of the pond or field, would bring a touch of good fortune, a moment of prosperity, ordinary people would make the gift straightaway. If gathering together at the family shrine would give a household a sense of protection or unity or purpose, gather it would gladly, night after night.

There is nothing mysterious about this tendency, which we find all over the world. Yet it reflects the highly mysterious fact that all human beings are vulnerable to ill fortune and the even more mysterious fact that all must die. Grappling with these realities, both prosaically and romantically, Chinese peasants moved in the traces of the "darker than any mystery," regardless of how clearly they realized it, for they had to sense, in their bare humanity, that their survival was a struggle of spirit, of meaning, as much as of body. So they prayed, celebrated, suffered, endured with more or less wonder that all this was necessary and more or less gratitude that it should be possible.

At a juncture between folk and literate religion we find such influential traditional Chinese practices as divination by means of the *I Ching (Book of Changes)*. This is a collection of hexagrams that entered the

Confucian canon, becoming the basis for commentaries on change, the future, and human nature. In a series of lectures sensitive to the psychological sources of traditional Chinese divination, Richard Wilhelm has spoken of the central concept, change, as follows:

> Change has no limiting form. Change transforms itself and becomes one; one transforms itself and becomes seven; seven transforms itself and becomes nine. Nine is the end point of this transformation. Change then transforms itself again, however, and becomes two; two transforms itself and becomes six; six transforms itself and becomes eight. One is the origin of the transforming form. What is pure and light rises up and becomes heaven; what is turgid and heavy sinks down and becomes earth. Things have an origin, a time of completeness, and an end. Therefore three lines form the trigram Ch'ien. When Ch'ien and K'un [another trigram] come together all life originates, for things are composed of yin and yang. Therefore it is doubled and the six lines form a hexagram.[20]

The "therefore" in this argument gives it away. The change that the text discusses is more metaphysical than physical. No empirical science controls observations of particular changes. The *I Ching* is not interested in the development of the fetus as such, or the storm, or the field of wheat. It is interested in the flux of change that runs through all such development and that seems to run through all creation. When the mind grapples with this flux, it tries to stop it, keep it around, tied down, tamed, long enough to make its acquaintance.

Such an effort through many generations brought popular Chinese religion to a stage where myriad people contemplated the patterns of divining sticks as expressions of the supernatural control of fate or the future or the shape of local dramas. The physical aspects of change blurred with the metaphysical, the psychological, and the magical. Ordinary people working the sticks were not practicing physics or psychology or even religion, at least not purely or with a clear method for determining what they wanted to know and testing whether they had found it. Rather, they were pitching in all their investigative faculties, from the most sober and reasoned to the most suggestible and emotional, trying to take a reading, offer a report, catch a breeze on the wing.

In the sixteenth-century C.E. folk novel *Monkey*, various magical (precise and ineffable) forces combine to power the hero through his endless adventures, which are both physical feats of derring-do and spiritual accomplishments. Notice, however, the overriding concern that Monkey brings forward when he has the chance to question the great patriarch Subodhi:

> "Take pity upon me and teach me the way of Long Life. I shall never forget your kindness."
>
> "You show a disposition," said the Patriarch. "You understood my se-

> cret signs. Come close and listen carefully. I am going to reveal to you the Secret of Long Life."
>
> Monkey beat his head on the floor to show his gratitude, washed his ears and attended closely, kneeling beside the bed. The Patriarch then recited:
>
> "To spare and tend the vital powers, this and nothing else, is the sum and total of all magic, secret and profane. All is comprised in these three, Spirit, Breath, and Soul; Guard them closely, screen them well; let there be no leak. Store them within the frame; that is all that can be learnt, and all that can be taught. I would have you mark the tortoise and the snake, locked in tight embrace. Locked in tight embrace, the vital powers are strong. Even in the midst of fierce flames the golden Lotus may be planted. The Five Elements compounded and transposed, and put to new use. When that is done, be what you please, Buddha or Immortal." [21]

Long life is the great treasure that Chinese adepts have always sought. By alchemy, breath control, good *karma,* or other means, Chinese magicians have tried to preserve their vital powers and ward off death. The common notion has been that a secret wisdom exists by which one can gain long life. Thus Buddhist masters and Taoist sages have loomed as possible saviors or sources of such wisdom. Here Monkey (human nature in its rapscallion, trickster side) beseeches the patriarch to let him in on the secret. What the patriarch chants for him is a potent brew.

There is a vital power in the body—breath, spirit, soul. We must spare and guard it, keep it strong, realize that it determines our density. The admonition to allow no leak suggests the physical overtones: we are discussing a bodily, somatopsychic force. The tortoise and the snake, like the female and the male or the slow and the quick, suggest yin and yang. Kept together properly, they make the vital spirit strong, fertile.

The golden lotus is the law of the Buddha. The flames are the passions, the realm of *samsara.* No matter how troubled one's times, the law can take root, for it is the law of light, the expression of reality. The five elements (water, fire, wood, metal, earth) change, combine and uncouple endlessly, as the *I Ching* suggests. Sages use their magical power to direct this change. That is the most they can do. Call them buddhas or (Taoist) immortals, it makes little difference. In folk religion they have been more alike than different.

A major reason for this is that what ordinary people have sought from folk religion is simply benefit—power against sickness, bad luck, despair, infertility, early death. Whoever, whatever, has showed signs of providing this benefit has been a treasure. Thus the only relevance of any mystical aspect of the patriarch's achievement was its role in his power. "Magic" has been a loose term, covering any capacity to work good, especially prodigious, effects. (We prescind from "black magic.") Unless direct perception of ultimate reality, for example, through meditation, increased a sage's power, such perception has been beside the point. For folk religion, a writhing demoniac could be more interesting than a restrained master of meditation.

Must mysticism become powerful, full of psychosomatic force, if we are finally to take it seriously? What vitality ought to flow from the flood of enlightenment? Stories abound in which mystics talk to the birds, command the winds, raise the ill from their pallets. Age after age, an all too human population longs for signs and wonders. So examining mystical claims inevitably raises questions for discernment. What is authentic power, vision, immortality, and what not? What is pure, holy, transforming force, and what only animal spirits?

Not even rapture, ecstasy to the seventh heaven, removes these hard questions. If only in the aftermath, when life on the ground resumes, we critics must exercise our minds, exorcise our hearts. Folk religion reminds us how tangled a web we weave, when first we set out to deceive *samsara* and enter upon ultimate reality. Nothing in the public realm is pure. Everything public, popular, folk begs discernment, understanding, goodness, faith.

Shinto

Although Confucianism, Taoism, and Buddhism developed distinctive shapes after coming to Japan, one may argue that the Japanese contribution to these traditions did not exceed, or was not more decisive than, that of China (or, in the case of Buddhism, India). In this section we focus on what was decisive for Japanese religious culture: the native naturalistic traditions. Our concern, as always, will be to orient and sharpen our questions about mysticism. For example, how has Shinto aligned the forces at work in the direct perception of reality? What does "the Way of the *kami*" do to the generic mystical sense that anything may become sacred if ultimacy chooses to manifest itself through it? Are there in Shinto any novel, noteworthy impressions of human spirituality that throw a fresh light on the mutual indwelling of being and beings, ultimacy and proximateness?

Let us plunge into Shinto, and the full ocean of Japanese folk religion, through the following observations of Ichiro Hori, a noted authority:

> From the introduction of Buddhism into Japan—or at least from the beginning of the Heian period [eighth century A.D.]—Buddhism and Shinto confronted each other overtly and covertly. This is seen symbolically in the frequent records of incidents in which, for example, the curse or anger of a kami [deity, sacred power] revealed itself in a thunderbolt or a mysterious fire after infringement on his divine territory by the felling of sacred trees in the shrine enclosure for the building of a Buddhist temple.

A shaman would announce the curse or anger of the deity. The mixture of Shinto and Buddhism meant that Buddhism gradually lowered its standards to accommodate the Shinto framework, specifically to cooperate with popular Japanese shamanism. One typical example of these tendencies is the belief in *goryo* [ghosts].

Belief in *goryo* arose at the end of the eighth century and flourished throughout the Heian period [794–1185]. It was originally a belief that the spirits of persons who had died as victims of political strife haunted their living antagonists in their lifetime, and was propagated through the mouths of popular shamanesses. Buddhist priests of the Tantric Tendai (*T'ien-t'ai* in Chinese) and Shingon *Mantrayana* (*Chen-yen* in Chinese) sects who practiced religious austerities and obtained magical virtues in the mountains, as well as powerful Yin-yang magicians, were invited to negotiate with and exorcise these revengeful spirits.

With the popularization of this belief, the possibility of becoming *gyoro,* formerly the privilege of nobles alone, was opened to the common people. Thus, the linking of popular shamanesses with the Buddhist mountain ascetics (*shugen-ja*) became more and more close. For example, during the Heian period almost every Buddhist priest utilized a shamaness or her substitute as a medium during his exorcism in order to know the names of the revengeful spirits and their complaints and curses. On the other hand, one of the earlier social functions of the Japanese popular Jodo (Pure Land) school seems also to have been to transfer these revengeful spirits of the dead into the merciful hands of Amida Buddha (Amitabha Buddha) by the repetition of his sacred name (*Nembutsu*), as well as to cause the believers themselves to arrive in Amitabha's Pure Land after death.[22]

The competition between Buddhism and Shinto was rawest at the edge separating the foreign from the domestic. Buddhism had come to Japan from without—Korea, China, India. After the advent of Buddhism, Shinto articulated itself, becoming partially systematic, but even then it remained embedded in immemorial Japanese traditions. The *kami* were the ancient gods, the 800,000 forces that had always been. No striking rock or tree or shore on the islands fell outside their movement. As incarnated in the sun goddess, the patron of the royal line, the *kami* furnished the foundation of the traditional Japanese social order. As practiced at numerous shrines throughout the land, Shinto conserved the old values by weaving the fabric of sacred meaning age after age. So, to fell trees in a Shinto grove for wood for a Buddhist temple could seem insulting. The old gods could flare at this upstart religious movement and prove unruly.

Hori stresses that both Buddhists and Shintoists accommodated, contributing to a dramatic popular religion, a folk stratum both lively and wide, in which, for example, shamanesses (important in Shinto divination) helped ghosts make their grievances known. He also stresses that some Buddhist schools incorporated such traditional shamanic divination, correlating it with either their own magic (Tantra) or their beliefs about an afterlife in the "pure land" (or Paradise as it is known in the west).

Two points, then, stand out. One is the syncretism, the amalgamation or running together, of Buddhism and Shinto in medieval Japanese religion. At the end of this development, only a few Japanese were pure Buddhists or Shintoists. The vast majority kept reality in boxes with two labels, overlapping loyalties. The mysticism in Japanese folk religion therefore presents few sharp-edged, purely Shinto or Buddhist perceptions of ultimate reality. The sense of sacredness that sustained or transported the average person was a hybrid.

Second, the use of shamanesses, female ecstatics, suggests how popular Japanese religion assumed that the *kami* (or, by extension, the *bodhisattvas*) operated. Crudely, they took over the human spirit, like a lover or a lord. The shamaness, who frequently was blind, yielded herself, her spirit, to make way for the *kami*. The *kami* came into her, using her body and voice for its revelations. When the relationship between the two was stable, one could speak of a marriage. The shamaness might pine while her *kami* was away. Her performance might resemble a homecoming or tryst. Few Japanese shamanic texts offer analyses of the interactions between *kami* and shamaness, force and vessel, illuminating their mutual enlightenment or empowerment. Most descriptions are extrinsic, at best poetic, but the mere fact that in the shamaness divine and human conjoined for illumination, meaning, and resolution is provocative. With or without Buddhist notions of intrinsic purity or divinity, many traditional Shinto rituals made the spirit of a consecrated woman the temple or sacrament of a *kami*.

In developing *Zen*, Japanese Buddhists both capitalized on Shinto naturalism and emptied it of its pluralism. In *Zen* hands, the 800,000 *kami* were condensed into a still small voice, though now the Buddha nature moved with the tides, was at home in the mountains. *Koans, haiku,* cherry blossoms, cups of tea and numerous other particulars gained space, sharper edging and etching, than what Shinto alone could furnish. Ultimacy and emptiness sharpened individual items to a one-pointedness reminiscent of successful meditation.

However, the late Soto *Zen* master Shunryu Suzuki has described the relativizing of meditation that, perhaps paradoxically, contact with Shinto and other native Japanese factors could encourage:

> Baso [1644–1694] was a famous Zen master called the Horse-master. He was the disciple of Nangaku, one of the Sixth [Ch'an] Patriarch's disciples. One day while he was studying under Nangaku, Baso was sitting, practicing zazen [meditation]. He was a man of large physical build; when he talked, his tongue reached to his nose; his voice was loud; and his zazen must have been very good. Nangaku saw him sitting like a great mountain or like a frog. Nangaku asked, "What are you doing?" "I am practicing

> zazen," Baso replied. "Why are you practicing zazen?" "I want to attain enlightenment; I want to be a Buddha," the disciple said. Do you know what the teacher did? He picked up a tile, and he started to polish it. In Japan, after taking a tile from the kiln, we polish it to give it a beautiful finish. So Nangaku picked up a tile and started to polish it. Baso, his disciple, asked, "What are you doing?" "I want to make this tile into a jewel," Nangaku said. "How is it possible to make a tile into a jewel?" Baso asked. "How is it possible to become a Buddha by practicing zazen?" Nangaku replied. "Do you want to attain Buddhahood? There is no Buddhahood besides your ordinary mind. When a cart does not go, which do you whip, the cart or the horse?" the master asked.[23]

The master saw in the disciple an overestimation of meditation. Baso was placing too many of his eggs in that basket. Meditation was only a means to transform the ordinary mind so that it would realize its intrinsic Buddha nature. To use meditation, or any other means, well, one had to estimate it correctly. Any method, practice, attitude, or emotion that eased the emergence of one's intrinsic Buddha nature would do, but no method would help much if it did not bear on the horse, or the crucial factor: the mind itself (the integral spirit). Because of this story, Baso became known as the Horse-master. When he won enlightenment, people smiled to think that Nangaku had stimulated him well.

Baso (also known as Basho) is one of the most famous writers of *haiku*. The following specimen suggests how he condensed the gist of illuminated existence into simple images, moments both physical and mental:

> On a withered branch
> A crow has alighted;
> Nightfall in autumn.[24]

The withered branch evokes the decay of the year in autumn. The crow spreads this mood. The fall of night is dark, cold, forbidding, yet perhaps also sobering. If all things go into night (death, decay, nothingness), we should bestir ourselves to find light and warmth. If ordinary existence is suffering, we should get on with the task of liberation. So, any arresting moment (poets see more than the rest of us) can retrigger the call for liberation. Any crow alighting on a withered branch can caw, "Remember death."

As the description from Hori suggested, popular Japanese piety blended *Zen* sensibilities, such as those of Baso, with millennial Shinto feelings. Note in the following extract from the diary of a fourteenth-century Buddhist priest-pilgrim, Saka, the acceptability of Shinto sentiments. Saka is recording his feelings while at Ise, the greatest of the Shinto shrines and the imperial home of the sun goddess established early in the first century C.E.:

> While on the way to these shrines one does not feel like an ordinary person any longer but as though reborn in another world. How solemn is the unearthly shadow of huge groves of ancient pines and chamaecyprais, and there is a delicate pathos in the few rare flowers that have withstood the winter frosts so gaily. The cross-beam of the Torii or Shinto gateway is without any curve, symbolizing by its straightness the sincerity of the direct beam of the Divine Promise. The shrine-fence is not painted red nor is the Shrine itself roofed with cedar shingles. The eaves, with their rough reed-thatch, recall memories of the ancient days when roofs were not trimmed. So did they spare expense out of compassion for the hardships of people . . . it is the deeply rooted custom at this Shrine that we should bring no Buddhist rosary or offering, or any special petition in our hearts and this is called "Inner Purity." Washing in sea water and keeping the body free of all defilement is called "Outer Purity." And when both these Purities are attained there is then no barrier between our mind and that of the Deity. And if we feel to become thus one with the Divine, what more do we need and what is there to pray for? When I heard that this was the true way of worshipping at the Shrine, I could not refrain from tears of gratitude.[25]

Purity is a leading concept in Shinto, symbolized often by salt. Purity relates to simplicity, lack of adornment. The roots of Shinto lie deep in natural processes. The *kami* mediate the power and beauty of nature to human beings, forming traditional Japanese culture. What is plain, basic, elementary rates higher than what is gilded or full of artifice. For example, the movement of animals, direct and unself-conscious, is more enlightened than the movement of reflexive, conflicted human beings.

Saka thinks that purity, both physical and mental, is the key to union with divinity. Remove defilements—physical dirt, mental soiling—and the gods will draw near, seem intimate. This thought moves him to tears. Divinity, holiness, or ultimate reality is not problematic or obscure. We need only actualize our best potential, remove our worst vices, and we can find the world sacred. The tall trees mold the religious spirit to reverence. The ever-pure, self-renewing sea shows us how freely and unconstrainedly ultimacy moves. On pilgrimage we can break free of the illusions that press daily living out of shape, letting our spirits return to these better realities of sacredness, these more solid truths.

Inasmuch as Japanese folk religion gained ballast, solid anchoring, in Shinto and Buddhist purity such as Saka's, it did not lose its way in low magic. The *kami* remained forces to venerate and serve, not just powers to manipulate. Nature stood apart, in sovereign holiness, beyond the conjuring of diviners and soothsayers. More sensitive people bowed low before nature, bathed themselves in the astringent sea, because they experienced that this was good for their spirits, their best humanity—healing, purifying, challenging, consoling.

Summary

We have surveyed the general history and worldview of traditional China and then examined the major religious traditions: Confucian, Taoist, Buddhist, and folk. We concluded by studying Shinto, the major catalyst for the transformation of these Chinese traditions in Japan, as well as an indigenous tradition interesting in its own right. What are the major points that we should underscore in the file as we store it away, and what are the major implications for our ongoing examination of mysticism?

The overwhelming fact about China, both historical and present-day, is the immensity of the population. Literally billions of human beings have lived their lives and sought meaning through the traditions that we have surveyed. Japan has been a much smaller realm, geographically and demographically, but even Japanese culture has formed hundreds of millions of people. The formative Chinese culture, dominant for well over 2,500 years, was a blend of peasant (folk), Confucian, and philosophical Taoist ideas. The great majority of the people have always been peasants, scratching their sustenance from the land, while since the Han Dynasty an articulate public philosophy, more indebted to Confucius than anyone else, has provided both the governing classes and the merchants with their sense of tradition, hierarchical order, etiquette, and ritual. Religious Taoism, concerned with yin–yang rituals, divination, exorcism, geomancy, astrology, spells, and other colorful concerns dear to the uneducated classes, interacted constantly with peasant concerns for fertility, the turn of the seasons, and protection against the vagaries of nature. Philosophical Taoism, loving the poetry of Lao-tzu and Chuang-tzu, furnished artists and the literati with inspiration for private life, the zone of retirement.

Buddhism brought China a more sophisticated, analytical philosophy than what the native traditions had generated and a clearer ascetic tradition. Even though the Chinese accommodated Buddhism to their own practical bent, they were drawn into deeper waters. Meditation, perhaps especially that developed by Ch'an masters, strengthened the retirement attractive to the refined, educated classes. Philosophical schools such as Hua-yen took Taoist reflections on nature to new levels of dialectical delicacy. Certainly, Buddhist priests developed rituals attractive to a wide range of the population, especially the impressive funeral rites. The monastic *sangha* enjoyed considerable success, so much so that several times government officials feared its power and launched bloody persecutions of it. Buddhism always retained the stigma of being something foreign, even after hundreds of years in Chinese garb, but it adapted much better than Christianity, which after 400 years could boast of only modest numerical success.[26]

Confucian, Taoist, and Buddhist traditions emigrated to Japan, sharpening native Japanese mores, approaches to nature, views of enlightenment, ultimacy, and meditation. The *sangha* flourished in Japan in many different schools, and the Buddhist priesthood became influential. The major interlocutor with these foreign influences was Shinto, the indigenous naturalistic religion later articulated as the Way of the *kami*. Shinto differed from Chinese folk traditions mainly because of the peculiar characteristics of the Japanese islands. The dominant combination of coastal and mountainous areas generated a concern for purity, a response to astringency, that was more muted in China, where landscapes varied more widely. Certainly mountainous China inspired wonderful landscape paintings, some of them virtual essays on emptiness, especially during the Sung Dynasty. However, Japanese moss, flowers, bamboo, and martial arts grew in hybrid, Shinto–*Zen* soil to a stronger discipline, a clearer one-pointedness, than their Chinese counterparts.

We have also noted the significance of Shinto shamanic traditions in Japan and their syncretism with Buddhist views of the *bodhisattvas*. The imperial mythology developed from the Shinto chronicles became the basis for chauvinistic Japanese ideologies, and throughout nationalistic periods Shinto rode high, dominating the most colorful and official annual ceremonies. The note of simplicity, indeed primitivism, that runs through classical Shinto statements and the centralizing of Shinto aesthetics at the imperial shrine at Ise have kept even modern Japan rooted to a perhaps romantic view of the earliest times when the *kami* were closer because the people were purer. Much of the appeal and the undeniable strength of Japanese higher culture, and much of the secret to Japanese discipline, lies in the marriage of Shinto and *Zen* sensibilities. When the warrior classes took *Zen* over for their spiritual rationale, they put a touch of steel into traditional Shinto simplicity. The result could be a striking, elegant refinement, spare and arresting spiritually.

If these paragraphs serve as a quick review of the high points of the traditions that we have examined in this chapter, what ought we to say in retrospect about their contributions to our study of mysticism?

First, Confucianism does not strike the typical comparativist as a formidably mystical tradition. It has thrived long enough and enrolled enough talent to have generated insights and practices that a comparativist can correlate with mystical tendencies seen more clearly elsewhere, but the general tenor of the most influential texts in the Confucian canon is worldly, pragmatic, prudential, political, or moral rather than mystical. The *Tao* that the Confucians venerate is the *Tao* of nature, certainly, but it is also the *Tao* of the ancient sages, the humanistic way to order the people toward peace.

Confucians loved rituals, both naturalistic and social. They thought that their realm below required the blessing and mandate of heaven, the mysterious, ultimate realm above, but heaven was inscrutable. Therefore, the usual occupation of the sage was with the affairs of earth. Confucius became happy whenever he heard the voice of the Way in the morning, but he nonetheless lamented the fact that he had never found a philosopher-king willing and able to carry his teachings into political practice. The most influential Neo-Confucian philosopher, Chu Hsi, responded to Buddhist metaphysical developments by developing what we might call metaphysical and epistemological analyses of the realms of being and mind, but these show little indication of having come from meditation, let alone from seizures of spiritual rapture.

Second, the philosophical Taoists are made of more impressive mystical stuff, though the mold in which they cast it tends to be poetic, paradoxical, and parabolic rather than analytical–epistemological. Lao-tzu and Chuang-tzu clearly were highly sensitive spirits for whom the *Tao* was a personal force, a soulful interlocutor. The Great Clod and ten thousand things that the *Tao* turned were full of wonders, a delight as well as an ultimate context against whom only the fool contended. Wisdom consisted in going with the grain, the tide, the breath of the *Tao,* learning to ride things the way they were. Of paramount importance was the perception that all things were mortal and vulnerable. Success was in great measure surviving, lying low, avoiding notice, not making waves. The spiritual heroes tended to live apart, in solitude, as iconoclasts. They tended to love the old days, the old ways, when people did not get lost in speculations or machinations, when life moved along close to nature. Before there were hundreds of laws, crime was negligible. Before rabble-rousers put in people's heads wild images of gain, most people were content. Lao-tzu and Chuang-tzu intuited with many other mystics that contentment, human fulfillment, is a matter of being and seeing more than having, more even than enjoying, if "enjoying" connotes material pleasures.

This is a mysticism of minimalism and process. The sage, whose wisdom comes from union with the *Tao,* has a rich spiritual life because of a poor material life, a poverty of desires. The sage is alert, flexible, nimble. Sages can move with shifting fortunes, attune themselves to new seasons, both natural and spiritual. When Lao-tzu tells us that he alone is dark and that his spirituality has made him idiosyncratic, dysfunctional, we further realize that the classical unknowing, even the classical desolation, of Western mysticism probably had its counterparts in the souls of the most sensitive Taoists. One could not make the *Tao,* the ultimate mystery darker than any mystery, one's beloved and not feel singular, lonely, alien. One had to pass through stages of muteness, feelings of idiocy, before the concerns of the hoi polloi lost their threats. After enlightenment, rocks again became simply rocks, worldlings again became simply worldlings. During the transitions to wisdom, though, rocks could loom as impenetrable strangenesses, weird impasses on the mountain trail. Worldlings could seem outright crazy, mouths ripped open in insane laughter, raging passions both mindless and useless.

Articulate, reasoned, truly analytical treatments of enlightenment and being occur only in Chinese Buddhist texts (which is not to say that Confucians or Taoists had no experiences of enlightenment or mystical fulfillment). There Ch'an and other masters advance the Indian traditions of *dharma* analysis, often with characteristically Chinese concreteness. The enlightenment experience of Hui-neng, the sixth Ch'an patriarch, occurs in what we can call a mystical text proper. The mutual indwelling of human spirit and objective Buddha nature recalls the prime mystical site, the operative mystical metaxy, that we have seen in Indian texts, both Hindu and Buddhist. The Chinese convictions about the suddenness of enlightenment and the intrinsic purity of the mind carried over into *Zen* circles, but even without the further polish that Japan gave them they produced impressive results. Silence surrounds the intrinsically enlightened spirit. *Koans* and other mental apparatuses tend to fall by the way, reduced to the order of means. The simpler the sitting, the moving, the thinking, the better. The more clearly the focus falls on the central light of all things, the empty ultimacy, the better oriented, the wiser, the vision seems to be. The way of practice tends to prevail over the way of speculation.

Mysticism is the direct perception of the empty ultimacy, which is also the *pleroma*, the comprehensive fullness of being and meaning. This ultimate is impersonal, personal, and beyond such distinctions. It is what it is, and its being is the source of all other being. Nothing can provide us with its clarification or exegesis. It must communicate itself, give its light or speak its word, as it sees fit. Generally, it does this inasmuch as our human spirits receive or generate meaning, light, or significance. The fact that we have ideas, move into possibilities, can make things both material and conceptual, ties us into Buddha nature. If stones live in intelligibility, we are hard-pressed to find the refractions. If water sings songs, it must do so at frequencies for which we have no ear. The same with plants and trees. Animals, with their more palpable intelligence, are a different story, but once again we are largely locked out. Probably we read more into the inner lives of animals, projecting from our own psyches, than we draw out certain significances, innate perceptions, or narratives. The fact remains that mysticism is a human phenomenon, even though sometimes (for example, in certain yogic forms) it seems hell-bent on trying to evacuate from consciousness everything sensible, imaginary, and so indebted to the animal part of our humanity. Mysticism is a fact of mind, awareness, inner light.

The engine of mysticism is the perception of ultimate reality, the being struck by the power of God or flooded with the light of Buddha nature. How this happens mystics babble endlessly to explain, finally wearying their way to ineffability. The fact, the experience, is more cogent than most explanations and, once appreciated, seems primordial, the fact-of-facts, the one thing necessary and directive. Let someone (for example, a Blaise Pascal) once meet or receive the living God, and everything changes; thereafter nothing is the same. Let a monk taste *nirvana,* cross to the "wisdom that has gone beyond," and everything limited, mortal, compounded of elements other than purity and light, seems at best provisional, drawing its bearability from the sustaining presence in it of what is absolute.

We may say, then, that mysticism is a function of spiritual realization. The direct perception of ultimate reality shows us that we are spiritual, something other than the matter of our bodies, even the fantasies of our minds that we draw from matter. We cannot define this spirit adequately, this spark in our clod, but we know that it gives us our wonder and pathos. Through it, in it, we sense, now and then even meet, what may be unconditioned—spirit with no debts to matter, meaning so pure that it just is, beyond and below and through all particulars, all meanings that fix and contain and sacramentalize it, perhaps beautifully but still achingly inadequately.

This relates mysticism to death. Perhaps through death the ache and inadequacy will pass and something whole, fully healthy, completely satisfying will come to be, abiding permanently, stabilizing us in light, being, love. Paradoxically, the agony of death may be the defeat of *samsara,* our exit from the ten thousand things, the turn of the Great Clod that delivers us not simply back to the earth to nourish the worms and the next cycle of mulching evolution but also to heavenly freedom, the ride of the spirit that need not end, the mountains that need no peaks but stretch on without term.

Neither Chinese folk religion nor Shinto traditions added significantly to the high traditions of East Asian mysticism. In our view, Buddhist and Taoist sources continued to furnish the most impressive mystical voices. *Zen* and Japanese developments of Mahayana metaphysics have formed many impressive masters of meditation in Japan, and they continue to do so today.[27] The general impression they give is of solid, occasionally deep, and especially pointed appropriations of emptiness. The mysticism of light and darkness, lack of own-being and so presence of ultimate reality as dancing flux, continues to prove attractive to many, all the more so as it gives soul to lovely ceremonies, a rich array of aesthetic gifts in architecture, sculpture, classical theater, flower arrangement, gardening, and more.

Mysticism in the generalized, popular sense of a special sensitivity to the natural environment, and often a delicate awareness of interior, psychological movements from an empty core of light, is alive and well in upper Japanese culture. Perhaps the best place to look for its analogue in contemporary Chinese culture is in poetry. This sort of mysticism, coeval with human refinement, never goes out of style. Perversion and repression, however horrible, never defeat it fully. It is the light of ultimacy shining in the darkness, which never overcomes it. No Chinese brutality or Japanese materialism has snuffed it out, as has no Nazism or Stalinism or American trashiness. It is ourselves coming to ourselves enough to recognize that we are prodigals, wastrels, living among the pigs on husks, and so finally crying bitterly and turning for home. With that turning we head back into the mystery of our own humanity, where it seems we have been made to dwell, and we find that the mere stop of our *exitus* and the bare first steps of our *reditus* begin our healing.

NOTES

1. Arthur Waley, trans., *The Analects of Confucius* (New York: Viking [1938]), 88. See also 3:13 and 5:12.
2. Ibid., 4:8, 103.
3. Ibid., 11:11, 155. A very illuminating essay on Confucian etiquette is Herbert Fingarette's *Confucius: The Secular as Sacred* (New York: Harper Torchbooks, 1972). *The Doctrine of the Mean,* a text that entered the Confucian canon, may offer more openings to the mystical influence of spirits.
4. D. C. Lau, trans., *Mencius* (Baltimore: Penguin Books, 1970), 120. We have changed the translation slightly, to make a clearer logic. A comparison with W. C. H. Dobson, trans., *Mencius* (Toronto: University of Toronto Press, 1963), 176, did not resolve the problem. There ought to be two senses to the key words "smaller" and "greater," but neither translator makes it clear what they are. Without such a dualism, the passage makes no movement, the Way seems arbitrary.
5. Lau, trans., *Mencius,* 7B.25, 7–8, and 2A. 9–16.
6. *Hsun Tzu,* chap. 23, in *Masters of Chinese Political Thought,* ed. Sebastian de Grazia (New York: Viking Press, 1973), 176–177.
7. Chu Hsi, "Complete Works," 11, in *A Source Book in Chinese Philosophy,* trans. Wing-Tsit Chan (Princeton, N.J.: Princeton University Press, 1963), 630–631, 638–639.
8. Arthur Waley, trans., *The Way and Its Power* (New York: Evergreen, 1958), 144.
9. Ibid., 168–169.
10. See the fine studies by Holmes Welch, *Taoism: The Parting of the Way* (Boston: Beacon, 1966), and Livia Kohn ed., *The Taoist Experience: An Anthology* (Albany: State University of New York Press, 1993).
11. Burton Watson, trans., *Chuang Tzu: Basic Writings* (New York: Columbia University Press, 1964), 27–28.
12. Thomas Merton, *The Way of Chuang Tzu* (New York: New Directions, 1965), 45–47.
13. For one overview, see Arthur F. Wright, *Buddhism in Chinese History* (Stanford, Calif.: Stanford University Press, 1971).
14. See James Whitehead, "Vimalakirti," in *Abingdon Dictionary of Living Religions,* ed. Keith Crim (Nashville, Tenn.: Abingdon, 1981), 795.

15. Philip B. Yampolsky, *The Platform Sutra of the Sixth Patriarch* (New York: Columbia University Press, 1967), 130.

16. Ibid., 132.

17. Francis Cook, *Huan-yen Buddhism: The Jewel Net of Indra* (University Park: Pennsylvania State University Press, 1977), 114–115.

18. Marcel Granet, *The Religion of the Chinese People* (New York: Harper Torchbooks, 1977), 41–42.

19. Max Weber, *The Religion of China* (New York: Free Press, 1951), 21–22. See also C. K. Yang, *Religion in Chinese Society* (Berkeley: University of California Press, 1961).

20. Richard Wilhelm, *Change: Eight Lectures on the I Ching* (Princeton, N.J.: Princeton University Press/Bollingen, 1973), 82–83.

21. Wu Ch'eng-en, *Monkey: Folk Novel of China,* trans. Arthur Waley (New York: Grove Press, 1958), 23–24.

22. Ichiro Hori, *Folk Religion in Japan* (Chicago: University of Chicago Press, 1968), 199–200.

23. Shunryu Suzuki, *Zen Mind, Beginner's Mind* (New York: Weatherhill, 1970), 80–81.

24. *Encyclopaedia Britannica,* vol. 1, p. 935.

25. H. Byron Earhart, *Religion in the Japanese Experience: Sources and Interpretation* (Belmont, Calif.: Dickenson, 1974), 15–16.

26. *1994 Britannica Book of the Year* (Chicago: Encyclopaedia Britannica, 1994), 584, lists 0.2% of Chinese as Christians.

27. See Frederick Franck, ed., *The Buddha Eye: An Anthology of the Kyoto School* (New York: Crossroad, 1982).

5

Jewish Traditions

General Orientation

The historical extent of Judaism depends on how we understand Abraham and Sarah, the best candidates for the title "progenitors," and where we place them chronologically. If we assume that they were historical figures, rather than literary constructs, we talk about the beginning of a people, a line of blood and culture. The most probable date for the first generations of that people, who become self-conscious as tribes derived from Jacob ("Israel"), the grandson of Abraham and Sarah, is the (Middle Bronze) period (2000–1500 B.C.E). Archaeological research suggests that the customs described in the cycle of stories about Abraham and Sarah in Genesis appear about that time. Later (perhaps 1200 B.C.E.) comes the exodus from Egypt under Moses, and later still comes the reign of King David (ca. 1013–973 B.C.E.).

Because of Genesis we associate Abraham and Sarah with the experience of faith in an other-worldly, truly unique God. The stories about Abraham stress his difference from his idolatrous father. They also stress the promise of the living, uncapturable God that Sarah and Abraham will become the font of a vast progeny, like the stars in the heavens or the grains of sand along the beach. This is mystical but probably only in the sense that all religious faith is a response to the mystery of God, a mode of access to the unknowable primacy of the divine and a mode of suffering from that primacy.

Jews vary in the degree to which they take up the stories of biblical Israel as immediately formative of their current culture. For secular Jews biblical Israel may be only a historical curio, but there are good reasons for considering the biblical heritage as part of a relatively smooth historical continuum and for making the Bible (Hebrew or Greek Septuagint) and the Talmud the twin classics of written Jewish revelation.

By the first centuries of the Common Era, after the fall of the Jerusalem Temple (70 C.E.), the focus of Jewish religious observance and interest had become the Torah (guidance, law) preoccupying the rabbis. Increasingly, Judaism became more interested in mastering Talmudic law for fully devout, observant living than in developing the prophetic aspects of its biblical heritage—the poetic declamation and political witness we now associate with Isaiah, Jeremiah, Amos, and Hosea. For most of the Common Era, the Talmud, in early (Mishnaic) or more developed (Babylonian, perhaps Jerusalem) form, has commanded the attention of Jewish religious specialists.

Thus Jewish mysticism has a long, rich, and symbolic heritage, with roots in both the Bible and the Talmud. A primary symbol, such as the chariot (*merkavah*) described in the first chapter of Ezekiel, may begin in the Bible, but its operation in the millennium 200 to 1200 or so C.E. is quite different from what one finds in Ezekiel. Rabbinic commentary had been developing at a steady pace, often offering brilliant readings of both biblical and Mishnaic texts. Although Jewish mysticism generally has furnished a counterweight to legal (Halakhic) interests, many of the Jewish mystics have been masters of Jewish law, and virtually all have been formed by it.

The 1994 Britannica *Book of the Year* lists a worldwide total of 18,153,000 Jews. (Comparable figures include Christians, 1,869,751,000; Muslims, 1,014,372,000; nonreligious people, 912,874,000; Hindus, 751,360,000; and Buddhists, 334,002,000.) By continental area, this listing finds 359,000 Jews in Africa; 6,264,000 in Asia; 1,475,000 in Europe; 1,132,000 in Latin America; 6,850,000 in North America; 100,000 in Oceania; and 1,973,000 in Eurasia. (The two major centers of population are the country of Israel, which in this listing falls into "Asia," and the United States.) Jews live in 134 of the 272 countries of the world, and they comprise about 0.3 percent of the world's population (5,575,954,000). All statistics such as these are perilous, but generally the proportions that a conscientious tally presents are trustworthy. Jews may be closer to 20 or 22 million worldwide, as Muslims may be closer to 1.25 billion, but the 1:50 or so proportion is probably trustworthy.

In his overview of Judaism for the *Encyclopedia of Religion,*[1] Eugene Borowitz describes the movement from the Bible to rabbinic Judaism that perhaps began in the first half of the first century C.E.:

> The rabbis themselves affirmed an unbroken transmission of authoritative tradition, of Torah in the broad sense, from Moses to Joshua to the elders, the prophets, and thence to the immediate predecessors of the rabbis (*Avot.* 1.1). By this they meant that along with the written Torah (the first five books of the Bible, also known as the Pentateuch or Law) Moses delivered the oral Torah, or oral law, which contained substantive teaching (legal and nonlegal) as well as the proper methods for the further development of the Torah tradition. As inheritors and students of oral (and written) law, the rabbis knew themselves to be the authoritative developers of Judaism.
>
> Modern critical scholarship universally rejects this view. For one thing, the Bible makes no mention of oral law. Then, too, it is reasonable to think of Torah as undergoing historical development. When, over the centuries, Judaism grew and changed, later generations validated this unconscious process by introducing, retroactively, the doctrine of oral law.[2]

The significance of this situation for our study of Jewish mysticism is that virtually always postbiblical Jewish mystics were immersed in, even pillars of, the communities that the rabbis were developing authoritatively. Many of them were authoritative rabbis, or masters, in their own right. Mysticism enjoyed and suffered from a beguiling, fearful existence along the border of Talmudic Judaism. It could seem both to support and to challenge the central concern of the rabbis: to interpret Torah so as to sanctify all of practical life.

Complicating the historial overview still further is the influence of *aggadah*, the less formal and legal, more folk and storied traditions of popular Jewish culture. (All traditional, pre-Enlightenment Jewish life is religious, close to what we now call Orthodox Judaism.) Rabbis could avail themselves of the stories, songs, and dances associated with *aggadah*. They could develop or curtail the religious imagination at work there. Some of that religious imagination ran in tandem with mystical concern for the *merkavah*, while mystical concerns of such later groups as the Kabbalists and Hasidim made for a different coupling. The lines between (*1*) sober legal concern, (*2*) concern that we might call popular or folkloric, and (*3*) mystical concern proper have not been hard and fast in Jewish history. (In our interpretation, not every warming of the pious heart qualifies as mysticism.)

Among the key assumptions and central theological positions shaping the predominant Jewish worldview through the postbiblical period (virtually, the Common Era) have been the notions of Torah, already indicated, and the covenant. From the time of Abraham, with important renewals with Jacob, Moses, and David, Jewish history has been time spent with and shaped by God. The Mosaic convenant, described most crucially in Exodus and Deuteronomy, has been the most important. The saving of the people by leading them out of Egypt and the formation of them into a nation with a new destiny (to live in the promised land) is the backbone of Jewish biblical history. The people whom the traditional rabbis served lived in the covenant struck with Moses on Sinai. To their minds, the laws given to Moses were the constitution of the covenantal relationship. Later law, both written and oral, developed the Mosaic, biblical beginnings. Talmudic legislation was largely commentary on either scriptural legal texts or earlier commentaries. Central to the entire enterprise was the conviction that Jews were God's people, living still in the intimate, privileged relationship God had given Moses.

Moreover, the nature or character of the Jewish God has been significant. He (biblical and rabbinic Judaism have both been patriarchal) is personal and sole, a center of understanding and will responsible for creation and brooking no rivals. The most important biblical text on this score is the *shema:* "Hear, O Israel! The Lord is our God, the Lord alone. You shall love the Lord your God with all your heart and with all your soul and with all your might" (Deut. 6:4–5).[3]

W. Gunther Plaut explains this translation:

> [T]wo affirmations are made: one, that the Divinity is Israel's God, and two, that it is He alone and no one else. Other translations render "the Lord our God, the Lord is One" (stressing the unity of God) or "the Lord our God is one Lord" (that is, neither divisible nor coupled with other deities, like Zeus with Jupiter).[4]

The monotheistic sensibility of biblical Israel, discernible in the texts about the call of Abraham as well as the texts about the fight of Moses to keep religion pure during the trek through the wilderness, came to rabbinic Jews as a foregone conclusion. The prevailing interest during the Common Era has not been theology—trying to understand the divine reality, either its simplicity or its goodness or its right to be loved—but religious practice: doing the deeds, the commandments, the legal obligations of the covenantal relationship with the Lord. To execute this interest well, some members of the community had to study the law, the Torah, carefully. In fact, in most centuries most male members of the community longed to study the Torah. (Women seldom could study Torah formally, though women might know the provisions of Torah for home life—kosher diet, observance of the Sabbath—better than their husbands.)

Traditionally, religious prestige, pride, sense of being fully a Jew focused on a mastery of Torah. Torah has been the form in which the will and the instruction of God has been available most clearly, fully, and trustworthily. To meditate on the Torah has been the height of mainstream religious desire. Psalm 1, set by the rabbis at the head of their Bible's songbook, makes this clear: "Happy are those who do not follow the advice of the wicked or take the path that sinners tread. But their delight is in the law of the Lord, and on his law they meditate day and night" (Ps. 1:1–3, NRSV).

Plaut suggests the significance of the *shema* in the following description of its place in Jewish prayer:

> It is recited at evening and morning services on various occasions, on retiring at night and at the end of Yom Kippur [the Day of Atonement], where it is almost at once followed by the sevenfold exclamation: "The Lord is God!" (I Kings 18:39). It is a custom among Orthodox Jews to cover the eyes while saying the words, in order to increase one's concentration.[5]

What might a properly concentrated appreciation of the *shema* produce by way of personal insight and testimony? The response of Alexander Suskind of Grodno (d. 1793), a Lithuanian Kabbalist, suggests how Jewish mystics often thought:

> I believe with perfect faith, pure and true, that You are one and unique and that you have created all worlds, upper and lower, without end, and You are in past, present, and future. I make You King over each of my limbs that it might keep and perform the precepts of Your holy Torah and I make You King over my children and children's children to the end of time.[6]

When this sort of faith came to climax in a direct experience of the King, mysticism proper flowered.

During the centuries after the fall of the Temple in Jerusalem, when the rabbis were developing what became the talmudic literature, two leading centers of interpretation arose. One was the area outside of Jerusalem, along the southeastern edge of the Mediterranean. The other was in Babylon, around the Tigris and Euphrates rivers, including Baghdad. When Jews emigrated to other lands, they took with them the style of rabbinic interpretation dominant in their parent group in one or the other of these primal centers. Thus Judaism in the Middle East and North Africa favored Babylonian models, until the expulsion of Jews from Spain in 1492. Jews in Asia and northeastern Africa probably emigrated from Iran, where, again, Babylonian traditions prevailed.

In southern Europe, as in the Middle East, from the seventh century C.E. Jews had to contend with Islam as well as Christianity. The medieval crusades often became harassments to Jews as well as Muslims. Northern and Eastern European Jewish communities served the surrounding Christian communities as merchants and bankers, usually an ambiguous status. The medieval European period, then, was a mixed experience. Mystics, as ordinary folk, might live in places such as Toledo, where Christians, Muslims, and Jews got along peaceably, or they might live in Eastern European communities, where spasms of violence (*pogroms*) were all too common.

Until its encounter with modernity, which for many Jewish communities did not occur until the end of the nineteenth century, most of Judaism retained its traditional focus on the Talmud. Reform and Conservative movements developed as a response to modernity (the call, spearheaded by the eighteenth-century Enlightenment, for critical assessment of traditional culture and for the rule of reason over faith). Today, Jews in both Israel and the United States divide into secular and religious camps. They may form common ethnic units or cultural systems yet go separate ways when it comes to religious observance. For mysticism the full implications of this relatively recent phenomenon are not yet clear. Probably, traditional symbols favored by Kabbalists and Hasidim will come in for critical study. However, if most of those interested in mysticism come from orthodox circles, such symbols may continue to rivet the Jewish imagination, as though modern history had changed nothing essential to Jewish piety.

The Biblical Period

In this chapter we treat Jewish mysticism historically, beginning with the biblical period; moving through the rabbinic, medieval, and post-medieval periods; and ending with modernity. In each period our focus continues to be what it has been when dealing with other religions: the dialectical interplay between our working definition of mysticism (direct experience of ultimate reality) and representative instances that might embody this definition, challenge it, and refine it.

Introducing a collection of studies of Jewish spirituality, Arthur Green has begun with Psalm 27:8–9, which has the plaintive cry, "Seek my face" ("In Your behalf my heart says, 'Seek My face!' O Lord, I seek Your face. Do not hide Your face from me."). Green takes this cry as a leitmotiv that has moved all ages of Jewish seekers, with which all would agree:

> Seeking the face of God, striving to live in His presence and to fashion the life of holiness appropriate to God's presence—these have ever been the core of that religious civilization known to the world as Judaism, the collective religious expression of the people Israel. Such a statement of supreme value—aside from questions of how precisely it is to be defined and how it is achieved—could win the assent of biblical priest and prophet, of Pharisee and Essene sectarian, of Hellenistic contemplative and law-centered rabbi, of philosopher, Kabbalist, *hasid*, and even of moderns who seek to walk in their footsteps.[7]

Obviously, mystical attainment occurs when a person finds or is shown God's face.

If we place biblical Israel in the context of its Near Eastern neighbors, the distinctive note of its religious culture is a tie to history. Whereas the surrounding cultures developed myths and rituals to celebrate the fertility of nature, Israel remembered Abraham, Isaac, Jacob, Moses, Joshua, and David. Because of the surrounding cultures, the Israelite cult was in danger of slipping into a mythological mentality and pursuing fertility, as, for example, the Canaanites did. That was the great fear of the classical biblical prophets. Along with social justice, purity of cult preoccupied Isaiah, Jeremiah, and Ezekiel, at times obsessed them. Why? Probably because they feared that in venerating deities of nature their people would lose the solidity, the actuality, of what had happened to make them who they were, what had given them a special identity. The God of their mystics would always be the Lord of their history.

Indeed, who were they if not the people whom God had led out of Egypt under Moses, whom God had joined to Himself on Sinai in a distinctive covenant with special laws? They had entered upon their land through the gracious help of their Lord. Their prosperity, in the sense of their retaining their identity and growing into its full promise, depended on their remembering the Mosaic events (the most crucial paradigmatic happenings) and on their keeping faith with the Lord to whom they were bound by their covenant.

In the tenth century B.C.E., under Solomon, the son of David, the biblical Israelites built a temple in Jerusalem to give their Lord a proper housing. As the biblical writers depict this development, it did not have unanimous support. As with the earlier move from leadership by charismatic "judges" to a king, the more astute among the observers could see both pluses and minuses. In the earlier case, a king would make Israel like the other nations and would unify the tribes. However, a king could also represent a decline from faith in God, from living by God's provision of charismatic leaders as the need arose, to rule through an office, something institutional, a form not necessarily vital spiritually. Similarly, a temple would formalize the Israelite cult and honor the Lord as divine sovereign over all creation. However, it might also imply that God could be contained, enticed to make a given edifice the divine throne, a concept especially problematic for mystics.

From the time that Moses had received no clear answer to his request for God's name, biblical Israel had nourished a strong iconoclasm. One could not represent God and usually it was idolatrous to try. A temple could pander to the idolatry latent in most human hearts. It could fan the desire to make God less terrible, more congenial. Whereas the tent of meeting that God had used in the wilderness was, like the charismatic judge, an up and down thing, dependent on the moment, geared to the inspiration of a living God, the temple that replaced the tent symbolically was fixed, independent of occasional inspiration. That meant that the God of the temple might be less alive or dynamic than the God of the tent. So, at least, did the detractors or doubters argue. The supporters of the temple, reaching back as editors to the traditions about the tent, made even the tent an abode of God that drained away some of the divine unpredictability.

Thus Jon Levinson has written of Leviticus 9:23–24, where the Lord issues fire to consume the offering laid out in the tent, as follows:

> At last, the basis has been laid for insuring the eternal availability of that elusive "presence" (*kavod*, usually rendered "glory"), which all the people saw. The visitation of God to Israel, the vision of God in Israel, need no longer be episodic or arbitrary. An enduring means of access of YHWH to Israel and of Israel to YHWH has been inaugurated, service without end, "an eternal ordinance for all generations" (Lev. 10:9). From now on, all that remains is the issue of whether or not Israel is observing the commandments, moral, ritual, and both, which entitle her to be graced by the electrifying divine presence manifest in the Tent and to remain in the presence of the fire that burned on Mount Sinai, wherever she may go.[8]

Here we see the spiritual move that leads to, or tries to justify retrospectively, the full elaboration of Torah characteristic of both priestly biblical Israel and rabbinic Judaism. God is present in the tent/temple; the two meld, as type and antitype, forerunner and fulfillment. The fire that burned on Mount Sinai, when God gave Moses the commandments and law, stays as long as Israel keeps the covenant.

To a mystical mind, this is a perilous move. The fascination with the fire of God (all the blazing holiness, power, creativity of the divine) is familiar. Everywhere, the mystic seeks God's face, that direct experience of ultimate reality, though fearfully, dreading what the ultimate may require. However, the mystic soon learns, through personal experience, that God comes and goes as God chooses. There is no containing God, no assuring that God will be present, except God's own promising. God is hidden, not because God is a tease but because the nature of God exceeds human perception and understanding. So any effort to formalize, regularize, or guarantee relations with God is likely to provoke a countermovement in defense of the divine freedom. As the mystic sees and feels things, no legal or sacramental system can take priority over God. Even the most reliable mystical defender of orthodox, law-abiding religion has to assert regularly the freedom of God to shape the covenant, the friendship, as God sees best. Otherwise, the mystic has forfeited what makes a mystic: direct awareness that God holds all priority in being and holiness.

Perhaps the first place that one should look for an indication of such awareness in Judaism is the formative experience of such great prophets as Isaiah and Ezekiel. The vision that Isaiah reports launched his prophetic mission. Note that the vision assumes that the Temple in Jerusalem is the special abode of God and that to serve the holy Lord the prophet has to be cleansed:

> In the year that King Uzziah died, I saw the Lord sitting on a throne, high and lofty; and the hem of his robe filled the temple. Seraphs were in attendance above him; each had six wings: with two they covered their faces, and with two they covered their feet, and with two they flew. And one called to another and said: "Holy, holy, holy is the Lord of hosts; the whole earth is full of his glory." The pivots on the thresholds shook at the voices of those who called, and the house filled with smoke. And I said, "Woe is me! I am lost, for I am a man of unclean lips; and I live among a people of unclean lips; yet my eyes have seen the King, the Lord of hosts!" Then one of the seraphs flew to me, holding a live coal that had been taken from the altar with a pair of tongs. The seraph touched my lips with it and said: "Now that this has touched your lips, your guilt has departed and your sin is blotted out." Then I heard the voice of the Lord saying, "Whom shall I send, and who will go for us?" And I said, "Here am I; send me!" (Isa. 6:1–8, NRSV)

Here the Lord (ultimate reality for biblical Israelites) comes in quite definite form: He is a great king. A considerable history of images precedes this vision, or the version of it that the editors of the canonical text have produced. The seraphs witness to the holiness of the Lord, as does the motif of smoke and fire. The prophet needs clean lips to announce the word of a clean, utterly holy God.

The prophet feels burdened with sin, something both like the ignorance that Eastern mystics battle and unlike it. Sin is culpable distance from God, uncleanliness that makes the creature as far below God morally as ontologically. Sin relates closely to a willful, law-giving God. People sin by opposing the divine will, disobeying the divine laws. As this text implies, though, below this operational sin lies a constitutional human unworthiness. Nothing can deal with God intimately unless God makes it worthy. Here, God makes the prophet worthy of the divine commission, which turns out to be charging the Israelite people with infidelity and threatening to fix them in moral blindness, by having the seraph burn away his dross.

The text from Isaiah offers a good example of the mystical dimension in biblical prophecy. The authority of the prophet comes from God's giving him a word that he must proclaim. Whether the giving is dramatic or quiet, it involves an unusual, extraordinary experience of the divine. Even more influential in Jewish mysticism are the first chapters of Ezekiel, where another prophet, exiled to Babylon, receives his call from the holy God. Isaiah saw the Lord in the Temple in Jerusalem. Ezekiel saw the Lord far away from Jerusalem, with the implication that no temple contained the Lord, that the Lord would go wherever he wished, wherever his people might be. The chariot (*merkavah*) that Ezekiel saw both symbolized this divine motility and became the favored focus of Jewish mystical contemplation. Here is an excerpt from the long description of the prophet's vision:

> As I looked, a stormy wind came out of the north: a great cloud with brightness around it and fire flashing forth continually, and in the middle of the fire, something like gleaming amber. . . . As I looked at the living creatures [who had human form, varied faces, and wings], I saw a wheel on the earth beside the living creatures, one for each of the four of them. . . . Wherever the spirit would go, they went, and the wheels rose along with them, for the spirit of the living creatures was in the wheels . . . and above the dome over their heads there was something like a throne, in appearance like sapphire, and seated above the likeness of a throne was something that seemed like a human form. Upward from what appeared like the loins I saw something like gleaming amber, something that looked like fire enclosed all around . . . and there was a splendor all around. Like the bow in a cloud on a rainy day, such was the appearance of the splendor all around. This was the appearance of the likeness of the glory of God. (Ezek. 1:4–5, 15, 20, 26–28, NRSV)

The repetition of the word "like" alerts us to the prophet's inability to describe his vision exactly. As well, this keeps him safe from idolatry since he does not claim to represent God precisely as God is. However, this text is kin to the text from Isaiah: both stress the sovereignty of God, who is a great king, and the splendor, which reflects the divine holiness. The angels, the four creatures, the throne, and the fire suggest a heavenly court. What went on in the Temple in Jerusalem to honor God formed the expectation of even the most visionary biblical prophets. Reaching back to Moses on Mount Sinai, a tradition of regal symbols presented the Lord in power like that of thunder and lightning, in speed like that of the most nimble chariot. The reality at the center of Ezekiel's vision is both human and nonhuman. It is both form and fire.

Probably, the experiences that underlie these two famous texts were shocking, hair-raising, incitements to both praise and dread. Ezekiel falls on his face, capable of no response other than complete abjection before the divine majesty. However, as we have them in these texts, the experiences bespeak a considerable mediation. The symbols of this visionary, mystical theology are consistent. The divinity described is continuous with the divinity Moses met on Sinai, the pillar of fire that led the wandering people by night in the desert, the fire that blazed from the tent, the cult of smoking sacrifices that the priests oversaw in the Temple in Jerusalem. Moreover, after their visionary experiences, the majestic Lord sends both Isaiah and Ezekiel to reform their people, to bring them back from their infidelity, which they have expressed by breaking the law of the convenant. Tradition is more than alive and well in these famous accounts. Tradition forms them through and through.

These texts from Isaiah and Ezekiel are not the only indications of direct experiences of ultimate reality at the center of Israelite, biblical experience. For example, after Moses has asked God to be with the Israelites and make them His own special people, God responds affirmatively and grants Moses a vision of the divine glory:

> And he said, "I will make all my goodness pass before you, and will proclaim before you the name, 'The Lord'; and I will be gracious to whom I will be gracious, and will show mercy on whom I will show mercy." But, he said, "you cannot see my face and live." And the Lord continued, "See, there is a place by me where you shall stand on the rock; and while my glory passes by I will put you in the cleft of the rock, and I will cover you with my hand until I have passed by; then I will take away my hand and you shall see my back; but my face may not be seen." (Deut. 33:19–23)

Other texts, much less than a full inventory, from which one might elaborate in detail the mysticism of the Hebrew Bible, include the experience of Elijah after the storm (I Kings 19), where he hears a still small voice (NRSV: "a sound of sheer silence"). Since this voice comes after the wind, the earthquake, and the fire that demonstrated the divine might, we might construe it as a better medium for God's communication than the prodigies of nature in which the nations delighted. Also, enough of the psalms describe emotional swings in the spiritual life to have inspired people aflame for God to take heart, persevere in their direct experience of God, wait for the return of comfort after affliction. Because of this, the psalms became a guide to ecstatic, if not mystical, experience. Finally, many rabbis interpreted the Song of Songs as an allegory for the love affair between God and Israel, offering a reading of Jewish history that made it a romance, a search for nuptial union with the divine, despite its many seasons of suffering.

Generally, the understanding of relations with God that emerges from key portions of scripture such as these is that God wanted to make the people holy as he was holy. This involved giving the people a law that could purify them so that direct intercourse, intimate dealing with their God, would become habitual. Thus the Book of Job depicts a man serious enough about God to fight for answers to his questions about divine justice. The response of God, out of the whirlwind (Chapter 38), ends by praising Job for his dogged persistence (Chapter 42), even as it forces him to broaden his perspective by reminding him that he was not there when the Creator laid out the proportions of the universe and gave all creatures their laws. Last, even so satiric a figure as the prophet Jonah has dealings with God that show God's desire to be with Jonah and to prod him gently to let go of his bitterness and take to heart some of God's generosity.

We should probably think of the conversations between God and Jonah, Job, and other biblical characters as the productions of the writers' imaginations. The Bible does not specialize in direct reports by the principals themselves. Still, the overall impression we get from these imagined conversations tells us volumes about the sense of God that dominated the biblical imagination. God is personal, intelligent, and full of will. People can engage with God, talk with Him as a lord, a parent, a friend, even a lover. They can be angry or overjoyed or pining away. The only thing they cannot be and please the biblical authors is indifferent.

The Lord at the center of Israelite history and identity is a name for the great reality with which all people have to come to terms if they are to live lives that are sane, balanced, and realistic. Regardless of whether ordinary Israelites made growing closer to God the passion of their lives or whether communities arose dedicated to this goal as a group enterprise, the authorities who assembled the canonical biblical text made the experience of exceptional individuals such as Abraham, Sarah, Jacob, Moses, Isaiah, Jeremiah, Ezekiel, and Job paradigms for later Jews to revere. That experience set direct dealings with God, in either full imagination or darkness and silence, at the core of human significance. Certainly, no people, not even the holiest of Jews, ought to separate themselves from the community, as Rabbi Hillel, a first-century C.E. master, made plain; but the community was never an anthill where everyone had to follow rigid patterns. Individuals remained free to follow the Song of Songs or the new covenant described in Jeremiah 31 toward such intimacy with God as the spirit of God moved them to seek.

We do not know whether the schools from which various priestly, prophetic, and sapiential writings may have come offered their students formal help with seeking God's face. We can say, however, that we should not dismiss the visions of Isaiah and Ezekiel as marginal to the biblical message. Although after the fall of the temple and the rise of a rabbinic Judaism, focused on what eventually became the Talmud, a prophet such as Ezekiel would not have been the role model for most students, Ezekiel and the other great prophets can strike us latter-day interpreters as representatives of a golden mystical time when God came and went in fire rather than letters and ink.

We may suspect that an influence of tradition similar to that in Judaism works in other accounts of mystical experience, for example, Buddhist reports, even though there the consistency of the dominant images may not be so clear, in part because there is no "Bible" so canonical. We may suspect, that is, that few mystics escape the influence of what fidelity to their religious tradition has led them to expect of God. Nonetheless, in these prophetic examples, the positive, kataphatic theology remains striking. Although the Lord of the burning bush and Sinai is hidden, mysterious, dark, the preference of the prophets and the later Jewish mystics is to use strong images: fire, imperial thrones, chariots, imperative commands. The God of the beginnings of Jewish mysticism is less the *neti, neti* (not this, not that) of the Hindu mystics than what the biblical prohibition against representing God might lead us to expect. Positive imagination remains more central in the prophetic experience than an iconoclasm championed because of an overwhelming experience that nothing can stand in for God.

As a gross, orientational judgment, we may say that the Western mysticisms based on the Bible (that is, Jewish, Christian, and Muslim mysticisms) make more of imagination and traditional symbols than do the Eastern mysticisms. Through Torah, the Incarnation, sacrament, and the Koran, Western mystics have tended to mediate and partially form their experiences of the divine. The directness of their encounters with ultimate reality may therefore be less than that of a Nagarjuna or Chuang-tzu, whose prime analogue is not a historical religious tradition but the power of the divine in nature. Inasmuch as the Mosaic covenant and the Torah have been the heartbeat of traditional Judaism, even its mystics have tended to speak in somewhat predictable terms. We ought to take this predictability as a caution not to base our definition of mysticism too strongly on Eastern, negative, impersonal symbols. Not all mystics want to evacuate the imagination or think that to meet the divine directly you have to go into a cloud of unknowing and chant in the face of positive creation a countervailing "nada, nada."

Certainly, many mystics find that their experiences of the divine relativize or recreate the symbols that they have inherited. Even if Isaiah says old words or invokes old sanctions, his experience gives them new urgency and fresh force. What the experience of Jewish mystics seldom does is overthrow the old words, sanctions, and symbols. The hold of the historicity that makes the Bible distinctive, the ties to paradigmatic events like the Exodus and the raising of the Temple, stays strong throughout later centuries.

Many Jewish mystics are well aware of the distance between God and any symbolic representations of God, as we found Ezekiel's tentative speech to suggest that he may have been. Indeed, a significant number of biblical texts make this point about the distance of God precisely, including a famous one from Isaiah (55:9, see also 45:15). Overall, justifying a perilous yet useful hypothesis, the heavy symbolism that we find in the visions of Isaiah 6 and Ezekiel 1 continues in the Kabbalists, the premier mystics of later Judaism.[9]

Rabbinic Judaism

We are working with a misnomer in this section inasmuch as since the fall of the Temple in 70 C.E., all Judaism has been rabbinic, that is, dominated by the rabbis and their study of Torah. However, the term makes a point: after the biblical period, in the Diaspora from Jerusalem, the preeminence of the rabbi and his study, which had not been obvious as little as 200 years earlier, became established quickly. By 600 C.E., when the Talmud was virtually complete, rabbinic Judaism had become the unquestioned form and the rabbi the incarnation of Torah. Prophecy was not a significant force. There were no priests and no cult in the Temple. The tradition of wisdom that one finds in the third portion of the Hebrew Bible (*Tanak*) had become an enspirited Talmudic learning, the ability to find the living presence of God among the many letters of the rabbis' reflections on Torah.

Two amplifications merit attention. One is the infusion of Hellenistic learning—philosophy, allegorical literary interpretation—that Philo of Alexandria (20 B.C.E.–50 C.E) represents most brilliantly. The other is the mystical interests of the rabbis, which tended to focus on the *merkavah*.

Philo represents the influence of Greek philosophy, especially the Platonic line that stressed the primacy of the soul and the ascent of the soul toward the heavenly source of intelligibility. This is a contemplative philosophy, convinced that *theoria* is the highest human activity. It can become mystical, inasmuch as the ascent of the soul can be ecstatic (a movement out of its ordinary status as the principle of life for the body, to union with God in a relatively independent status). Whether human beings can achieve this union through their own efforts or must await a divine initiative was a matter of debate. Most mystics, including those indebted to Plato and Plotinus (the foremost Neoplatonist), have reported that their transports, their spiritual consummations in visions of heavenly intelligibility, have come to them as grace—unpredictable, unmerited gifts.

Philo sought to harmonize this Greek philosophical tradition of contemplation with the Mosaic law serving as the constitution of his people's religious and ethical lives. He sought to contemplate the heavenly source of intelligibility without giving up the Torah, indeed while making what Moses experienced in receiving the Torah paradigmatic for the contemplative life. Of this effort and the mysticism that it created David Winston has said:

> In his allegorical interpretation of the divine voice as the projection of a special "rational soul full of clearness and distinctness" making unmediated contact with the inspired mind that "makes the first advance," it is not difficult to discern a reference to the activation of man's intuitive intellect. . . . In Philo's hermeneutical prophecy, then, we may detect the union of the human mind with the divine mind, or, in Dodd's terms, a psychic ascent rather than a supernatural descent.
>
> Philo's mystical passages contain most of the characteristic earmarks of mystical experience: knowledge of God as man's supreme bliss and sep-

> aration from him as the greatest of evils; the soul's intense yearning for the divine; its recognition of its nothingness and of its need to go out of itself; attachment to God; the realization that it is God alone who acts; a preference for contemplative prayer; a timeless union with the All and the resulting serenity; the suddenness with which the vision appears; the experience of sober intoxication; and, finally, the ebb and flow of mystical experience. Philo was thus, at the very least, an intellectual, if not practicing, mystic.[10]

This analysis may require some exegesis. First, Philo practiced the allegorical interpretation of texts popular in his native Alexandria. The assumption behind this interpretation was that texts carry several layers of meaning. Behind the literal, denotative import of the bare words could lie references to historical happenings or, more significantly, permanent truths of the spiritual life. Hellenistic philosophy stressed reason rather than myth or unrestrained imagination. In interpreting the divine voice that spoke to Moses, Philo was leery of magic and miracle. Winston's description of the projection of the rational human soul as the explanation for revelatory experiences may not square with his description of Philo's mysticism. The judgment that Philo tended to speak of a psychic ascent to God, where the accent would fall on human effort or at least human experience, may shortchange the mystical experience and conviction that primacy in the divine–human encounter always lodges with God, with the divine descent or condescension.

The description of Philo the mystic addresses his concern for union with ultimate reality, the Mosaic Lord who is also the first principle sought by Hellenistic philosophers (the God or the One), in Greek theoretical garb. Here, divinity is less the Lord of the covenant than the beloved of the soul, the light enticing the mind. For this reason, the mainstream of Jewish piety, led by rabbis immersed in a Talmudic view of Torah, never embraced Philo as a champion. His view of revelation and Torah could seem reductionist, one that subordinated rabbinic concerns to a more perennial quest of the human soul for union with the divine All. Certainly, the biblical Lord was All, the sole deity, creator, savior; but the predilection of the rabbis was to ponder the laws given by this Lord when He formed Israel historically. However worked over by the editors responsible for the final, canonical version of *Tanak*, the accounts of the Exodus and the raising of the temple refer to quite embodied events that occurred in time. Any transports of the soul accompanying these events are secondary. The rabbis are most interested in the legal, regulative consequences of such historical acts. Movements of the human soul outside of time in contemplative address of the Lord responsible for all of history and all of creation can remain at the edge of rabbinic interest, not necessarily rejected but certainly marginalized.

Of the preeminence of the symbolism of the chariot in the first millennium C.E., Louis Jacobs has written:

> From the period of the Talmudic rabbis (the first and second centuries) for roughly a thousand years, the Jewish mystical tradition was centered on contemplation of the vision of the *Merkavah*, the heavenly chariot described in the first chapter of the book of Ezekiel. The contemplatives of this lengthy period were known as the "Riders of the Chariot," that is, those who engaged in soul ascents to the heavenly halls where they saw God and his holy angels. It is interesting to observe that the majority of these "Riders" were scholars well-versed in the Law. Jewish mysticism, on the whole, far from being antinomian, generally seeks to infuse new life into the practical observances of the Torah.[11]

If we inquire about the mechanics of "soul ascents," we find ourselves in the realm that Philo worked over. Regardless of whether or not one accepts his Platonic anatomy of the soul, one has to deal with the same dynamics as he did. Reason, both mundane and ecstatic, comes in for study, as do emotion and will. A thorough analysis will consider imagination, what our picturings of the divine are worth. Mystics vary in the degree to which they interest themselves in these analytical, even technical matters. So do the rabbis who took up esoteric pursuits and tried to travel out to the *merkavah* or contemplate the Kabbalists' images of the beginning.

The difference between mystical practitioners and mere theoreticians has always been significant. It is one thing to read a text bearing on the *merkavah*, even to read it carefully and with considerable learning, so that one knows how traditional masters have interpreted it. It is another thing to move spiritually to a contemplation of the *merkavah* sufficiently intense to transport one to the heavenly court. Just as traditional rituals (for example, the *kachina* dances of southwestern Native Americans) have invited participants to lose themselves in the characters they are representing or the story they are dancing out, so traditional mystical exercises have invited those they have intrigued, adepts, to lose themselves in contemplation.

Contemplation is a relatively simple, holistic gaze, usually directed at an icon or a mental tableau such as the *merkavah* complex of images, occasionally directed at the divine darkness that has no apparent form or voice. The gaze in question here may begin with the eyes in *iconic* traditions (those, such as the Eastern Orthodox Christian, that consider icons valid, perhaps even necessary, media for contacting God, as opposed to iconoclastic traditions, which are leery of sensible representations). In all contemplative traditions, however, a quiet, peaceful gaze becomes the focus and preoccupation of the inner spirit. Whether maintaining relatively ordinary external perception or going deeply within ("shutting the doors," in the language of Lao-tzu), the contemplative tries to appreciate the simple reality, even the splendor, of the divine.

If successful, this contemplation makes the divine as real, actual, potent as anything seen, heard, or smelt. Mystics may debate about the different stages along the way to a consummating seeing or feeling of the divine. They may speak of a contemplation that can be acquired by human effort, in contrast to a contemplation that is a passive experience of divine action. They agree, however, that the contemplative experience itself is precious and that without such contemplation they would come unhinged. Habitual contemplative prayer, regular effort to ride in the chariot, is a positive addiction. When Philo speaks about the longing of the soul and its sufferings from the absence of God its beloved, he reveals how practice forms the contemplative's being. Nothing in the world of space and time can satisfy the practiced religious contemplative. The end of Torah, in the sense of the experience that might bring a full appreciation of its wonders as the Word of the Lord, lies in a vivid, more or less direct encounter with the source of the Torah, the living Lord who speaks forth all the meaning of Torah.

This brings us to the development in rabbinic Judaism of a less markedly mystical but still highly spiritual understanding of rabbinic, Talmudic work. Although a given rabbi may not have tried to ride the chariot, he could not avoid the central conviction of mature rabbinic Judaism that study of the Torah is the highest human activity, possibly even the prayer or devotion that God likes best. To merit so high a designation, study had to become a vehicle for contacting God. It also had to become a means of moral perfection, the ripening of human character.

The rabbis have tended to favor performance over intention, action over thought. When it came to basic orientation regarding the Torah, they taught the common people that the first priority was to do what they had been commanded, to keep the plain letter of the law. In their own reflections on study, however, the rabbis went beyond such a minimum. The following passage from the Talmudic tractate *Shabbat* illustrates the rabbinic concern for intention:

> Said R. Jeremiah in the name of R. Eleazar: When two scholars sharpen each other's minds in the study of the law—the Holy One, praised is He, will prosper them as it is written: "And in your majesty prosper" (Ps. 45:5)—the word for "majesty," *vahadarkha*, may be read as *va-hiddedkha*, which means "your sharpening." Moreover, they will rise to greatness, as it says [in the conclusion of the previously quoted phrase]: "prosper, ride forth." But one might assume that this is so even if their studies are not sincerely motivated. Therefore the previously quoted verse continues, "for the sake of truth." One might assume that this is so even if they became conceited [because of their knowledge]. Therefore does the verse finally add, "and meekness and righteousness." [12]

The commentarial style is as interesting as the content. Rabbi Jeremiah speaks in the tradition of Rabbi Eleazar, an eminent predecessor.

The whole discussion boils down to how rabbis ought to study the Torah. Traditionally, they have studied in pairs, reading and arguing out loud, matching wits in interpretation. The goal would be an encyclopedic memory of the entire Talmud, such that one could bring to bear on the interpretation of a given text the full resources of the Talmudic treasury. The reference to Psalm 45 and the general willingness to twist words to draw forth new possible meanings exemplify the rabbis' concern to make their interpretation of Talmudic passages also an exegesis of scriptural verses. More authoritative even than the Talmud has been the *Tanak.* Talmudic scholars see themselves as interpreters of scripture, given the task of making the whole Torah serve the sanctification of their people. The reason for the 613 principal laws, or privileged obligations, has been to form a people worthy of the Lord of the covenant, who is holy. The jots and tittles do not exist for their own sake but to form people in the ways of the Lord, the behavior and character that the holiness inscribed in Torah ought to educe.

This text promises to the student of Torah a singular greatness. The task of studying Torah is the highest imaginable. That does not mean, however, that human attitudes, intentions, virtues, and vices need not be a worry. The ideal is to study sincerely, for the sake of truth (rather than mystical union). The ideal is to study humbly, without conceit, with meekness and righteousness.

The study of Torah has been so intense that rabbis with any gifts for reflection have been bound to think about what it asked of their spirits and gave back to them. What it asked of their minds was clear enough: intelligence, attention, clear focus, daunting powers of memory, and imaginative ingenuity (to find new possible meanings in the established textual words). Our text suggests what the study of Torah asked of the student's character. The dialectical clash of study could make people arrogant, sharp, and impatient, even superficial. It could also incite a love of one's own opinion, even a temptation to bend interpretations to win debating points. Not so for the best rabbinic study, which, our text says, pursues only the truth and seeks only the guidance that the text wants to offer. To do this, the student has not only to learn the interpretative opinions of the classical commentators but also to offer the text a docile, limpid hearing. Without prejudgment or any agenda, the student has to be to the text as soft wax waiting for an impression.

How large is the step from this moral maturity, this virtuous character, to an immersion in and a reverence for the text that merits the name mystical? Estimates vary. However, the more deeply one takes the effort to wait upon the text in purity of mind and heart, the closer one will come to the mystical dimension. The voice behind the text, it bears remembering, is the voice of the divine Lawgiver. Moses is not the source of the law, only the mouthpiece. So in hearing the law, taking Torah to heart, one could be hearing the voice of God, opening oneself to the divine will. Inasmuch as this modality took over a given student's being and became the love of a given rabbi's life, the study of Torah could become a trysting, a meeting with the beloved Holy One.

The Medieval Period

The historical breakdown that we are using is rough. If we place the end of the biblical period definitively at 70 C.E., tentatively perhaps as much as 150 years earlier, and make 70 C.E. the beginning of rabbinic Judaism, we can defend a characterization of the first six centuries C.E. as generally rabbinic, that is, preoccupied with the style and material content that made up the Talmud. The medieval period arises out of our rabbinic period, with relatively few hiatuses. Christian developments continued to wash upon Jewish shores, but new were influences from Islam.

Among the main features of medieval Jewish mysticism, as we can see it now from a considerable distance, were the influence of allegorical exegesis (further developed from the time of Philo); the move of many Jews into Europe and the development there of a distinctive Ashkenazic (German, European) piety; the restraining influence of a rationalist philosophy, especially that of the polymath Maimonides; the effort to penetrate more deeply the spiritual significance of the commandments, or the *mizwot,* that summarized the practical implications of Torah; and the gathering force of an esoteric Jewish spiritual movement, Kabbalism, that sought secret knowledge of the Torah, sometimes by decoding the Hebrew letters in key scriptural words (a process known as *gematria*). Individual letters could carry specific numerical values, and their total in a given word or sentence could express a significance known only to the insiders who possessed the cipher.

These movements overlapped inasmuch as allegorists shared a nuanced, more-than-literal approach to the (scriptural) text with Kabbalists or as rabbis seeking the deeper meaning of the *mizwot* ran in tandem with rabbis seeking the deeper meaning of a biblical or talmudic text. The philosophers never became a dominant force, but they kept alive in Jewish intellectualism the Greek tradition of rational analysis and contemplative regard of ideal forms and heavenly sources of intelligibility.

The *Encyclopedia Judaica* defines allegory as "an extended metaphor, usually in the form of a narrative, portraying abstract ideas in symbolic guise." After discussing instances of allegory in the Bible, it goes on to characterize the use of allegory in Talmudic and medieval literature:

> Allegory was used in the talmudic period, and especially in the medieval period, in three types of literature, each using allegory in its own, different

> way: (a) homiletical literature used allegory in trying to translate facts and ideas known to the public into ethical teaching, by discovering the hidden meaning behind the well-known phenomena; allegorical interpretation of Scripture was frequently used in this literary type; (b) fiction, both poetry and prose, used allegory in order to develop a multilevel story or poem; (c) theological literature, especially medieval philosophy and Kabbalah, used allegory as a means to express the idea that the phenomena which are revealed to the senses are but a superficial and sometimes false part of the divine truth, whereas allegory can penetrate to deeper and truer levels. . . . The philosophers used allegory not only to explain away the physical attributes of God in the Bible and the talmudic literature. They interpreted whole biblical stories as allegory.[13]

Allegory makes a comparison, point for point, between one series of things and another series. It is a contrived, rational approach, not an overflow of mystical enthusiasm. Jesus' parable of the sower who reaps different measures of harvest (Matt. 12, Mark 4) is an allegory inasmuch as after giving it as an extended metaphor Jesus (or the evangelical editor) explains it point by point. The conviction that many scriptural texts merit, even cry out for, such a transposition could give the allegorists constant employment.

It is understandable that each generation of Jews reading the *Tanak* would want to find its relevance to their own times. To a modern mind, however, the first significance that a text carries is historical or literary, depending on its genre. The first significance of the account of Moses' receiving the tablets of the law on Mount Sinai, for example, is some sort of claim that the events depicted are real in whole or part. Alternatively, the first significance is the meaning that the biblical author achieves by placing this text where it is, shaping it to make the law appear in a certain guise, presenting Moses in this way or that, and so forth. For a modern reader, the first significance is not *esoteric* (a hidden meaning that only allegorists or symbolists can elucidate).

This suggests some of the distance between a modern mind and a medieval one. Physical science, begetting the scientific method, has shaped the modern mind so heavily that it tends to relegate nonempirical, not fully rational interests or modes of thought to second-class status, not thinking them as real or important as historical or literary–critical modes of thought. If the mystics of a given religious tradition or historical period respond passionately to such nonempirical, imaginatively visionary interests and modes of thought, moderns are likely to suspect their realism.

The pietism developed by Ashkenazic Jews who had moved into European lands gave rise to the prayer of the *hasid*. This word, quite ancient, connoted a person (usually male) of prayer, or devotion, with a desire to see God's face. The language of the Ashkenazic communities was Yiddish, an ingenious marriage of Hebrew and German. The pietistic literature produced in Yiddish indicates that what some scholars call the *virtuosi*, the rabbis esteemed for their spiritual attainments, sometimes pursued esoteric, Kabbalistic interests. They might read a text for the secret meaning contained in the numerical values of the letters in key lines, making this meaning the main concern of their prayer. Less esoteric Ashkenazim tended toward a prayer that both focused the mind and gave the heart a means to pour out its need, love, hope, and so forth. The more famous Hasidim, who arose after the Baal Shem Tov (1700–1760), continued the pietism of the earlier devout Ashkenazim, stressing even more than the medievals the ardor of the heart for its Lord and the Torah.

Of Maimonides (ca. 1135–1204) Isadore Twersky has written in the following lavish terms:

> . . . distinguished Talmudist, philosopher, and physician, and one of the most illustrious figures in Jewish history. He had a profound and pervasive impact on Jewish life and thought, and his commanding influence has been widely recognized by non-Jews as well as Jews. His epoch-making works in the central areas of Jewish law (*halakhah*) and religious philosophy are considered to be unique by virtue of their unprecedented comprehensiveness, massive erudition, and remarkable originality and profundity. Their extraordinary conjunction of halakhic authority and philosophic prestige has been widely acknowledged. While the generation before the age of Maimonides produced philosophically trained Talmudists—scholars well versed in both Greek science and rabbinic lore—the extent to which Maimonides thoroughly and creatively amalgamated these disciplines and commitments is most striking. Many people of different ideological inclinations throughout successive generations tend to find in or elicit from his great oeuvre a kind of *philosophia perennis*.[14]

Note that Twersky does not describe Maimonides as a mystic and does not lament the lack of mysticism in his work. A *philosophia perennis* is a love of wisdom, a layout of reality, that holds always and everywhere. This was the assumption, as well as the ideal, of classical Greek philosophy, such as that of Plato and Aristotle. The medievals continued to think in this way, in good measure because of the influence of classical thinkers whom Arab scholars made available in translation. The classical and medieval view of human nature, indeed of the nature of anything in creation, stressed its constancy. Nothing like the modern and contemporary appreciation of diversity, change, or the statistical character of many phenomena (the laws of genetics, for example) guided medieval thinkers such as Maimonides. In both his exegeses of Talmudic texts and his speculative works on philosophy, Maimonides sought eternal, unchanging verities. He was convinced that the world offered such verities, since neither God nor nature could be untruthful or unreliable.

Maimonides did his mature work in Cairo, becoming the unofficial head of a Jewish community existing in the midst of a predominantly Muslim culture. He earned his living as a physician, but apparently his real love was study of the Torah and philosophy. The contemplative life (*theoria*) that we found in Philo, the earlier thinker most his equal, attracted Maimonides strongly; but he thought that scholars ought not to rely on the charity of the community and so led an active life, earning his own keep. As well, he spent much time and energy attending to the welfare of the local Jewish community, of which his son Avraham became the official head (*nagid*).

Maimonides stands for the alternative to a mystical Judaism insofar as he followed Aristotle and the Talmudic sages to present as fully rational an explanation of Jewish beliefs as possible. His fourteen volumes of commentary on the Talmud and his *Guide for the Perplexed* present a blend of Talmudic and philosophical learning. In these and his other principal writings his goal was the instruction of people eager to find the heart of given laws, practices, and beliefs. However, he presented this heart through rational analysis, not mystical visions. One might say that his devotion to study worked in his life as a *mystique*, but this did not make him a mystic.

Even though Maimonides admitted an esoteric dimension in Torah, he tended to explicate this dimension through allegory, not mystical revelation. (In his *Guide for the Perplexed, Part III* he does discuss the influence of the divine on the mind of the prophet, but it stretches terms to consider this a tract on mysticism.) Maimonides commented on each of the 613 principal *mizwot*, and his list of the thirteen articles of Jewish faith entered into the Jewish prayerbook. Briefly, they state that God exists, is one, is incorporeal, is eternal, prohibits idols, gave the Torah, and sanctioned prophecy; that the Mosaic prophecy is unique, both the written and oral laws; that the law is immutable and eternal; that God is omniscient and will mete out reward and punishment; that the Messiah will come; and that there will be a resurrection of the dead. None of this is mystical, though all of it could become food for mystical contemplation. Thus inasmuch as the foremost Jewish philosopher apparently was not greatly interested in mysticism, we find mysticism relativized, or made less central than it is in Hinduism or Buddhism.

Although the method of Maimonides in explaining the commandments was to lay out the reasons of the divine mind for such laws, other medieval students of the *mizwot*, such as the Kabbalists Issac of Akko and Joseph ben Todros Abulafia, rejected what they considered the rationalism of Maimonides and other medieval philosophers, arguing that it detracted from faith. Such highly influential Kabbalistic works as the *Zohar* sought an esoteric meaning in the commandments, as well as in the words of scripture. For example, why should Jews have been commanded to make sacrifices of animals? The *Zohar* deals with this question rabbinically by moving in a somewhat rhetorical rhythm of questions and answers:

> What is the reason for sacrificing an animal? It would be better for a person to break his spirit and engage in repentance. Why should he slaughter an animal, burn it on the altar? But it is a mystery. . . . The mystery of the sacrifices includes many mysteries, which can be revealed only to the truly righteous, from whom the mystery of their Master is not concealed. (*Zohar* 3: 240b)[15]

Maimonides seldom recurs to mystery. His advice, both explicit and implicit, to Jews, is to exert their minds, become learned, and see the harmony of Jewish law and traditional philosophy. The *Zohar* takes the various commandments as semaphors, signs given by the Lord, flags waving down the spirit and telling it to attend devoutly. In the above text Moses de Leon (1240–1305), the most likely author of the *Zohar*, refers to the theme of the biblical prophets that God wants mercy rather than sacrifice. For him the purification of the spirit, in part through mastery of its untoward inclinations and sacrifice of its vices, is more important than the outward ritual of killing animals. Still, the cult of the Temple established by God deserves respect. Many of the leading rabbis loved to contemplate the rituals of the Temple, even though those rituals had ceased many centuries earlier. For the *Zohar* a devout spirit, a righteous person, might find in rituals such as the sacrifice of animals the mystery of their Lord. Anything made by God, commanded by God, spoken by God could be revelatory. Indeed, the more devout the spirit, the more likely that the person would be living in a world of wonders and revelatory signs of God's presence and intentions.

The *Zohar* marks the consummation of the early, formative Kabbalistic movement. The Kabbalah arose in the thirteenth century among Spanish Jewish seekers whose goal was to find the inmost heartbeat of their tradition and live purified lives by it. (*Kabbalah* means "tradition.") Introducing the early Kabbalah, Joseph Dan has written:

> The Kabbalah is only one of many forms of Jewish mysticism during its nearly two millennia of development. Since the thirteenth century it has emerged as the most important current, and in subsequent centuries all Jewish mystical expressions were made, with few exceptions, through the symbolism provided by the Kabbalah. In the period of the development of the early Kabbalah it was not the only Jewish mystical system; it achieved this status only after the *Zohar* became the authoritative text of Jewish mysticism. . . .
>
> The most characteristic and recognizable symbol of the Kabbalah is that of the ten *sefirot*. . . . This strange and untranslatable term first appears in the *Sefer Yesirah* (Book of Creation), a short cosmological and cosmogonical work probably written during the fourth century C.E. Some of the terms used in this work are closely related to the *heikhalot* [early rabbinic]

> and *merkavah* literature. All later theologians undoubtedly drew the term *sefirah*, as well as many other terms that became central to Jewish philosophical and mystical speculation, from this short tract.[16]

The *heikhalot* literature deals with the ascent to the divine chariot but also with themes of the Song of Songs that anthropomorphize God. As we have seen, the *merkavah* literature ponders the mystery at the center of Ezekiel's vision of the divine chariot, wanting to ride there, with the fire of God. The Kabbalistic concern with the *sefiroth* takes another focus. Kabbalists want to know how things were in the very beginning, when creation came into being. No doubt, they believe that if they can gain a vision of that time, or that moment out of time, they will gain a rich understanding of what divinity is in itself and also of how divinity has related the world to its own creativity.

An early Kabbalistic text, "The Fountain of Wisdom," speaks of the original moment as follows:

> For before the celestial world—known as the 377 compartments of the Holy One, blessed be He—was revealed, and before mist, electrum, curtain, throne, angel, seraph, wheel, animal, star, constellation, and firmament—the rectangle from which water springs—were made; and before water, springs, lakes, rivers, and streams were created; and before the creation of animals, beasts, fowl, fish, creeping things, insects, reptiles, man, demons, specters, night demons, spirits, and all kinds of ether—before all these things there was an ether, an essence from which sprang a primordial light refined from myriads of luminaries; a light, which, since it is the essence, is called the Holy Spirit.
>
> Know and comprehend that before all the above-mentioned entities there was nothing but this ether. And this ether darkened because of two things, each having different sources. The first issued an infinite, inexhaustible and immeasurable light. The gushing forth was sudden, not unlike the sparks which fly and burst forth when the craftsman forges with a hammer.[17]

We shall have further occasion to comment on Kabbalistic symbols such as these. The point to note here is the movement beyond anything that Genesis, the book of *Tanak* dealing with creation, has enumerated. An esoteric tradition, seeking by allegory or imaginative contemplation the inner meaning of texts in Genesis and the Talmud on the beginning, moves toward an original light, flame, emanation. One can sense correlations with the flame of the *merkavah*, but the symbolism of the chariot is not central. The flame, the light, here the Holy Spirit, is the source of created realities, that from which they issue, that which always contains their ultimate explanation inasmuch as it presides over the mystery of their coming into being.

In the figures of the original light and fire, the *sefiroth*, and other aspects of the first moment, the original being, that drew Kabbalists like moths, a mystical mind could find much to love. Going out to these symbols, this raiment of the divine, mystics could think they were encountering ultimate reality. How direct their encounter was is more our question than theirs. These Kabbalists moved in a tradition that assumed that such symbols of the divine originality conveyed the reality of what they depicted sufficiently to anchor a devout life. The imaginative journeys of the Kabbalists, like those of the riders of the chariot, were movements of their spirits, trips of their souls, that took them into zones more real than anything sensible or earthly. So they treasured their journeys and the symbols that both motivated and guided them. They thought of the *sefiroth* as aspects of the divine construction of reality that they were born to contemplate.

The Kabbalists spawned many different interpretations of the basic notion of the divine spheres. Prior to the expulsion of Jews from Spain in 1492, debates about the proper understanding of the *Zohar* divided Castilian interpreters into conservative and liberal, usually ecstatic, camps. Conservatives wanted to retain the understanding of the Kabbalistic emanations that had kept this mystical movement sober. Ecstatics such as the late thirteenth-century interpreter Avraham Abulafia sought a more personal experience, one concerned less with verifying the divine spheres (as the form of divinity) than with helping the devout person cleave to God.

In the sixteenth century, the charismatic figure Isaac Luria became the center of a community in Safed, northern Galilee, that made Kabbalistic themes the marrow of its religious life. Indeed, during the seventeenth century the thought of Luria dominated many Jewish religious circles. With Luria, Kabbalistic study became the means to dramatic sanctity. For example, he developed the personal reputation of being able to discern the states of soul of his disciples, to talk with animals, and so on. Supposedly he knew the different stages through which souls had passed in their transmigrations, and the symbolic reading that he gave to the emanations of the divine energy explained the traumatic exile of the Jews from Spain.

Important to Luria was the teaching that God had contracted himself to make space for creation. In a second phase, God had reemerged, filling vessels with the divine light. Some of these vessels broke, putting into creation sparks of the divine light. The greatest task of human beings was to liberate these sparks through devotion. *Tikkun* ("mending") became a summary of the work that prayer and good deeds could accomplish. It was not simply social but cosmic, repairing the world, the creation that had suffered a mishap. Eventually all of the sparks could return to God, separated from the world of matter.

If one studies the dispersion of Jews after the destruction of the Temple in 70 C.E., the overall pattern is that for the first 1,000 years the various geographic centers present a religious life directed by a halakhic interpretation of the covenant. For such an interpretation the main task was not to gather up sparks of divine light but to keep the law as the rabbis had kept codifying it in talmudic fashion. Certainly such communities did not advance in lockstep, forced to uniformity, but for the first millennium of Diaspora mystical movements remained subordinate to fairly straightforward legal acceptance of the Torah. As we have seen, Maimonides gave little play to mysticism. Indeed, his love of the worldly, empirical Aristotle inclined him to avoid the flights of the Neoplatonists, Christian and Gnostic alike, who wanted to escape from the material world.

After the seventh century, many Jews were living under Muslim rule, where a sober outer appearance served them well. The great stimulus to Kabbalism was the explusion from Spain in 1492. Certainly during the later medieval period the Kabbalists had begun a devotional and speculative movement that augmented what people might find in the Talmud, but the new emotional energies generated after 1492 gave this movement its great power. Nonetheless, Kabbalism remained an interest tied to Iberian origins in particular and European traditions in general.

Jews who emigrated to North Africa after 1492 sometimes took over magical (not strictly mystical) practices of Muslims devoted to saints, amulets, and the like. Kabbalistic studies continued in the Maghreb (Algeria and Morocco). Spanish Jews also emigrated to the Ottoman Empire, setting up communities such as that of Isaac Luria in Safed. In the seventeenth century the so-called false messiah Shabbetai Tsevi, born in Smyrna, connected Kabbalistic fervor with a message of messianic restoration, attracting many followers. Even his conversion to Islam did not dampen their hopes completely. Just as Jews in Iberia had felt pressures to convert to Christianity, so in the Ottoman Empire they felt pressures to convert to Islam. Shabbetai Tsevi succumbed, and sober Talmudists saw a lesson.

Jews who ended up in Asia and northeast Africa developed in such isolation from the mainstream that their practices became peculiar. Thus the Ethiopian Jews, known as Falashas, and both the White and Black Jews of Cochin, India, created a mixture of halakhic and idiosyncratic traditions that eventually made their Jewishness seem deviant. The same for the Indian Jews known as the Bene Israel. The point for our purposes is that while European communities often developed mystical groups and practices indebted to the Kabbalah, few communities outside Europe did. This includes the communities of expatriates to the Western hemisphere, who took hold from the seventeenth century.

The Postmedieval Period

Once again our chronological designation is loose, assuming that periods and schools of Jewish mysticism overlapped and seldom arose or passed away neatly: the postmedieval era carries forward concerns of our medieval period, and its own concerns continue into the modern period. As noted earlier, modernity, a critical mentality owing most to the eighteenth-century Enlightenment, came late to the majority of Jews, certainly to the culture of the *shtetl,* the small Eastern European village which prevailed, dominated by orthodox patterns, into World War II. Overall though, our adjective "postmedieval" applies most strongly to the period of the fifteenth through the eighteenth centuries.

The two primary mystical developments in our postmedieval period were the elaboration of Kabbalistic themes and the development of a new Hasidism, centered in Eastern Europe, that made the rabbi the devotional center of the community. The Hasidic rabbi, following in the footsteps of the charismatic Baal Shem Tov, was faithful to *halakha* but more naturally a master of *aggadah*—a teller of stories and a maker of *midrash* (devotional commentary on biblical texts). In addition, he was an example of ardent prayer, going before the Lord to plead the cause of his community—protection from their enemies and prosperity through fidelity to the *mizwot.* The rabbinic courts that developed made the head rabbi a prince and drew the devout from miles around. The Hasidic communities one finds nowadays in Williamsburg (Brooklyn) and Israel descend from the eighteenth-century beginnings in Eastern Europe.

The greatest recent authority on the Kabbalah, Gershom Scholem, has described the mythical character of Kabbalistic theology as follows:

> The mythical character of Kabbalistic "theology" is most clearly manifested in the doctrine of the ten *sefiroth,* the potencies and modes of action of the living God. The Kabbalistic doctrine of the dynamic unity of God, as it appears in the Spanish Kabbalists, describes a theogonic process in which God emerges from His hiddenness and ineffable being, to stand before us as the Creator. The stages of this process can be followed in an infinite abundance of images and symbols, each relating to a particular aspect of God. But these images in which God is manifested are nothing other than the primordial structures of all being. What constitutes the special mythical structure of the Kabbalistic complex of symbols is the restriction of the infinitely many aspects under which God can be known to ten fundamental categories, or whatever we wish to call the conception underlying the notion of the *sefiroth.* In the *Book of Creation,* where the term originates, it means the ten archetypal numbers (from *safar* = to count), taken as the fundamental powers of all being, though in this early work each *sefirah* is not yet correlated with a vast number of symbols relating it to other archetypal images to form a special structure. This step

> was first taken by the *Bahir* and the medieval theosophy of the Kabbalah, reviving gnostic exegeses concerning the world of the eons and going far beyond them.[18]

In explication of this pregnant analysis of Scholem, let us note the following. First, this authority sees the Kabbalah as contemplating the emergence of the godhead, the substance of God that human beings can appreciate, through the emanation of the *sefiroth*. In speaking of a "theogonic" process, he implies that God becomes what God is through this emanation. True, God was hidden and ineffable before the production of the *sefiroth*, but God only became real for us, a factor in human awareness, or our God, by processing forward through creation.

The special note of the Kabbalistic doctrine of the ten *sefiroth* is that it narrows down the primary modes of the theogony. The God whose attributes could be as extensive as the range of creatures he produces can be grasped adequately through ten primary categories or archetypal numbers that specify his ways of appearing. The world has a structure, and this greatly simplifies our task of understanding how God appears in the world, where God is to be found, and how God is to be served.

In calling the medieval development of this mythical theology a theosophical enterprise, Scholem tips his own hand. His evaluation is ambiguous, if not negative, because "theosophical" is seldom a term of praise. It connotes imaginative flights, often excesses, unanchored in tradition, ignorant or contemptuous of orthodox consensuses about the matters it treats. The powerhouse of theosophical speculation involves mystical transports in which the soul expands, driving the imagination to identify oceanic psychological states with divine ultimacy. This is "gnostic" inasmuch as theosophists claim that they possess a special, esoteric, privileged understanding of God and ultimate secrets unknown to the common run of people, even believers.

The Gnostics of the early centuries of the Common Era developed elaborate schemes detailing the eons and cosmic levels into which creation and the divine plan of salvation emerged. They spoke of falls from original grace and sparks of the divine lingering in the depths of human awareness and waiting rekindling. Often these Gnostics urged an ascetic abstinence from sex and more than necessary food, under the general conviction that matter held people back from the work of rekindling the spark of divinity in their depths and remounting, through the various cosmic levels, to their originally much higher status, close to the divine source of the emanative process of creation.

The Kabbalists and Hasidim took over the notion of divine sparks, purifying it of its worst Gnostic excesses (variations of pantheism) and making it into a view of the world as shot through with divine goodness waiting for devout Jews to capitalize upon it as an energy for hallowing all space and time. The Kabbalists also shared the notion that their inmost concerns were secret, a knowledge not available to the masses, indeed dangerous for any but the most mature. Many rabbinic schools insisted that one should only study the Kabbalah after a long apprenticeship to the ordinary interpretations of the Talmud, lest one drown in mysteries way over one's head.

Even with such cautions, however, serious questions remain. What are the controls on the Kabbalistic imagination? For example, how valid or trustworthy can a view of creation be if it goes so far beyond the plain descriptions of Genesis as to find little warrant there? Do the commentaries on Genesis that one finds in the Talmud justify the doctrine of the *sefiroth?* If not, is the only authority for the Kabbalistic theogony the beauty or power of the imagery one finds in the most beloved Kabbalistic texts?

The God of mainstream Jewish theology does not become divine through an emanative process. The God of mainstream Jewish theology may be hidden and well beyond human comprehension, but the manifestation of this Lord through creation is completely his own doing. The world appears because God says, "Let there be." Creation is a function of the divine word and will. There is no indication in the Bible that God becomes through the arising of the world. There is no indication that God plus the world is more than God alone or that God needs the world for any divine completion or fulfillment. In corraling the divine creative activity into ten categories or numbers, Kabbalistic theology risks a peculiar idolatry. What gives human imagination the right to limit the divine? The argument would have to be that human imagination is only presenting a self-limitation that God has chosen to put on his own creative activity or modes of appearance in the world. However, where is the basis for this argument? What texts or experiences justify the claim that God has limited the work of creation to the categories of the *sefiroth?*

The Kabbalists certainly are mystics in the sense that they seek ultimate reality, the divine as it is in itself. They may also be mystics in the sense that many of them thought they found what they had sought, in their pictures of the *sefiroth,* their experiences of the unfolding of the divine fiery center through their ten numbered emanative categories. When they lingered at the hinge of the process, the moment when the fire emerged from the ineffable divine darkness into the first emanative number, they stood at the imaginative equivalent of the place where divine consciousness seemed to flow into human consciousness and constitute its awareness.

The divine light illumines human spirituality, making it reflective and so capable of realizing both particular meanings and its nature as competent in principle to grasp any meaning. When human intelligence realizes that whatever exists is intelligible, that being and meaning are interchangeable, it enters the idealistic territory claimed so vigorously by such schools as the Buddhist Yogacara ("mind-only"). It senses a priority of mind over being, idea over materialized fact, inasmuch as mind and idea are the sources of what makes something identifiable.

The Kabbalists do not take this ontological turn. Their speculations are more theosophical, to stay with Scholem's analysis, more imaginative than ruthlessly analytic in an ontological mode. In part this is because the end for which Kabbalistic exercises serve as means is the worship of a personal God. In contrast to an impersonal Buddha nature or Suchness or *nirvana*, the Jewish Lord is an unlimited personal center of knowing, loving, understanding, and willing. How directly the chariot vision of Ezekiel shapes the transposition of this personal character (so clear in Genesis and the formation of the Mosaic convenant) toward the less personal imagery of fire and emanation in the *sefiroth* is for historians of the textual chain of influences to trace. The Hebrew Bible certainly has sufficient indications of the hiddenness of God, the divine darkness, to justify limiting the divine personalism; but it does not justify the imaginative picture of the emanation of the *sefiroth* upon which the Kabbalists lavished so much contemplative attention. That came from other sources, the authority of which, as we mentioned, seems suspect.

In distinguishing between theosophical and ecstatic Kabbalah, Moshe Idel has argued that, whereas some texts focus on the imagery of the divine (the *sefiroth*) that we have analyzed and Scholem makes central, others are preoccupied with the experience of the individual mystic ("ecstatic").[19] Idel's analysis of this second group of texts shows that the ecstatics manifest many of the characteristics of mystics in other religious traditions. Although the imagery and conception that they have inherited are uniquely Jewish, they struggle with the paradoxes of the divine hiddenness and manifestation, nothingness and totality, that mystics of the East, for example, long to understand.

Still, Idel's study of Kabbalistic sources does little to suggest that Jewish mysticism has been significantly apophatic. Unknowing, darkness, nothingness, and the other principal negative symbols play only a limited role. The divine nothingness, in fact, is usually a retraction of the divinity due to human sins, a withdrawal of the presence previously vouchsafed through the *sefiroth*. It is not a permanent condition imposed on the ecstatic (an inevitability in the experience of the mystic) because of the divine infinity. A few texts speak of the annihilation of the ecstatic in union with the divinity, but these do not represent the mainstream. Idel makes the case against Scholem that Jewish monotheism did not mean that the tradition ruled out mystical union with the divinity from the start (that is, because the very idea would have seemed blasphemous). Many ecstatics did, in fact, claim to find God's face or be swallowed up by the divinity, but the predominant Kabbalistic imagery never became yogic, in the sense of an anti-imagery born of a desire to evacuate the imagination and the intellect because these faculties seemed superficial compared to the deep human spirit that communed with divinity ontologically, being to being.

Now, the question of the authority of claimed mystical experiences, both those that seem to square with tradition and those (such as the Kabbalistic experiences that go far beyond *Tanak*) that seem not to square, is not unique to Judaism. All mystics in all traditions face this question, principally because mystical experience is private, a happening in the consciousness of individuals that most of their contemporaries do not share. Convinced of the validity of the experience as the individual mystic may be, the community can remain dubious if what the individual claims seems novel or unorthodox. (In mysticism strictly so called, where the individual experiences an extraordinary force passively, mystics report regularly that they can have no doubts that the action of the ultimate or God on them is indubitable. They may have second thoughts later, when they are not held by the divine action and are trying to analyze it, but during the rapture or quiet holding itself they cannot deny that what is happening is not their doing but the work of another, a power or holiness more than human.)

Most traditions solve this problem, to the extent that it is solvable, by insisting that mystical claims remain less authoritative than mainstream doctrine and therefore optional, that is, opinions that bind only those who claim to have derived them from personal indubitable experiences. Thus when a mystic such as the Muslim al-Hallaj claims to have become God, Muslim orthodoxy gets upset. When he will not renounce his claim, even in the face of a firestorm of criticism accusing him of blasphemy (there is no god but God, and God does not take partners, does not share the divine soleness), the Muslim community feels entitled, indeed obliged, to put him to death as a menace to orthodox faith, a denier of the straight path that the majority must be able to walk untroubled if Islam is to guide the many to salvation, to the Garden rather than the Fire.

Christian struggles with mystical experience are analogous, as the operation of the Inquisition shows. The dangers from the side of a too zealous defense of orthodoxy, a too active pursuit of heresy verging on the paranoid, emerge in the Inquisition in sometimes horrible form: torture, burning at the stake. On the whole, Judaism has avoided the worst horrors of heretic hunting, focusing on performance and ethics more than on theory and opinion. If a mystic kept the *mizwot*, living as a solid citizen of the ordinary religious community, he or (rarely) she could think as seemed good. The rabbis debated points of law fiercely, and it was only a short step to allowing individual opinion to flourish in the area of mystical, Kabbalistic speculation. As noted, there was a tendency to limit Kabbalistic exercises to those considered mature, already well established in *halakhic* orthodoxy, and also a tendency to keep one's Kabbalistic work quiet, if not esoteric. However, enthusiasm for the burning of Kabbalistic texts could cause Kabbalists to break these conventional contraints and try to bring the Kabbalah closer to the mainstream of Jewish thought and practice.

The Hasidim who looked back to Israel ben Eliezer, the Baal Shem Tov, the "rabbi of the good name," as the founding figure of their modern era (we have seen that the *hasid,* the pious person, had been a stock figure since the beginning of the rabbinic period), manifested a distinctive, winning joy. In the collective profile of the many Hasidic masters who led *shtetl* communities, stories from the *aggadah* that encouraged people to find a devout life possible, beautiful, a thing of delight were more prominent than stern calls for fidelity to *halakha.* The Hasidic masters would dance with the Torah on the Sabbath. They would urge praying to the Master of the Universe with ardor, tears of need and joy. Ordinary people considered their rebbe the center of the world, the *axis mundi,* the incarnation of the Torah. Drawing on the doctrine of the sparks of divinity strewn throughout creation, they could look at their lives and work as sanctifying the world by cultivating these sparks through a zealous execution of the *mizwot.*

The first two entries in Martin Buber's collection of tales about the Hasidic masters deal with the preexistence of the soul of the Baal Shem Tov:

> They say that when all souls were gathered in Adam's soul, at the hour he stood beside the Tree of Knowledge, the soul of the Baal Shem Tov went away, and did not eat of the fruit of the tree.
>
> It is said that the soul of Rabbi Israel ben Eliezer refused to descend to the world below, for it dreaded the fiery serpents which flicker through every generation, and feared they would weaken its courage and destroy it. So he was given an escort of sixty heroes, like the sixty who stood around King Solomon's bed to guard him against the terrors of the night—sixty souls of zaddikim to guard his soul. And these were the disciples of the Baal Shem Tov.[20]

The notion of the preexistence of the soul of the Baal Shem Tov at the time of Adam squares with the Gnostic tradition elaborated by the Kabbalists. Who knows what happened in the primordial time of Adam and what is the authority for going beyond the account of Adam's eating from the Tree of Knowledge given in Genesis? Clearly, we have in the first story a pious elaboration of the meager biblical details, driven by the admiration for the Baal Shem Tov pulsing in the breasts of many of his disciples. They felt that his holiness, his religious purity, had to set him apart from the taint put in motion and passed down the line by father Adam. The Baal Shem Tov had to be an exception, one set aside by God and preserved in purity. (This is the reasoning behind the Christian exemption of both Jesus and Mary from original sin, the effects of the "fall" of Adam and Eve.) However, the source of this tradition is more a pious sense of what would be fitting than anything authenticated by the mainstream of Jewish tradition, the Bible or the Talmud.

The second story is touchingly human. The Baal Shem Tov senses the wickedness of the world and at least tacitly asks for protection against it. The symbolism of protection by sixty *zaddikim*, righteous Jews, recalls Solomon and suggests that the Baal Shem Tov possessed a wisdom like that of the archetypal sage of the Hebrew Bible. Less directly implied, though clear if one knows the biography of the Baal Shem Tov, is an anticipation that this great religious figure will do battle with the forces of evil on a regular basis. Whenever his people come into threat, from either Christian persecution or natural disaster, he is their first defender, champion, advocate before God.

The Hasidic grand rabbis went before God in prayer as the first mediators between the Holy One and their people. They bore the afflictions of the entire people, groaning and weighted down. Sometimes the people would attend their prayers, much like a shamanic community gathered round to witness the mystical flight of their leader to the god withholding the game or punishing them for breeches of taboo by sending sickness. The rabbis tended to be less dramatic than the shamans, less given to producing vivid accounts of the perils of their journeys (passing the guardian dogs, walking through walls of fire), but on occasion the people could deduce from their groans and prostrations and silences the fierceness of their struggles, which could be wrestlings with the Evil One.

The stories of the fabled Hasidic masters often stress their singularity, the paradoxes guiding their lives, because they lived closer to divinity than most people, not apart from the *halakha* and customs of the community but with considerable freedom to interpret both as suited their personal spiritual dispositions. A story about Rabbi Shalom Shakhna of Probishtch (d. 1803), a later master, somewhat removed from the immediate influence of the Baal Shem Tov, illustrates this point:

> Rabbi Shalom Shakhna, the son of Abraham the Angel, lost both his parents when he was very young, and grew up in the house of Rabbi Nahum of Tehernobil, who gave him his granddaughter to wife. However, some of his ways were different from Rabbi Nahum's and unpleasing to him. He seemed to be very fond of show, nor was he constant in his devotion to the teachings. The Hasidim kept urging Rabbi Nahum to force Rabbi Shalom to live more austerely.
>
> One year during the month of Elul, a time when everyone contemplates the turning to God and prepares for the Day of Judgment, Rabbi Shalom, instead of going to the House of Study with the others, would betake himself to the woods every morning and not come home until evening. Finally Rabbi Nahum sent for him and admonished him to learn a chapter

> of the Kabbalah every day, and to recite the psalms, as did the other young people at this season. Instead, he was idling and loafing in a way particularly ill becoming to one of his descent.
>
> Rabbi Shalom listened silently and attentively. Then he said, "It once happened that a duck's eggs were put into a hen's nest and she hatched them. The first time she went to the brook with the ducklings they plunged into the water and swam merrily out. The hen ran along the bank in great distress, clucking to the audacious youngsters to come back immediately lest they drown. 'Don't worry about us, mother,' called the ducklings. 'We needn't be afraid of the water. We know how to swim.' "[21]

Rabbi Shalom is claiming the right to go his own way, as the Spirit moves him. He finds solitary contemplation in the woods more nourishing than the standard study of Torah (and Kabbalah, which apparently had by his time become somewhat standard, open even to young students). Rabbi Shalom is not a monk. (Judaism developed no monastic life centered in celibacy. Marriage was the ordinary pathway and the rabbis looked down on the single life, in good measure because they took seriously the biblical command to be fruitful and multiply, but also because they thought that family life was a great source of balance and a good school for realism.) He is a contemplative, perhaps even a mystic, drawn to commune with God, ultimate reality, through nature. This sets him somewhat at odds with usual Jewish piety, which looked more to the social world of the community, and above all to its Talmudic collection of traditions, than to the natural world for its inspiration. Whether his natural parents would have supported this tendency more than his adoptive parents did is not clear. However, Rabbi Shalom certainly thinks that he is like a duckling brought up in the nest of a hen, and his rather unflattering suggestion is that Rabbi Nahum has assumed that only the ways of hens are normal. That he can swim with little effort in waters that chicks might find dangerous comes as a surprise to Rabbi Nahum and may frighten him, but Rabbi Shalom will be just fine.

The story is a call for freedom. The ways of God in guiding eminent rabbis may be more flexible, indeed more scandalous, than what the narrow-minded can imagine. As tamer versions of the "crazy wisdom" that we found in the Tibetan gurus living out in the Himalayan wastes, some of the Hasidim cultivated eccentricities designed to offset the potentially stifling influence of legal controls and community customs. To make the point that God is always greater than our traditions, assumptions, and understandings about God, they would go against the established grain, though seldom into anything that might be labeled sin.

So Rabbi Shalom's offenses, when studied in the clear light of day, boil down to going his own way and not marching in step to the *shul*, the study hall, that the other students frequented. He is in the collection of tales of later masters because his choice, or his following of the guidance of the Spirit within him, justified itself by a later demonstration of a wisdom and holiness both edifying and practically helpful to his people.

Modern Jewish Mysticism

Here, our main focus is the nineteenth and twentieth centuries, when Judaism began to move out of its postmedieval phase and reassessed its traditional, orthodox reliance on Talmudic law. The Reform Movement crystallized some of this transition and showed the main motivation for critical analyses of the Talmudic tradition: to adapt Judaism to the truly modern world, where first Newtonian and then post-Newtonian physics; Kantian and Hegelian philosophy; the American and French Revolutions; colonialism in Africa, Asia, and Latin America; Marxism; Darwinian biology; Freudianism; and other new developments were shaping a new human consciousness.

Before dealing with the implications of these modern changes for Jewish searches for direct experiences of ultimate reality, let us begin from one of the end points of this process, the spirituality of the late Rabbi Menachem Schneerson (d. 1994), Grand Rabbi of the Lubavitch Hasidim, perhaps the most influential of the groups living in Brooklyn and keeping a close watch on events in Israel. In 1964, when Rabbi Schneerson was sixty-two, the Jewish novelist Harvey Swados interviewed him but never published an account of their interaction. In 1994, Robin Swados, the daughter of the novelist, published this interview on the Op-Ed page of the *New York Times* (June 14, p. A15).

After Swados had finished with his questions, he found the rabbi turning the tables and asking him some questions about how he regarded his own Jewish heritage. These questions illustrate Rabbi Schneerson's appropriation of the traditional obligation of the Hasidic rebbe to challenge people to appreciate the work of God in their lives (the function of the "sparks," we might say). Here is a slice from the fare that the Grand Rebbe offered his new acquaintance late one night more than thirty years ago:

> You [Swados, novelist] have certain responsibilities which the ordinary man does not—your words affect not just your own family and friends but thousands of readers.
>
> I'm not sure I know what those responsibilities are.
>
> First, there is the responsibility to understand the past. Earlier, you asked me about the future of Judaism. Suppose I ask you how you explain the past, the survival of Judaism over three millennia.
>
> Well . . . the negative force of persecution has certainly driven people together who might otherwise disintegrate. I'm not certain that the disappearance of that persecution, whether through statehood in Israel or

through the extension of democracy in this country, wouldn't weaken or destroy what you think of as Jewishness.

Do you really think that only a negative force unites the little tailor in Melbourne and the Rothschild in Paris?

I wouldn't deny the positive aspects of Judaism.

Then suppose that scientific inquiry and historical research lead you to conclude that factors which you might regard as irrational [mystical?] have contributed to the continuity of Judaism. Wouldn't you feel logically bound to acknowledge the power of the irrational, even though you declined to embrace it?

Are you suggesting, Rebbe . . . that I should reexamine my writing, or my personal code and my private life.

Doesn't one relate to the other? Doesn't one imply the other?

That's a complicated question.

Yes. . . . It certainly is. . . . I warned you that I wouldn't be diplomatic, didn't I?

The Rebbe is stereotypically Jewish in preferring to move things along by questions rather than declamations. He wants Swados to think for himself, engage himself personally in the issues evolving between them. Jewish study of tradition has been an active affair, an intense dialogue with the talmudic text, at times even a fight. Jewish prayer has often been a questioning, even querulous encounter with the divine, guided by memories of Abraham haggling over the fate of Sodom and Gomorrah and of Job accusing God. The tradition has thought it better to risk offense, forwardness, whining than to risk detachment, indifference, carelessness. Passion could always turn around from complaining to ardent love. Indifference left one with no place to go, made one a stone rather than a fire. Certainly, reverence for the Lord and the desire always to bless the holy divine name bulk large in traditional Hasidic spirituality. However, personal engagement, challenging God to take seriously the history that He created as a covenant forming Israel into a people, bulks even larger.

Here, Rabbi Schneerson wants to move Swados, a rather secularized modern Jew, back into this history. He wants to remind Swados of the tradition, even the chromosomes and blood, from which he comes and on which he depends. There is a mystery to the survival of the Jewish people. There is an obligation to the past, to keeping faith with one's forebears, incumbent on all Jews who recognize this mystery and come to any significant sense of wonder at it. It is only a short step from such wonder to prayer, or outbreaks of praise, gratitude, complaint at the sufferings involved, or petition for the strength to keep going, keep extending the Jewish line. If the Rebbe can get the novelist to entertain any of this wonder at his roots, the living traditions that, probably more than Swados realizes, have formed his writing, indeed his very self, he will have fulfilled his obligation to this interviewer by giving him spiritual nourishment, challenge, and hope for his soul.

Rabbi Abraham Joshua Heschel (1907–1972), another esteemed exponent of the Hasidic tradition, for many years taught Jewish ethics and mysticism at Jewish Theological Seminary in New York. In one of his popular books, Heschel had the following to say in answer to the question, "Does God require anything of Man?" Notice how Heschel, like Schneerson, wants to spring his interlocutor free of a worldview dominated by rationalism and to open the interlocutor to the possibility that something more than our limited reason ought to determine how we think about the potential in life, especially in the life of Jews wanting to appropriate their tradition:

> From a rationalist point of view it does not seem plausible to assume that the infinite, ultimate supreme Being is concerned with my putting on Tefellin [boxes with scriptural texts; phylacteries] every day. It is, indeed, strange to believe that God should care whether a particular individual will eat leavened or unleavened bread during a particular season of the year. However, it is that paradox, namely, that the infinite God is intimately concerned with finite man and his finite deeds, that nothing is trite or irrelevant in the eyes of God, which is the very essence of the prophetic faith.
>
> There are people who are hesitant to take seriously the possibility of our knowing what the will of God demands of us. Yet we all wholeheartedly accept Micah's words: "He has showed you, O man, what is good, and what does the Lord require of you, but to do justice, and to love kindness, and to walk humbly with your God . . ." (Micah 6:8).
>
> If it is the word of Micah uttering the will of God that we believe in, and not a peg on which to hang views we derived from rationalist philosophies, then "to love justice" is just as much law as the prohibition of making a fire on the Seventh Day. If, however, all we can hear in those words are echoes of Western philosophy rather than the voice of Micah, does that not mean that the prophet has nothing to say to any of us?[22]

Rabbi Heschel would move his readers into the world of the biblical prophet Micah, where the Word of God echoes as the foremost reality. He would remind us of the consciousness that formed the relationship between God and Israel, an awareness of the will of God as the most precious reality in the people's lives. The reason why Torah is holy and precious is that it expresses the divine will. When the prophet summarizes what God wants of human beings as doing justice, loving kindness, and walking humbly with God, he is giving voice to a desire of God's heart, not to some pale summary of a good attitude derived from mediocre human reasoning.

There is an understated passion in these three precepts, a kindling ready to take flame. The prophet speaks gently, as though God were inviting an audience of children to take their first steps. However, justice, kindness, and walking humbly can become strong features of a truly admirable life, indeed of a life worthy of an eminent Hasidic teacher. Heschel is judicious in his choice of this text from Micah. He knows that contemplating it deeply can bring a reader close to the heart of the mysterious Lord of the covenant, who loves justice, moving kindly, and humble awareness that everything good about the covenant comes from divine grace.

While heirs of the Hasidic tradition such as Schneerson and Heschel strove to modernize its devotional outlook (including its affection for the Kabbalah), another reaction to modernity, the Reform Movement, which sought to reconcile traditional Judaism with modern rationalism, was at work in both Europe and the United States. The not-so-veiled attacks of the Hasidic masters on "rationalism" express the worry of the Orthodox that the Reform Movement would give away precious portions of the traditions by which Jews had survived for thousands of years. The Reform Movement sought an appropriation of modern science, philosophy, and political thought that shifted religion from law and mysticism to ethics. The hope of the Reformers was to draw from Judaism universal principles applicable to the moral lives of all human beings. If possible, they avoided criticism of *halakha,* but on occasion they questioned the relevance of the 613 *mizwot,* the laws for kosher and observance of the Sabbath that had been the backbone of *shtetl* life. As well, they distanced themselves from the Kabbalah, the interest of esoteric Jews in the symbolism of the Song of Songs as an expression of Israel's marriage to its Lord, the expectation of the coming of the Messiah, and other emotional preoccupations of the traditional religious life.

One of the most influential early statements of the Reform view was the Pittsburgh Platform produced by a conference in 1885 that reconciled European (largely German) Reformers with Americans. The eight principal points of this platform reflect an effort to mediate between Jewish tradition and a new, modern cultural context: (*1*) an acknowledgment that all religions attempt to grasp the infinite, but a conviction that the Jewish idea of God, as presented in scripture, represents the highest conception; (*2*) a conviction that the Bible gives the record of the consecration of the Jewish people as priestly and the highest source of religious and moral instruction in the world, along with an acknowledgment that the biblical modes of speech are not modern but ancient and inclined to stress the miraculous; (*3*) a key statement suggesting the infra-Jewish tensions between the Reformers and the traditional Talmudists: "We recognize in the Mosaic legislation a system of training the Jewish people for its mission during its national life in Palestine [during the biblical period], and today we accept as binding only the moral laws, and maintain only such ceremonies as elevate and sanctify our lives, but reject all such as are not adapted to the views and habits of modern civilization," a principle bound to offend the Talmudists inasmuch as it made "modern civilization" more authoritative than the Torah; (*4*) continuing the attack on *halakha* implied in principle three: "We hold that all such Mosaic and rabbinical laws as regulate diet, priestly purity, and dress, originated in ages and under the influence of ideas altogether foreign to our present mental and spiritual state. They fail to impress the modern Jew with a spirit of priestly holiness; their observance in our days is apt rather to obstruct than to further modern spiritual elevation"; (*5*) the Pittsburgh Platform affirms that the universal culture developing in the world connects it with the messianic yearning of Jews for an age of justice and peace, but Reform rejects the notion that Israel is a nation, plumps for a self-conception as a religious community, and rejects the goals of both a return to Israel (Zionism) and the restoration of sacrificial worship according to the traditions of the Aaronic priesthood (that is, rejects the reestablishment of a cult like that of the Temple in Jerusalem during the biblical period); (*6*) a description of Judaism as a progressive religion, striving to accord with principles of reason, mindful gratefully of the spread of its views through its daughter faiths, Christianity and Islam, especially the spread of its monotheism and high morality; (7) a reassertion of the immortality of the human soul but a rejection of the ideas of the resurrection of the body, heaven, and hell; (*8*) in the spirit of the Mosaic legislation, a commitment to social justice, regulating the relations between the rich and the poor toward a fairer sharing of the world's resources.[23]

These principles are mainly self-evident, though in 1885 many Jews would have found them radical, if not blasphemous. The most offensive aspect to the traditional Talmudists was the attack on *halakha* as outdated. To those who viewed the Torah, both biblical and Talmudic, as revelation given by God, any rejection of its authority in the name of modern needs or wisdoms must have seemed an inversion of right order. Who were human beings to alter the order that God had given from Sinai and through the labors of so many holy rabbis? It was this order, these traditions, that had held Jews together for three thousand years, as Rabbi Schneerson made clear to Swados. It was the prophetic voice (Moses was the premier prophet) expressed in the particulars of traditional Jewish observance (for example, donning *tefillin*) that had preserved the community during countless persecutions, as Heschel implied. Certainly, Heschel wanted Jews to move by the spirit of a prophet such as Micah, honoring the famous three expressions of the divine will found in Micah 6:8 at least as much as the laws about lighting fires on the Sabbath. Certainly, he was aware of the proper pressures for updating the tradition and accommodating to new times, but the reduction of the tradition implied in Reform Judaism did not sit well with Jews dedicated to Talmudic law. It seemed so strong a threat to Jewish identity that they attacked it bitterly.

A full history of the modern period of Judaism would deal with the rise of the Conservative and Reconstructionist movements, further attempts to mediate between Talmudic traditions and modern needs. Through all of this innovation, Orthodoxy came to a consciousness it had not previously possessed. Previously, Jews had simply gone their traditional ways, adhering to their traditional understandings of the *mizwot,* the ways to study Talmud, the place of the Kabbalah, and other mystical pursuits. Now, under the pressure of reforming movements, the traditionalists had to become more reflective and articulate about what they did, why they acted as they had so long, why they loved what they loved. Thus even when Jews continued the old ways they could not do so with the old consciousness. Whereas a century earlier, when the Enlightenment was transforming European consciousness, only the relatively few Jewish intellectuals living, studying, and working in the non-Jewish mainstream were affected deeply, when Reform brought modern, largely Enlightenment principles into the Jewish community proper, the majority of Jews had to take note.

The effect on mystical pursuits was predictable. Those who accepted the arguments about the need to make Judaism accord more with modern rationalism tended to back away from the esoteric aspects of the traditional, rabbinic regimes. Fearing to be thought irrational, mythological, and concerned with miracles, they let the modern rationalists' suspicion of mysticism, indeed disdain for mysticism, color their faith. Judaism, it was thought, ought to concentrate more on ethics, universal principles enlightening the good moral life incumbent on all human beings, than on the peculiarities of its *halakha* or the strange symbolism of the *Zohar.* It ought to enter the mainstream and not stay sequestered in the ghetto. The rabbi ought to be a teacher in touch with the demands of modern life, more than a transmitter of Talmudic traditions apparently far out of touch. All these biases echo in the concerns of Schneerson and Heschel to preserve the place of the irrational, the mystical, in Jewish piety.

In fact, in the postmodern era, which owes large debts to Einstein and Freud (two secular Jewish intellectuals), but also to the horrendous wars of the twentieth century, rationalism has gone out of style. That does not mean that mysticism has become the rage, but it does mean that any simple effort to make a narrow understanding of the methods of the physical sciences an adequate paradigm for human intelligence at its most creative meets with rejection nowadays. For example, the behaviorism of a B. F. Skinner tends to get laughed out of court. Human beings are more intuitive, complicated, irrational or pararational or suprarational than a behaviorist model allows. Reason is subtler, more imaginative, more poetic, symbolic, metaphorical than what a purely functional view of human intelligence (one that takes it as simply refining skills needed for biological, evolutionary survival) takes into account.

We may conclude our sketch of modern (now postmodern) trends affecting Jewish mysticism by looking briefly at two responses by eminent present-day Jewish scholars. At the end of her demanding work *Saints and Postmodernism,* Edith Wyschogrod speaks for intellectuals alienated from traditional religion but moved by recent moral heroes seeking a new holiness:

> I have also maintained that postmodern saintliness is not premodern saintliness, because inscribed in postmodern saintly life is the weight of recent history. Hagiography [writing about saints], when written in the idiolect [special speech] of the postmodern saint, bears the trace of the rational and egalitarian suppositions of the Enlightenment as well as the force of the Kantian moral law and the criticisms and appropriations of it in utilitarian and pragmatic ethics. If liberal theories of justice (Rawls), phenomenological ethics (Scheler), and contemporary Kantianism (Gerwirth) referred to in this study fail to persuade, it will not do to return nostalgically and uncritically to an older ethos. Nostalgia is amnesia, a wiping out of both the sea-changes brought about by recent history [for Jews, above all the Holocaust] and the sins of older communities such as slavery in Greece and the persecution of Jews, Moslems, and heretics by medieval Christians. This backward thrust is an example of what I called earlier the myth of the tabula rasa [clean slate] and leads to impossible dreams such as Alasdair MacIntyre's hope for a restoration of a monastic ethic or a return to an Aristotelian version of the good life as one governed by the classical virtues. . . .
>
> Borrowing the compassionate strands of the world's religious traditions, the absurdist gestures of recent modernist art and literature, and modern technologies [postmodern] saints try to fashion lives of compassion and generosity.[24]

Debated here are the influences of such ideas, or experiences, as "the Other" and the breakdown of tradition via deconstruction of premodern scriptures (exploration of the abritariness or historical conditioning of any text, even one claiming the authority of divine revelation and long granted such authority by a community dedicated to saintliness) on any postmodern quest for lives of compassion and generosity. Simply put, the postmodern intellectuals who interest Wyschogrod are those who challenge Ivan Karamazov's conviction that if God is dead, anything is permitted.

Alternately, what Nietzsche's proclamation that God is dead means does not seem patent. If there is an "Other" ingredient in human consciousness, formative of human consciousness, can "God" ever be dead? Much depends on definition and interpretation, the extent to which such an Other and God can coincide. Wyschogrod's strength lies in her insistence that we cannot deal with the question of human saintliness, the development of a generous and admirable humanity, apart from the recent history that has shaped us. We have to let the slaughters of the wars of the twentieth century, the brutalities of the recent totalitarianisms, and the breakdown of traditional world views and moral assumptions through developments in both the physical and social sciences make their full impact. Otherwise, we lapse into nostalgia, an indefensible forgetfulness.

For a student of mysticism, Jewish or other, this analysis is stimulating but not fully convincing. Apart from the few intellectuals who feel at home in the complicated, highly reflexive language ("idiolect") and conception to which Wyschogrod is inclined, most people interested in saintliness, even most intellectuals, seem to grant the traditional theistic conceptions more possibility than does postmodernism. The depravities of the twentieth century are not without precedent in prior centuries. The skepticism induced by postmodern deconstructionists is not so different from the negative theology prominent in many religious traditions. One cannot equate the Other with God unqualifiedly, but one can recall that the God of the high religious traditions, the sophisticated theologies elaborated from direct experience of ultimate reality, is always beyond capture by human reason, always more unlike than like what human beings can say about him or her or it.

Postmodern intellectuals do us the favor of underscoring the newness of the situation and the consciousness that recent history has brought on the scene. Often, however, they seem ahistorical in their own right, unable to discern the actual, century-by-century struggle of human beings to find meaning in a world bound finally to defeat them, to push in their faces a radical, irreducible mysteriousness. One might call the wisdom resulting from this struggle and defeat a defensible *philosophia perennis.* One might also note its impressive articulation in the mystical traditions of such intensely pursued and ongoing collaborative interpretations as the Jewish Talmudic, and suggest that it can serve postmodern quests for saintliness as a good counterbalance to their tendency to exaggerate the changes in human nature wrought by recent history—to underplay the constancy of such problems as sin, the need for grace, the inevitability of tradition, the function of hallowed traditional texts and symbols, the role of rituals mediating encounters with ultimate reality, the force of death, and the wonders of creativity and love.

Arthur Green, working from an interesting combination of Hasidic and Reconstructionist convictions, has reflected on the peak experience of Moses on Mount Sinai in ways that suggest how the mystical tradition has continued to influence Jewish theologians:

> Moses is the one who saw beyond the darkened glass, who looked into the brightness. What did he bring back from that indescribable moment? Was it something like the great piece of music, or the scientific breakthrough, a result of revelation creativity as we might understand it? Yes, but articulating it is not simple. All of our Torah, in the broadest sense, may be called an ongoing, stammering, and always inadequate attempt at this articulation.
>
> Out of Sinai comes Y-H-W-H, the reality and the word. Sinai offers Y-H-W-H as the singular divine presence that pervades all the world and reaches beyond it in ways we humans are not given to fully understand. This reality, Sinai tells us, is accessible to human beings at the greatest moments of their lives. The same ecstatic presence that filled the hearts of Israel as they walked proudly out of Egypt, the same presence that was to so fill the Tent of Meeting that no person was able to enter it, could be found in human life, both for individuals and for the nation, again and again in the future. *Ehyeh Asher Ehyeh,* "I shall be that I shall be," is interpreted by the rabbis to mean "I shall be with you again as I was with you then." The manifestation of Y-H-W-H that happened in Israel's minds and hearts at Sinai is an assurance that such manifestation does not happen then alone. Revelation reveals the *possibility* of revelation, not just that once, but whenever the human heart and mind are open to it.[25]

Here we see a more optimistic view of tradition, premodernity, and mysticism than what Wyschogrod offers us. Whether Green is indulging in nostalgia, forgetting the dislocations introduced by the twentieth century, and assuming that human consciousness is sufficiently a *tabula rasa* in each new child to allow a direct outreach to God or reception of a divine influx, depends on how he would interpret the specific postmodern changes that Wyschogrod has described. Clear from these lines, however, is the canonical status that he gives to the Mosaic experience. Because God disclosed the divine reality there, God can disclose the divine reality at any time. Because there human consciousness entered a new dimension, where it became aware of the bare primacy of the divine being and the utter freedom of God to be for human beings just as God chose to be, unutterably, for Moses, human consciousness may return to such a new dimension whenever its own needs or achievements combine with divine grace to repeat the Mosaic elevation.

This conviction amounts to an endorsement of what we have been calling mysticism. Green is saying that Jews believe that God can give them direct experiences of the holy divine ultimacy at any time in their history, on the model of what God gave to, and through, Moses at the origin of their covenant. Probably the relations between this possibility and keeping the *halakhic* traditions can never become absolutely clear, but the entire thrust of Green's book makes his theology a positive view of the Talmudic traditions, as long as we do not take them lifelessly.

Steeped in a continuing effort to make Jewish tradition vital, to transmit the ardor of the best past rabbis to a new generation, scholars of Jewish law and mysticism such as Green have become reformers of a new order, not rationalistic as the early leaders of Reform Judaism tended to be, but sufficiently shaped by irrational, postmodern experiences to defend a mystical view of the well-springs of human creativity and saintliness, which view they find to be both a faithful rendition of Jewish experience and an honest estimate of human potential.

Summary

We have followed the historical course of Judaism, studying the biblical, rabbinic, medieval, postmedieval, and modern periods into which we have divided it. To say the least, this course has been rich and full of important developments. The biblical origins of Israel have remained central to Judaism inasmuch as the Mosaic covenant associated with Sinai has shaped both the Jewish people's sense of God and the Torah through which they have tried to live with and be holy before God and therefore be worthy of election and blessing. Thus the symbolism of the tent and the temple (the biblical housing of God) has informed the postbiblical generations insofar as they read the Mosaic texts with a desire to find holy ultimacy. The symbolism in the great prophetic visions of Isaiah and Ezekiel has also borne down heavily on later generations, as the chariot that Ezekiel saw became another housing for God, the demesne most beloved of the early rabbinic mystics for hundreds of years.

Nonetheless, despite this continuity with such biblical beginnings, the move after the destruction of the Temple in Jerusalem in 70 C.E., from leadership by a sacrificial, ritualistically focused priesthood to leadership by teachers immersed in the study of Torah, changed Judaism significantly. The law became the consuming passion. Understanding the law and keeping it became the primary way of dealing with God. At times the mysticism of the rabbis of the first millennium C.E. seems to have flowed out of the law, as though devotion to *halakha* ought to make one a rider in the chariot. At other times their mysticism seems to have beckoned as a relief, an expatiation, for spirits risking cramp by legal precision.

The Kabbalistic developments that gathered full force in what we have called the medieval period may appear initially as more Gnostic, esoteric, and dubious than the mystical accents worked out of what Ezekiel reported, but further analysis can narrow the difference. Certainly, the symbolism of the *sefiroth,* and the *gematria* elaborating them to suggest that the Hebrew alphabet expressed the building blocks of creation, represents a vibrant, fantastic journey. However, the center of Ezekiel's vision, the fire of the moving chariot, invites an elaboration such as this, a further imagining.

Hasidic piety, gathering force in the postmedieval period, taking fuel from further persecutions and messianic hopes, introduced new emphases on stories, joy, and redemption. The Hasidic rabbis loved the Torah, danced with the Torah, and often loved the Kabbalah as well; but they displayed a countenance homelier than that of the Kabbalists, a mysticism more suited to an Eastern European village, thick with mud, than a sophisticated Spanish city. The sparks of redemption gave them great hope, while the sufferings of their people kept them busy before the Master of the Universe, pleading.

The recent period of Jewish spirituality, dominated by the rise of reform movements, has forced many Jews to look carefully at what had been for the majority a relatively simple, homogeneous, traditional way of life. "Orthodoxy" was a new category, born of the need to contrast traditional ways with challenging new ones. In addition to its veneration of the *halakha*, Orthodoxy continued to mine the mystical traditions of the Kabbalah, as well as other sources of "irrational" strength, joy, and a capacity for suffering that reason alone was often unable to provide. The contrasts between the readings of current times that we found in Edith Wyschogrod and Arthur Green indicate how Jewish quests for meaning tend to go nowadays. The reconciliation of traditional Jewish mysticism with such postmodern factors as the Holocaust is not easy, but clearly both poles of experience must receive their due.

When it comes to characterizing the distinctiveness of Judaism's direct experiences of ultimate reality, what does our survey suggest? First, it suggests that the biblical formation of Israel in the desert was momentous. Second, it suggests that Jewish mystical experience has been markedly kataphatic (imaginative, positive about symbols, confident that the effort to represent God is valuable). Let us elaborate these two propositions.

The God whom Jewish mystics have sought is the God of the ancestors Abraham, Isaac, Jacob, Moses, and David. This God is personal, engaged with the original tribes and their descendants, as with a people uniquely his own. The people think of themselves as formed through a unique covenant, and they think that no other people has received so much from the one God, the Maker of heaven and earth.

The Mosaic axis of this faith, which reflects the constitutional or paradigmatic character of the Exodus from Egypt and the establishment of the covenant (both personal bond and communitarian law) on Sinai, has been both priestly and sapiential. The consecration of the people through the observance of the Torah has been priestly, with or without explicit sacrificial rituals. The study of Torah for both the glorification of God's name and the religious welfare of God's people has been sapiential—a matter for teachers, rabbis, spiritual masters, sages.

From these Mosaic beginnings, Jewish mystics have seldom taken yogic, shamanistic, or even medicinal turns. While comparativists might find in descriptions of Jewish mystical consciousness elements of such turns, the basic thrusts have not been interior, toward *enstasis* (yogic), as they have not been toward "magical flight," the almost avian ecstatic tendencies of Central Asian and American shamans. Although some eminent rabbis have been miracle workers, most have not made healing their hallmark or watchword. The medicinal or therapeutic accent that one finds in the Buddha, who was concerned to get the poisons of *samsara* out, or even in Jesus, who documented the presence of the Kingdom of God by healing the sick, is stronger than what one finds typically among the Jewish mystical masters, where, of course, it is not entirely absent.

Conversely, the concern of the Jewish mystical masters for the Mosaic symbols and Torah has nothing like a full counterpart in Buddhism or even in Christianity. Christianity may reinterpret such symbols as the Exodus in light of the passover of Jesus from death to resurrection, but eventually the commitment of Christianity to the Incarnation of the divine word develops among its mystics a more individualistic accent. Similarly, no Buddhist law (*dharma*) makes a people as distinctive as their law has made Jews.

The monotheism of the biblical experience, expressed as clearly as it could be in the unnamable presence of the Mosaic Lord, brings the Jewish mystical beginnings into an orbit familiar to us from our study of Eastern ultimacy. Nothing can ever represent the Lord adequately; he is a hidden God. However precise the Torah becomes, it can never rightly remove the priority of the divine freedom. The Lord may have sworn fidelity to the covenant and may never repent of this swearing, but the covenant does not compel him. There is nothing prior to the Lord, no principle or force more elementary. That the Master of the Universe should have assumed historical relations with a people and joined himself to Israel in a covenant is the more wondrous for this. From Israel he never needs a thing. If he chooses to need the fidelity of Israel, even the affection, that is a further wonder. If he condescends to enter the tent or the temple, that is completely a gratuity.

The second mark of Jewish mysticism that we underscore, its affirmative, imaginative character, presents an intriguing challenge when one tries to correlate it with the stronger strata of biblical monotheism that we have just seen. On the one hand, the historical, institutional forms of religion, with which preoccupation with the Mosaic Torah and its stress on performance (ethical follow-through, *mizwot*) let the Jewish rabbis square their teachings, through a conviction that human reason, imagination, and will can serve divinity well, can descry divinity only partially. The biblical accounts of redemption (Exodus and entry into the Promised Land) and creation, along with the prophetic accounts of revelation and the rabbinic accounts of finding the divine will through the *halakha,* support a confidence that God is in the world, at least vestigially (as sparks of holiness) and that seeking God is not a waste time. On the other hand, the esoteric character of the Kabbalistic symbols, which sometimes the mystics elaborate with little critical control, indeed carelessly, heads in the direction of a mystical antinomianism or plain confusion, an intuition that because no laws or conceptual niceties or syllogisms can capture the divine mystery in its utterly simple primacy, what one says does not matter fully. Probably no Jewish mystical master would ever have signed a document with this proposition, but in practice the texts sometimes appear to be free-wheelings, sportings, exuberant larkings, or rocketings around the divine fire and *sefiroth*—cosmogonic dances at the primordial beginning that can never come under human control. Certainly, sometimes the Kabbalists, like the *halakhists,* stand on the verge of claiming control, and so of risking idolatry, but the better explanation for the imaginative flights of a classical Kabbalistic text such as the *Zohar* may be the freedom that an awareness of "defeat from the start" can create.

The God at the beginnings of the Jewish quest for direct experience of ultimate reality never leaves the further historical course of that quest. Through all its legal and symbolic permutations, Jewish spirituality stays coherent through fidelity to its beginnings on Sinai, its conviction that its God moves in its history, and its love of the divine fire, the furnace of holiness.

Naturally, the distinctiveness of the history, both happenings and symbols, that shaped their people gives the Jewish mystics most of their special features. The core of that history is the Torah, how it arose and how Jews formed themselves through it. The commonness of Jewish mysticism, what it shares with other traditions to justify our ranging it alongside Buddhism or Islam, stems from the depth that it reached. At the point where Jews confessed that Israel had to hear repeatedly and daily that the Lord is One, they could move away from any restriction that would have made their theology or mysticism simply tribal, ethnic, or provincial. Crucial for Jews as constituting themselves through the Torah might be, they could know from their mystics as well as their rabbis that the Master of the Universe was the source of all creatures and that that chariot of the Master could carry a great variety of riders.[26]

NOTES

1. Eugene Borowitz, "Judaism: An Overview," in *The Encyclopedia of Religion,* ed. Mircea Eliade (New York: Macmillan, 1987), 8:127–149.
2. Ibid., 129.

3. Translation from W. Gunther Plaut, ed., *The Torah: A Modern Commentary* (New York: Union of American Hebrew Congregations, 1981), 1366.

4. Ibid., n. 4.

5. Ibid., 1371.

6. Ibid., 1369.

7. Arthur Green, ed., *Jewish Spirituality* (New York: Crossroad, 1985), 1: xiii.

8. Jon Levinson, "The Jerusalem Temple in Devotional and Visionary Experience," in *Jewish Spirituality*, ed. Green, 1:35.

9. For other aspects of the biblical foundations of Jewish mysticism, see *Jewish Spirituality*, ed. Green, 1:1–165. Probably the standard overview of Jewish mysticism as a whole remains Gershom G. Scholem's *Major Trends in Jewish Mysticism* (New York: Schocken Books, 1941). On current approaches to the literary aspects of biblical texts, see Robert Alter and Frank Kermode, eds., *The Literary Guide to the Bible* (Cambridge, Mass.: Belknap/Harvard, 1987).

10. David Winston, "Philo Judaeus," in *Encyclopedia of Religion*, 11:289. See also Winston's study, "Philo and the Contemplative Life," in *Jewish Spirituality*, ed. Green, 1:198–231. The classical study is probably Harry A. Wolfson's *Philo* (Cambridge, Mass.: Harvard University Press, 1962).

11. Louis Jacobs, "Judaism," in *The Study of Spirituality*, ed. C. Jones, G. Wainwright, and E. Yarnold (New York: Oxford University Press, 1986), 493.

12. B. Z. Bokser and B. M. Bokser, eds., *The Talmud: Selected Writings* (New York: Paulist, 1989), 60–61.

13. "Allegory," in *Encyclopedia Judaica* (Jerusalem: Keter, 1972), 2:643.

14. Isadore Twersky, "Maimonides, Moses," in *Encyclopedia of Religion*, 9: 131.

15. See Daniel C. Matt, "The Mystic and the *Mizwot*," in *Jewish Spirituality*, ed. Green, 1:383.

16. Joseph Dan, "Introduction," in *The Early Kabbalah* (New York: Paulist, 1986), 7; also Mark Verman, *The Books of Contemplation* (Albany: State University of New York Press, 1992).

17. "The Fountain of Wisdom," in *Early Kabbalah*, 50.

18. Gershom G. Scholem, *On the Kaballah and Its Symbolism* (New York: Schocken, 1965), 100–101.

19. Moshe Idel, *Kabbalah: New Perspectives* (New Haven: Yale University Press, 1988).

20. Martin Buber, *Tales of the Hasidim* (New York: Schocken, 1947), 1:35. A good balance to Buber's presentation is Jerome R. Mintz's *Legends of the Hasidim* (Chicago: University of Chicago Press, 1968).

21. Ibid., 2:49.

22. Abraham J. Heschel, *Man's Quest for God* (New York: Scribner's, 1954), 103.

23. "Pittsburgh Platform," in *Encyclopedia Judaica*, 13:570–571.

24. Edith Wyschogrod, *Saints and Postmodernism* (Chicago: University of Chicago Press, 1990), 256–257.

25. Arthur Green, *Seek My Face, Speak My Name* (Northvale, N.J.: Jacob Aronson, 1992), 113.

26. For some of the further chapters and personalities that a full survey of Jewish mysticism might include, see Louis Jacobs, *Jewish Mystical Testimonies*

(New York: Schocken, 1977). On Rabbi Abraham Isaac Kook (1865–1935), an influential modern Zionist given to mysticism, see Arnold M. Eisen, "Secularization, 'Spirit,' and the Strategies of Modern Jewish Faith," in *Jewish Spirituality*, ed. Arthur Green (New York: Crossroad, 1987), 2:303–305. A good collection of the stories in which one might find the imagination that has nourished popular Jewish piety is Ellen Frankel's *The Classic Tales* (Northvale, N.J.: Jason Aronson, 1989).

6

Christian Traditions

General Orientation

Historically, Christianity is about 2,000 years old. It arose from the preaching, life, death, and claimed resurrection of Jesus of Nazareth, a first century C.E. Jewish religious figure (*hasid*). Jesus preached to his fellow Jews, both in his native area, Galilee, and in other parts of the land, including Jerusalem. His disciples remembered him as having worked cures, indeed miracles, for the sick, the dying, and the possessed. We deal with the message and person of Jesus more fully in the next section. Here, the main point is that Christianity began as a messianic, sectarian Jewish movement. In the generation before the destruction of the Jerusalem Temple (70 C.E.), the followers of Jesus competed for allegiance with such other groups as the Pharisees and the Sadducees, while after the destruction of the temple their movement competed with the rabbinic movement developing to the west of Jerusalem, at Jamnia and other sites along the Mediterranean.

The principal bone of contention between traditional Jews and the followers of Jesus concerned messiahship. Traditional Jews denied that Jesus had been the Messiah, the leader anointed by God to rescue them from foreign rule and spiritual neediness. The followers of Jesus claimed that he had fulfilled the biblical prophecies concerning the Messiah by showing in his person, message, works, and resurrection from the dead a full dossier of credentials. At times this disagreement grew intense, taking on characteristics of a family feud. The New Testament reflects this situation. Many of the New Testament authors are intent on showing the messianic character of Jesus, and sometimes their frustration with fellow-Jews who refused to accept Jesus as the Christ is palpable.

Within a generation of the death of Jesus, however, the Christian community had opened itself to Gentiles as well as Jews. Missionary experiences of the early Christian leaders Peter and Paul, especially, convinced the first churches that Jesus had broadened the covenant. No longer were Jewish heritage and circumcision necessary for membership in the people whom God had elected. Christianity could present itself as the fulfillment of pagan (Hellenistic) aspirations, as well as the fulfillment of Jewish biblical prophecy. Early Christians continued to think of *Tanak* as biblical revelation, but when they generated sacred writings of their own, the Hebrew Bible (and the Greek, Septuagint translation of it) assumed the status of an "old," venerable collection of writings requiring reinterpretation in light of the coming of Jesus, the Messiah, to fulfill its promises. (Islam dealt more radically with both the Old and the New Testaments, judging that the Koran had superseded them.)

By the end of the first century C.E., Christianity had developed an identity separate from that of Judaism and, arguably, was well on its way to becoming a religion significantly different. It owed many of its genetic ideas to the Jewishness of both Jesus himself and the first generations of disciples, but Greek ideas had become nearly equally important. (Indeed, many Jews had come under the influence of Hellenistic ideas.) The fusion of Jewish and Greek ideas in the first four centuries or so of Christianity made a religious amalgam discernibly different from the rabbinic religion, which was developing simultaneously what would become the Talmudic literature. Capital in this difference was the status of Torah. A simplistic summary might say that, for Christians, Jesus took the place of Torah (as, in Islam, the Koran took the place of Jesus).

The patristic era, which occupied perhaps the second through the sixth centuries C.E., featured church leaders (bishops) inclined to think about doctrine and discipline in Hellenistic terms. At a series of councils of these bishops, the ecumenical Christian community hammered out its responses to doctrinal issues, liturgical challenges, questions of proper moral behavior, and religious movements such as monasticism. Although earlier Christians had suffered persecution and produced some martyrs, from the fourth century, when the Roman emperor Constantine favored Christianity as the religion most useful for consolidating his realm, the churches enjoyed considerable freedom to proselytize. Indeed, in the eastern Empire (later, Byzantium) church leaders developed almost symbiotic ties with the rulers of the state.

At the outset of its medieval period Christianity suffered from the turmoils introduced by Germanic, Celtic, Frankish, and other tribes flourishing in the West after the fall of Rome (430 C.E.). The papacy (bishopric of Rome) became a key, perhaps the most powerful, medieval European institution because it stepped into the vacuum caused by the fall of Rome. Eventually, most of the previously pagan Western tribes became Christian, due in large part to the labors of missionary monks.

In the East the culture remained less divided, though steadily the Eastern and Western branches of Christianity grew farther apart. The use of Greek in the East and Latin in the West was one crucial factor. Another was the different cultural situations and so different doctrinal, disciplinary, and other challenges. By 1054 the two churches had anathematized one another, each condemning the other's doctrines or church polity. The Crusades that brought Western knights to Jerusalem and other parts of the East from the end of the eleventh century further damaged infra-Christian relations, while the rise and spread of Islam from the seventh century colored the lives of most Christians in both the Middle East and southern Europe. The scholastic synthesis in thirteenth century Europe marked a high point in Christian culture, but by the fourteenth century the old era had ended and disarray threatened.

Movements for reform dominated the postmedieval Christian period in the West. The "captivity" of the papacy from Rome to Avignon in the south of France during the fourteenth century (1309–1377) marked the nadir of the church's disarray, making the Black Plague epidemic seem a divine punishment. The sixteenth-century Protestant Reformation, spearheaded by Martin Luther and John Calvin, had begun smoldering with the disgust of such fifteenth-century religious leaders as Jan Hus, John Wycliffe, and, at the turn of the century, Erasmus. The rallying cry of most reformers was "back to the New Testament," along with a demand that the church strip itself of the wealth corrupting it and again become a servant of the common people. The reformers favored placing both the Bible and the liturgy in the vernacular so that the common people could understand them. They also attacked the monastic life as too influential, if not unbiblical. New modes of thought challenged the scholastic modes established in the medieval Christian West. Economic developments took many lay Christians away from feudalism toward trade and urban life.

Modernity arose in Europe during the eighteenth century, though naturally it came on the heels of many late medieval or Reformation developments. The principles of the Protestant Reformers that only Scripture ought to rule church life and that all Christians are competent to determine how God is speaking (through scripture) in their lives, paved the way for the giants of the Enlightenment (David Hume, Immanuel Kant) to advocate the autonomy of individual reason. The natural science that had arisen with Sir Isaac Newton, the opposition of the church in the affair of Galileo (the lingering force of the Inquisition), and the new political horizons opened by the American and French revolutions interacted to create a thoroughly new age in human consciousness.

During the nineteenth and twentieth centuries, this modernistic impression of newness only deepened. Colonialism, which had begun in the sixteenth century, reached its peak in Africa, India, and eastern Asia. Socialist thought became more and more important in Europe. The Industrial Revolution changed manufacturing and stimulated mass movements to the cities. Developments in biology, geology, and medicine changed both people's sense of the natural world and their prospects for longevity.

At the end of the twentieth century, the Christian intelligentsia stands much where we found Edith Wyschogrod placing postmodernity. After the slaughter entailed in two world wars, to say nothing of countless other conflicts, older theologies could seem irrelevant. With stunning developments in physics, astronomy, biology, and medicine occurring every decade, "nature" seems vastly different from the mechanistic system Newton had envisioned. Even Marxism seems passé, making necessary a new politics, one both competent economically and capable of moving people to altruism. The global character of world culture, foreseeable in the year 2000, preoccupies such Christian leaders as Pope John Paul II, who returns repeatedly to the theme of the third Christian millennium, asking how best to think about it and become the midwife of its promise.

If this is a quick sketch of Christian history, what is the equivalent for Christian faith, doctrine, understanding of reality, world view? First, Christian faith has had to deal with the same basic realities of the human situation that beg clarification, or at least commentary, from all the religious traditions, indeed from all human beings who think. That is to say, Christians have had to explain how they understand the natural world, human nature, human sociability, and ultimate reality (divinity).

In the course of meeting this obligation, Christian theologians have developed such distinctive doctrines as those of the Trinity, the Church, sin, grace, and creation. None of these doctrines stands without historical predecessors in the Hebrew Bible, the parent book to both Christianity and later Judaism. Moreover, for all of them we can find comparable teachings in the other world religions, including those originating in India and China. All of these doctrines, however, have a different feel, tone, or style in Christianity, as we ought to expect. All reflect the uniqueness of Christian history, as well as the bold Christian claim that only Jesus has been a full incarnation of God.

So, second, when we ask about Christian distinctiveness, where it comes from, where it lodges and centers, the most obvious answer is Jesus the Christ, the "founder" of Christianity, the personal historical beginning. As we see in the next section, one cannot separate Jesus from the New Testament, which is virtually the only record of his life. Prescinding for the moment from the complicated interpretational questions that this situation raises, what can we say orientationally about how confessing Jesus to be the Christ has shaped Christian faith and mysticism?

The key moment in the development of Christian teaching came when the followers of Jesus confessed him to be divine. Most interpreters of the New Testament say that at least the seeds of such a confession grow in both the gospels and the epistles. Indeed, the implications of calling Jesus the Messiah soon led to calling him divine, especially since the strongest argument put forward in defense of his messiahship was the resurrection, God's having raised Jesus (to new, heavenly life) after his death (by crucifixion, as a criminal).

The sense that Jesus was (is) fully divine, as well as fully human, gives Christian faith its distinctiveness. Always this "both/and" shapes how Christians have thought about reality. For example, when it came to estimating the natural world, Christians could begin with the account given in Genesis, but they had to correlate this account with their own conviction that, as the Logos, Jesus had existed with God prior to his conception as the son of Mary, indeed prior to the creation of the world. The first verses of John reflect this need to rethink Genesis. In John creation became a function of God's heavenly speech, of which (eternal utterance) Jesus was the incarnate Word. Creation became a revelation of God lesser than Jesus, who gained his privileged status by giving flesh to divinity and becoming the primary sacrament (holy material sign) of God's own nature.

From considerations such as these came the Christian conviction that divinity itself is a perfect community: Father–Son–Spirit. Without division (though how?), these three show the singleness of God to be brimming with a light, life, and love that include "personal" values (understanding, love). The Father generates the Son. The Son, incarnate in Jesus through Mary, expresses the Father fully. The Spirit is the love between Father and Son, as substantial as they are. (After the Ascension of Jesus, his return to the Father, the Spirit took over most of his roles on behalf of believers.) The textual grounds for this conception of God lie in the New Testament, especially the Johannine writings, though often they are more latent than developed. The elaboration of this trinitarian conception through the patristic age, which relied strongly on Greek philosophical categories, took the Christian sense of divinity some distance apart from its Jewish beginnings (though Word and Spirit are important realities in Jewish biblical theology).

At least as distinctive as the trinitarian view of divinity, however, was the source of such a new theology: Jesus himself, Christology. As soon as it denied that Jesus was simply another creature, insisting rather that as the Word he had existed with God (the Father) and as God from "the beginning" (the state out of time peculiar to God as God), Christian doctrine began to reconfigure the sense of reality it had inherited from Judaism. All things had to hold together in Jesus the Christ because he had put into flesh the endless purely spiritual knowing of God.

The Christian view of human nature stems from pondering the implications of how Jesus treated his fellow human beings. Certainly, this view reflects the conception of God that runs throughout the Hebrew Bible: Creator and Lord (of a people covenanted to him through historical experience). However, the New Testament presents Jesus more prominently as a healer and savior, one bent upon righting a wronged, sickened human nature. He is not so much the Creator of his fellow human beings as the divinity who reknits their bones and takes away their guilt.

Sin is the capital Christian word for the wrongness from which human beings suffer, while *grace* is the capital Christian word for God's unmerited response to sin, God's righting of humanity's wrongness. Jesus announces a time of making right, salvation: the dawning of the Kingdom of God. Subsequent Christian theology has said that Jesus achieved in principle the reworking of human nature, a new creation. Like a new Adam, he took charge of a new race. Although still burdened by sin, this new race was no longer alienated from God or the object of God's wrath. Where sin, leading to death and wrath, abounded, grace has abounded the more, leading to Christian hopes for *glory* (complete fulfillment with God).

The Christian hope that the Christ will return to consummate history and usher in the moment of glory stems from a sense that what Jesus accomplished in principle ought to take hold fully and manifest its power totally to effect a complete meeting of human nature with God that would make human nature like God, deathless and filled with bliss.

The Christian understanding of human sociability also flows from the incarnational center that we have elaborated. The Christian community is the body of Christ, the branches of his vine. The spiritual life coursing there is the life of Christ himself, what he enjoys with the Father and the Spirit as the incarnate Word. The sacramental rituals that the Christian community celebrates mediate this life, while the scriptures that it venerates, in which it finds the most truthworthy revelation of God's being and purpose, take their central interpretation from the event of Christ and the conviction that Jesus is the definitive revelation of God.

In all ways, then, the centrality of Christian faith in Jesus the Christ makes the difference. If Jesus was divine, the Word of God incarnate, the Christian worldview is cogent. If he was not divine, the Christian worldview is merely interesting myth and speculation. There are parallels to this situation in the religions founded by other charismatic figures (the Buddha is the best parallel, Muhammad and Confucius are lesser ones), but such parallels do not take away the distinctiveness of the Christian world view that the doctrine of the Incarnation has sponsored. One of the major matters for us to study, therefore, is the impact of Christian incarnationalism on Christian mysticism. Has the Christian commitment to the conjoint divinity and humanity of Jesus determined the mainstream understanding of mysticism in Christianity, or have other factors carried equal, if not greater, weight?[1]

The New Testament

The first Christians were Jews who sought God through the Torah. The books of the Hebrew Bible, *Tanak,* were their scripture, but from virtually the moment of the death and resurrection of Jesus, Christian disciples began to generate memories and interpretations of their master. The earliest written expressions of such memories and interpretations to gain general approval were letters sent by the Apostle (authorized representative) Paul to (largely Gentile) communities that he had founded in Asia Minor. A generation or so after the missionary activity of Paul, several communities apparently put together different though overlapping collections of both the teachings of Jesus and memories of striking things he had done.

What we now call the synoptic gospels (Matthew, Mark, and Luke) represent mature forms of such collections. They are *synoptic* because at a single glance one can take in both their likenesses (considerable) and their differences (provocative). They are *gospels* because the key to their rhetorical intent is a proclamation of glad tidings, or good news. Indeed, they represent Jesus as having himself proclaimed such good news.

Specifically, the Jesus of the synoptic gospels says that God has commissioned him to announce a time of salvation, a new (messianic) eon when the reign of God will become so manifest that justice and peace will flower. Jesus eventually dies because of his fidelity to the mission God gave him, the charge to make such a proclamation up and down the land. The strange constellation of events (the resurrection) that the synoptic authors place after the death of Jesus is in part the ratification by God of what Jesus had proclaimed.

Before moving to an analysis of what the religious experience of Jesus himself probably was, let us fill out our sketch of the New Testament by describing the Johannine view of Jesus. If the letters of Paul place the death and resurrection of Jesus in the center, arguing that through those events God opened the covenant to the Gentiles, and the synoptic gospels place the proclamation of the reign of God in the center, arguing that the preaching, works, death, and resurrection of Jesus show him to have been the Messiah, then the Johannine writings (gospel, epistles, and with qualifications, Revelation) present a Jesus incarnating divinity, serving as the central sign of what God is like, offering not only peace and justice but a sharing in what God alone is: deathless, eternal, holy.

The Jesus who walks through the first half of the gospel of John is a maker of "signs": words and deeds conveying the sublimity of his heavenly origin, the power of his revelation of the Father. The Jesus of the second half of John speaks with the Father as though he were already in heaven or had never left heaven, as though he were the divine Word only lately come into flesh. He dies on the cross, but the nimbus of his glory makes even his death seem a foregone triumph. From the outset, the moment when he took flesh, the Johannine Word sanctified the world, giving all who would believe in him a sacramental view of creation. Seen in faith, the world is a cornucopia of the signs, a great treasure-trove of testimonies to the divine glory. The Word orchestrates this sacramental view of creation, not denying the evils that deface it but defeating them. The light that shines from the Word made flesh is stronger than all darkness. The darkness has never grasped it (John 1:5).

These are but a few of the themes that the New Testament weaves around the central figure of Jesus. If we ask what the religious experience of Jesus himself was and whether we ought to call that experience mystical, we enter the complicated question of how the New Testament relates to the historical Jesus. Do the texts that we have in the New Testament report accurately and dependably what Jesus said and did? Can we trust that the moods, sentiments, and experiences that they assign to Jesus actually occurred?

The consensus of scholars of the New Testament is that we cannot. The New Testament offers no detached, simply objective views of Jesus. Always what it presents reflects the faith of its authors that Jesus was the Christ. In other words, always the texts are interpretations prejudiced in favor of Jesus, shaped by people who had given him their hearts. There is no "historical Jesus" available to us, no Jesus uninterpreted. Always, therefore, we have to add the caveat that such and such seems to be so, based on the ways that the New Testament authors have presented the data.

Now, this situation is not unique to the New Testament. There are no simply historical scriptural presentations of Moses, David, the Buddha, Confucius, or Muhammad. As we use the words nowadays, "scriptural" and "historical" can be antonymns. *Scriptural* entails a commitment in faith. *Historical* entails a scholarly detachment from a commitment in faith, scholarly efforts at scientific objectivity. (Whether in fact there can be such a scholarly detachment or objectivity in humanistic studies has become a matter of some debate. *Hermeneutics,* the study of interpretation, is a complicated business. All texts reflect the historical and cultural circumstances from which they come. All texts also invite the engagement of their readers, who bring the biases of their own historical times and cultural circumstances. There is no text, no written source of meaning, that is not mediated, or interpreted at both its origin and its reception. The degree of interpretation may be greatest in scriptural texts because they present claims about ultimate reality, but scriptural texts are not *sui generis.* They are more like other humanistic texts than unlike them.)

To return to the matter of the mystical experience that Jesus himself enjoyed, let us focus on a story in Mark, the synoptic gospel that the majority of scholars judge to be the oldest:

> Six days later, Jesus took with him Peter and James and John, and led them up a high mountain apart, by themselves. And he was transfigured before them, and his clothes became dazzling white, such as no one on earth could bleach them. And there appeared to them Elijah with Moses, who were talking with Jesus. Then Peter said to Jesus, "Rabbi, it is good for us to be here; let us make three dwellings, one for you, one for Moses, and one for Elijah." He did not know what to say, for they were terrified. Then a cloud overshadowed them, and from the cloud there came a voice, "This is my Son, the Beloved; listen to him!" Suddenly when they looked around, they saw no one with them any more, only Jesus. (Mark 9:2–8, NRSV)

A lot is going on in this text. Jesus takes the leaders of his little band of disciples up a mountain. Often mountains are places thought to be especially holy to God (as deserts can be), and Jews of Jesus' day were likely to associate any mountain with Sinai, the site of the Mosaic covenant. Jesus is "transfigured": a radiance like that of the glorious divine godhead, the *Shekinah,* comes forth from him. He is light, presumably from the holy divine light, the fire of the Lord who moves by chariot.

Elijah, standing at the head of the line of great biblical prophets, and Moses, the great law-receiver, speak with Jesus. At the least, Jesus is the equal of Elijah and Moses, their peer. The message of Jesus, his work, must cohere with the prophetic work of Elijah, who fought the evil Queen Jezebel and heard the still small voice of God at the mouth of his mountain cave. The message of Jesus must also cohere with the work of Moses, with the Torah established in the covenant at Sinai. The New Testament authors are not subtle in saying that they see Jesus as a new Elijah (legend said that Elijah would reappear to finish history) and a new Moses.

The babbling of Peter stands for the inadequacy of human reason before the divine. The divine brings a stupor to the human mind, draws forth from human limitation a radical incompetence. Peter would make booths for the three heroes, as though they were celebrating the annual festival of *Sukkoth.* All this goes by the boards, however, for from the cloud (as on Sinai) a voice declares Jesus to be the Son of God, the beloved. The consequence of this declaration is an imperative, a command: Listen to him! He is now the foremost prophet of God, spokesperson for God. He is now the divine Word. After this declaration, Jesus holds the holy field alone. Only he is necessary for the divine revelation, the divine work of salvation and sanctification.

Bernard Cooke, dealing with Jesus as God's beloved, treats the transfiguration under the heading of what Israelite prophets seem to have experienced, and he links the transfiguration to the scene of the baptism of Jesus (Mark 1:9–11), where a divine voice also rings out:

> The prophetic elements of the baptismal scene find reiteration and strengthening in the account of the transfiguration. There Jesus is depicted as the realization of what was foreshadowed in Moses and Elijah, the law and the prophets. The theophany [manifestation of the divine] is again of Abba [Father] and Spirit (this time in the manifestation of the cloud, the *kabod* Yahweh). Jesus' teaching receives divine approbation, "Hear him," and the scene marks the beginning of Jesus' journey to Jerusalem where he was to undertake the mysterious restoration of Israel through the paradox of death as the Servant. Though it is disputed whether the title "Servant" can be traced to Jesus himself, from this point onward Jesus works to help his disciples understand that his Messiahship is to be interpreted in terms of the figure of the Servant of Isaiah 52–53. What lends credibility to Jesus' actual attribution of "Servant" to himself is the narrative's account of how completely alien and puzzling this notion was to the disciples' understanding prior to the resurrection experience.[2]

When we contemplate the overall portrait of Jesus that the New Testament authors paint, what stands out as most mystical is his intimacy with the God he called Father. The fact that the voice from the cloud calls Jesus "Son" and "Beloved" squares with the fact that the evangelists have Jesus comport himself as a trusting child of God. The great test of his trust is the moment in the garden of Gethsemane, when he asks his Father to spare him the coming trials (crucifixion and death; see Mark 14:32–42). The Father does not spare him, but Jesus finishes his life commending his spirit into the Father's hands.

When Christian theology became ontological, probing the relations between the human being of Jesus and the divine, it established that the center of the existence of Jesus was the divine Word. One psychological translation of this ontological analysis might be that the intimacy with God that Jesus displays reflects the oneness of the Word with God outside of time. This is not a translation that the authors of the New Testament would have worked out deliberately, but their sketches of Jesus do make unique his identification with, trust in, and constant reference of his person and work to the Father. Indeed, the Johannine Jesus utters a series of "I am" statements that are probably deliberate evocations of the divinity whom Moses experienced at the burning bush (Exod. 3).

The mystical point would be that orthodox Christians would interpret the intimacy between Jesus and God as a function of the divinity that the two shared. Certainly, the holy man Jesus, fully human and fully a Jew of his times, would have worshiped God wholeheartedly, in the spirit of the *shema*. However, the depths or fullness of that worship, the direct experience of ultimate reality to which that worship would have led Jesus, would have expressed the oneness with God that the Word always had because always he was the Word of the Father, the eternal Son.

True, the evangelists stressed the humanity of Jesus, the man they or their sources in faith had known personally, but the more they set Jesus apart from the ordinary run of human beings, making him the Messiah or Savior or divine Son, the more their depiction of his relations with God approached the intimacy that Christ-our-God, a title beloved by the Orthodox and used in their liturgies, would have had with the Father-our-God. Ultimately, Christian trinitarian theology came to say that the only difference between the Son and the Father was where they stood in their relationship: begotten and begetter, originated and origin, spoken and speaker, known and knower, and so forth. Apart from such relational differences, the Son was all that the Father was.

The intimacy with the Father bestowed on the incarnate Son, the Word become flesh, was bound to reflect this identity. Certainly, one could call the intimacy of the man Jesus, his full identification with his Father, mystical, if only because it seems to have gone far beyond the ordinary unions with God that human beings have reported, into a directly experiential union with ultimate reality. However, one ought in the case of the mysticism of Jesus to trim away connotations of transports or ecstasies. The Word was always intimate with the Father, was never not experiencing the Father directly. What happened at the transfiguration was less the transport of Jesus to a heavenly station where he could commune with Elijah and Moses and call forth the divine voice from the cloud than a manifestation for the sake of the disciples of who Jesus was, what status he had, always, everywhere. If so, then largely unaware the evangelists present Jesus as a paramount mystic, one never not perceiving holy ultimate reality directly.

The Patristic Period

While Christian scriptural texts such as Revelation reflect contemporary (ca. 95 C.E.) Jewish spiritual or interpretational movements, such as *apocalyptic* (concern with a divine revelation of what will happen soon during the imminent end of history), Christian mysticism as a whole owes more to Hellenistic than Jewish conceptions (admitting that many Jews themselves were to some degree Hellenistic). The great qualification to this statement is the constant reliance of Christian mystics on the Bible, which is thoroughly Jewish. Apart from this qualification, however, we find that such Christian patristic masters as Origen and Augustine draw their outlook, their ontological and psychological orientations in reality, their analytical tools for interpreting mystical experience, from the Hellenistic cultural world rather than from the world of rabbinic Judaism. They do not share the passion for the commentaries of previously eminent rabbis on Torah that led the Jewish leaders of the first centuries C.E. to create the Talmud. The Talmud remains foreign to the Christian masters, and they produce no comparable masterwork.

The greatest treatises of the Christian fathers are commentaries on scripture, however, and often their literary method is allegorical, with priority of interpretational place reserved for the spiritual sense that a text can bear, its somewhat Gnostic revelation of what God may be doing in the mind and heart and soul of the sincere seeker. (Whereas the rabbis tended to inquire about what God was doing with, or wanted of, the community, Israel the people, the fathers tended to inquire about the experiences, the needs and transformations, of the individual seeker.)

In the desert (which they assimilated to the forty years of wandering when Israel had been closest to God) the heroes of the early Christian monastic movement (Judaism has never produced a significant monastic tradition) provided the speculative fathers ample evidence of how a life spent pursuing union with God wholeheartedly might unfold. So, for example, Athanasius, bishop of Alexandria and leader of what became the triumphant orthodox party at the council of Nicaea (325 C.E.), wrote a life of Antony, the first truly famous desert solitary, in which Antony became a paradigm of orthodox faith.

Hearing a call to sell all that he had and strive for perfection, Antony went into the wastes of the Egyptian desert to seek God in silence. He found what he sought so fully that at the end of a long life he seemed completely whole and healthy. Along the way Antony had wrestled with demons, but Athanasius looked upon the sanctification of Antony as the restoration of what was natural to human beings. Health and intimacy with God were natural, what human beings ought to enjoy. Sin was the unnatural deprivation or cessation of these qualities. By nature, human beings were made to commune with God, find signs and powers of God everywhere. A Greek instinct that being itself is good pervades the spiritual analyses of the early fathers. In the desert one had to practice purgation of vice, but the being revealed as vice progressively fell away was naturally ordered to God, apt for direct experience of God.

Such classical collections of testimonies to the experience of the desert fathers as the Eastern Church's *Ladder of Perfection* by John Climacus and its *Philokalia*, and the Western Church's *Conferences* by John Cassian, suggest that the early saints whom the Spirit of God formed in the desert were minimalists, deeply in love with God but fiercely committed to humility. If the collections of their sayings are faithful witnesses, the early desert saints focused on vices and virtues more than stages of prayer. They thought they ought to pray constantly, and they considered prayer a great source of purification, along with fasting and other ascetic practices. However, they were not the psychologists that we find coming to the fore in modernity. Their interest in feelings, images, consolations, desolations, and the like was oblique rather than direct. Their God was demanding, holy, a constant prod to confession of sin and warfare against the self. Their God wanted their preoccupation with vice and virtue to result in a new being, a recreation of all that they were in the holiness of divine life.

As they matured, the desert heroes came to trust in their God, so we find lovely stories of masters assuring disciples that God had forgiven the disciples their sins. Even in warm maturity, however, the desert fathers and mothers remain minimalists. Seldom are they not reticent when it comes to displaying their mystical experiences or gifts. Not for them the narcissism coloring so much present-day Western spirituality.

The speculative mystical theologians involved themselves more deeply with questions of human constitution and how the indwelling divine Spirit worked than did the desert masters. Thus Origen (ca. 185–254), commenting on the Song of Songs writes of the sensitive capacities that mysticism can develop:

> "We hurry after you, after the fragrance of your ointments" (cf. Cant 1:3,4). And this happens, as we said, only when his fragrance has been perceived. What do you think they will do when the Word of God takes over their hearing and sight and touch and taste?—and when he gives to each of their senses the powers of which they are naturally capable?—so that the eye, once able to see "his glory, glory as of the only Son from the Father" (Jn 1:14), no longer wants to see anything else, nor the hearing want to hear anything other than the "Word of life" (1 John 1:1), The Touch will not touch anything else that is material, fragile, and subject to decay, nor will the taste, once it has "tasted the goodness of the Word of God" (Hebrews 6:5) and his flesh and the "bread which has come down from heaven" (Jn 6:33, 52–58), be willing to taste anything else after this.[3]

Origen takes the lovers in the Song of Songs as pursuers of God, figures moving along the mystical journey. He is interpreting the text on the inmost, spiritual ("anagogical") level. So the reference to the fragrance of the beloved, which is but one of the many intoxicants that love uses, stimulates Origen to contemplate the transformation of human sensation that contact with divinity, intimate sharing with divinity, change into a share of the divine life (*theosis*) tends to produce.

The five outer senses take on a new inner capacity. They become sensoria of the divine, ways for divinity to anoint human flesh and bring out its naturally sacramental potential. Note that the priority in this process lies with God, specifically with the Word of God. For Origen the "Word of God" has several connotations. It is the biblical revelation, the holy books of the covenants old and new. It is more primordially the divine Word, the Son living in the bosom of the Father from eternity. Also, Origen assumes that this Word has taken flesh in Jesus the Christ and so operates in the sacred history narrated in the two covenants. Prefigured in the Old Testament, and walking across the pages of the New, the incarnate Word presents the beauty of divinity, as well as its wisdom and salvific intent. The divine lover whom the beloved of the Song of Songs typifies gains the fullest humanity he (Origen assumed that in mystical intercourse divinity would be masculine and the human soul feminine) can through the incarnation of the Word that makes Jesus the Christ. So, the Word to whom Origen refers so steadily in this exegetical passage is the Lord whom the disciples of Jesus met during the transfiguration (but also lived with day by day less dramatically).

Origen began his career as a *catechist,* one commissioned to teach Christian neophytes their new faith. He lived in Alexandria, the city of Philo Judaeus, the city of Clement of Alexandria, an influential Christian catechetical figure himself. As Origen matured, he became one of the greatest biblical theologians that Christianity has ever produced.

Note in the present passage that Origen assumes that the development of the human senses so that they can exult in their divine lover is natural. This is a Hellenistic assumption characterizing the East more than the West. Although Augustine was thoroughly Hellenistic (Neoplatonic) in philosophical outlook, his experiences with the (dualistic) Manichean heresy and with his own turbulently sensual nature ("Give me chastity, but not yet") pushed him toward a different anthropology than that of Origen, one more negative in its assessment of the human nature that we find typically at the beginning of the mystical quest.

Augustine could refer to the human race as a *massa damnata,* a horde fated for hellfire. He argued so forcefully for the complete gratuity of salvation that he made Pelagius, a monk advocating much ascetic assertion of human free will, seem an arch-heretic. Augustine's views of original sin deeply influenced Martin Luther, who had been a monk in the order that claimed Augustine as its founder, and John Calvin. They so insisted on salvation's coming only through faith and grace that the Catholics whom they fought doctrinally during the sixteenth century counterasserted the significance of human works.

At any rate, here we find Origen sketching a transformation of the human senses through the work of the Word of God that makes mysticism markedly incarnational. The end product, the people sanctified by romantic love relations with God, will not quit the material world, stamp the dust of creation from their feet. Rather, they will sense the world anew, finding divinity everywhere, learning how the presence of God takes away from material creation its mortality, making it deathless like its divine source.

Gregory of Nyssa (ca. 335–395), commenting on the phrase "watchful sleep" in Song 5:2, suggests a greater love of darkness or at least imagelessness than what we find in Origen. Origen loves light, associating it with the peaks of direct experience of God. Gregory makes a fuller place for spiritual simplicity, convinced that divinity is bound to overshadow, or overwhelm, the limited human light:

> Sleep usually follows drinking, and in this way the banqueters can promote their good health by allowing time for digestion. Thus after her banquet the bride, too, is overcome with sleep. But this is a strange sleep and foreign to nature's custom. In natural sleep the sleeper is not wide awake, and he who is wide awake is not sleeping. . . . But in this case there is strange and contradictory fusion of opposites in the same state. For *I sleep,* she says, and *my heart watcheth* (Cant. 5.2). . . .
>
> From what we have said, then, when the bride proudly declares, I *sleep and my heart watcheth* (Cant. 5.2) we learn that she has risen higher than ever before. . . .
>
> The contemplation of the good makes us despise all these [created] things; and so the eye of the body sleeps. Anything that the eye reveals does not attract the perfect soul, because by reason it looks only to those things which transcend the visible universe. . . .
>
> Thus the soul, enjoying the contemplation of Being, will not awake for anything that arouses sensual pleasure. After lulling to sleep every bodily motion, it reserves the vision of God in a divine wakefulness with pure and naked intuition.[4]

This orientation is Platonic. The physical senses guide us through the material world, which is lower than the realer world of intelligibilities, or intellectual forms. When we have realized that the material world is farther down the scale of participation in God than the world of intelligibilities, we may fall asleep to the world of sensations but be watching carefully in the depths of our souls for intelligibilities, gathering ourselves together in a simple gaze upon Being. Here Gregory does not bother to correlate Being with the biblical God, but it is clear that Being is the Creator of all things and so the God whom Genesis makes the source of the world and whom Jesus calls his Father.

The Platonists instinctively thought of mystical progress toward direct experiences of God as an ascent toward the light. In the presence of the light, which gave one's mind understanding, one might find that the light was too bright for human comprehension and so one might enter upon a great darkness. This would be progress, nonetheless, because the intellectualist slant of Platonic philosophy took the "upper" portions of the human complex, the contemplative mind especially, as the site and engine of human transformation. If the upper portions turned out to be darkness as well as light, sleeping for the senses but watchfulness for the spirit, so be it. That apparently, experientially, was how God chose to bring the mind to perfection.

One might speak of the heart, as the Bible does, when a more holistic language became appropriate. However, among the fathers shaped by Platonism (the great majority), movement upward, away from the earth and toward the heavens, prevails over movement inward to the heart, let alone incarnational movement downward, along the lymph or blood circulating from head to foot.

In studying the influence of a great Neoplatonist, Augustine, whom he calls the "founding father" of Western mysticism, Bernard McGinn has located several texts commenting on the Psalms that express Augustine's view of how the soul knows God. For example, in his commentary on Psalm 99.5, Augustine says, in the name of God, "You will draw near to the likeness [the image of God in which human beings have been created] by the measure you advance in charity, and to the same degree you will begin to perceive God (*sentire Deum*)" (Hom. on Ps. 99.5 [PL 37:1274]).

McGinn comments as follows:

> What does it mean to "sentire Deum"? It is not that God comes to us as if he had been absent, or even that we "go" to him. God is always present to us and to all things; it is that we, like blind persons, do not have the eyes to see him: "The one you wish to see is not far from you. . . ." (99.5). We need to become like him in goodness and "loving in thought" . . . (99.6) in order to grasp him. Note that Augustine's language here abandons the spiritual sense of sight to emphasize the other spiritual senses. Only such experiential contact with God can produce the higher form of knowledge that perceives that we can really say nothing about him.[5]
>
> Continuing his commentary on Ps. 99, at verse 6 Augustine also says, "When, as someone like him [made in God's image, but having to clarify or educe this through spiritual progress], you begin to draw near and to become fully conscious of God (*persentiscere Deum*), you will experience what to say and what not to say insofar as love grows in you, because "God is love" (I John 4:8). Before you had the experience, you used to

think you could speak of God. You begin to have the experience, and there you experience that you cannot say what you experience.[6]

When we actually experience the divine, whether through vision or love, we find that we are dealing with a magnitude and a simplicity which, together, defeat us. We cannot say what we experience because what we experience is everything, no thing, too thick, too subtle, above all too elementary, primary, foundational for us human beings to render it. There is nothing before God from which we may derive God. There are no parts in God (except as we put them there for our convenience) and so no sensible or conceptual handles. If we fashion attributes for God, physical or spiritual or moral, we must remember at least now and then that they came from us, not from divinity itself. No matter what we say about God, our saying is more unlike what whole divinity is in itself than like whole divinity because what we say comes from our own partiality.

Augustine is sufficiently an experient of God, a man of significant contemplative attainment, to know that even his most poetic achievements, his best rhetorical flourishes, fall far short of the reality he is trying to describe. That reality stands before him afresh each day, soliciting new attempts to describe it, if only because engaging his mind and heart with it is his chief delight, as it is the chief delight of religious contemplatives everywhere.

However, at the end of the day the divine reality remains as far off, as inadequately described, as it was at the beginning. Augustine the human worker may have grown in the process of working, struggling to describe and understand, but much of such growth will have been negative, realizing more fully why he can never know what God is. At best, Augustine the theologian will have thought or prayed or written himself into a sharper, perhaps even an ecstatic, state of appreciation. The mysteriousness of the divine fullness or the power of the divine priority or the stretching forth of the divine endlessness may have taken him out of his usual egocentricity toward the theocentricity that actually obtains in creation, that actually gives reality its shape. Few human beings, even saintly human beings, are able to maintain a vivid sense of this theocentricity, however. Most human beings have to struggle afresh every day to make God the operative center of their reality.

Perhaps Augustine became the father of Western mysticism because he struggled so diligently for so many days. There were enough mystical moments, times when the theocentricity bound to be mysterious came into focus, in his enormous corpus of writings that as they became the foremost patrimony of Western Christian intellectualism, a Neoplatonic desire for the light became part of the Western spiritual mainstream.

The faith that Augustine sought to understand committed him to a lightsome God who was the source of the world that Genesis described.

This God was also the Father of Jesus the Christ and so was the trinitarian divinity that the relations of Jesus the Christ with his Father and Spirit implied. This God spoke and loved, all the while remaining hidden in light inaccessible. The Incarnation and Redemption did not remove the hiddenness of God. What they disclosed about God only deepened the divine mystery, showing that its morality, its goodness and love, were as profound as its unlimited being.

Certainly, the scriptures and the sacraments were precious embodiments of the divine love and grace. Certainly, they extended the wonderful courtesy of God, the humaneness of God, in dealing materially with creatures of blood and bone. However, this, too, did not bring God under human control. None of this gave human beings the name of God so that they might control the divine being. Human beings never gain an understanding of God that makes God obvious. Ingredient in the mystical tradition that Augustine fathered is a recognition, encoded in the very moment of peak experience of ultimate reality, that human speech, indeed human totality, cannot express ultimate reality adequately. The more intensely we know God, by God's gracious drawing of our minds and hearts into the divine mystery, the more profoundly we realize that God simply is, outside all categories, and that God's expressing what we are is more central to any mystical encounter than our realizing or expressing what God is.

The Medieval Period

The chronological framework of the developments to which we turn now is flexible, perhaps targeting the period from the seventh to the early fifteenth centuries. Augustine remained the great theological master in the West. In the East, Origen, the Cappadocians, and the desert fathers exerted great influence. Bridging the patristic and medieval periods was the significant body of works attributed to the Neoplatonist Dionysius the Areopagite. These works managed both to sketch a highly ordered view of the celestial and ecclesiastical hierarchies and to inculcate a strongly apophatic, or negative, view of divinity: we deal with the living God in darkness, unknowing, plunged into something far beyond our reckoning. Two paragraphs from an introductory description by Paul Rorem give the gist of recent scholarly estimates of the provenance of the Dionysian works:

> Although the author of these works was influenced by the Alexandrian–Cappadocian Christianity of Gregory of Nyssa, for example, and by the later Neoplatonism of Iamblichus and especially Proclus, these debts were so artfully concealed that his pseudonym managed to survive even the penetrating criticisms of Renaissance humanists. It was conclusively ex-

> posed only in 1895. Twentieth-century scholarship has identified the author's general context as a late-fifth century mixture of Syrian Christianity and Athenian Neoplatonism. None of the various attempts at a more specific identification of the author has yet commanded a scholarly consensus. The corpus may never be persuasively tied to a specific historical figure. There is, therefore, no historical basis for an investigation of the spirituality of this author except for his pseudonymous writings, which are intentionally misleading about the original context and community.
>
> The Areopagite's literary remains present a peculiar but coherent synthesis of Neoplatonic metaphysics, biblical exegesis, and liturgical interpretation. The Neoplatonic framework of "procession and return," of descent and ascent, charts the Dionysian path of spiritual "uplifting," an "anagogical" ascent to the divine summit itself. Guidance for this upward path comes through the interpretation of Christian symbols—both biblical and liturgical—according to a systematic program of negation. The Dionysian interpretation of scriptural and liturgical symbolism incorporates negation within the anagogical or uplifting journey, for such interpretation must negate and thus transcend the superficial appearances of the symbols in order to rise up to their higher, conceptual meanings. Then even the loftiest interpretation must also be negated and abandoned in the final, silent approach to the ineffable and transcendent God beyond all speech and thought.[7]

The pseudonymous character of this writing reflects a time when writers tried to appropriate the prestige of an eminent authority for their own work. Acts 17:34 names a Dionysius as a convert of the apostle Paul at the Areopagus (public square) in Athens. The late-fifth-century Syrian writer presents his work as the wisdom of that early Christian convert. The implication would be that from earliest times the view of the mystical pathway to God that he presents held sway in (at least Pauline) Christianity. Actually, what the mystical theology of the Areopagite reflects is nearly five centuries of interaction between Christianity and its Hellenistic environment. Neoplatonists such as Plotinus (ca. 205–270) had developed an impressive mysticism of their own, seeking the One beyond all beings. Christian seekers after the one God often found such Neoplatonic speculation attractive, as the work of Dionysius suggests.

The key question for Christian interpreters has always been whether the substance of this work is more Christian than Neoplatonic. How fully do the Christian symbols and convictions of faith structure the vision? The complicated mode of predication, which both uses traditional biblical and liturgical symbols and puts them aside, testifies to a keen awareness of the limits of all language in the face of ultimate reality. The retention of hierarchical symbols related to both the biblical orders of angels and the ecclesiastical orders of ministerial ranks helps to root this theology in the Christian world. To what extent the author retains a directive force from the humanity of Jesus is unclear. The divinity of Jesus gives him license to move beyond the materiality of even Christological symbols, but when God becomes so ineffable that we have to wonder why there is a historical order of salvation, the precisely Christian character of the mysticism is dubious. Nonetheless, there is no denying that Dionysius had enormous influence, shaping such diverse authorities as the Celtic master Scotus Erigena (ca. 810–877) and the high medieval master Thomas Aquinas (1225–1274).

We leave with a specimen of the language of Pseudo-Dionysius, from a reflection on the divine darkness in his *Mystical Theology:*

> O Trinity, beyond essence and beyond divinity and beyond goodness, guide of Christians in divine wisdom, direct us toward mysticism's heights, beyond unknowing, beyond light, beyond limit, there where the unmixed and unfettered and unchangeable mysteries of theology in the dazzling dark of the welcoming silence lie hidden, in the intensity of their darkness all brilliance outshining, our intellects, blinded—overwhelming with the intangible and with the invisible and with the illimitable. Such is my prayer. And you, beloved Timothy, in the earnest exercise of mystical contemplation abandon all sensation and all intellection and all objects or sensed or seen and all being and all nonbeing and in unknowing as much as may be, be one with the beyond being and knowing. By the ceaseless and limitless going out of yourself and out of all things else you will be led in utter pureness rejecting all and released from all, aloft to the flashing forth, beyond all being, of the divine dark.[8]

This radically negative theology has alternated with an affirmative theology, based on the Incarnation, to give Christian mysticism a zigzagging appearance. Inasmuch as they committed themselves to an orthodox Christology, according to which Jesus the Christ is both fully human and fully divine, all Christian theologians had to be both affirmative and negative. Jesus was "like us [fellow human beings] in all things except sin," yet he was also wholly beyond us, the unknowable Logos of God. How a given mystical writer experienced prayer or taught others to approach the Christian God came more from that writer's own temperament than from anything imposed *a priori* by Christian doctrine.

So, for example, Bernard of Clairvaux (1090–1153) developed in his exegesis of the Song of Songs a highly romantic, nuptial symbolism. Without denying that mystical experience transcends both sensation and intelligibility, he probed the human experience of love to produce a contemplative atmosphere quite different from that of Denis:

> Love is a great reality, but there are degrees to it. The bride stands at the highest. Children love their father, but they are thinking of their inheritance, and as long as they have any fear of losing it, they honor more than they love the one from whom they expect to inherit. I suspect the love which seems to be founded on some hope of gain. It is weak, for if the hope is removed it may be extinguished, or at least diminished. It is not

> pure, as it desires some return. Pure love has no self-interest. Pure love does not gain strength through expectation, nor is it weakened by distrust. This is the love of the bride, for this is the bride—with all that means. Love is the being and the hope of a bride. She is full of it, and the bridegroom is contented with it. He asks nothing else, and she has nothing else to give. That is why he is the bridegroom and she the bride; this love is the property only of the couple. No one else can share it, not even a son.[9]

When we recall that Christ is the bridegroom (see, for example, Matt. 9), we realize that Bernard is transposing the Song into a script for Christian mystical progress. The Christian soul, considered feminine (receptive, if not passive) in relation to God, prospers best by loving wholeheartedly, ardently, as though complete devotion were the core of her being. This outlook works with a stereotype of feminine emotionalism, as though all women were born *apassionarias*. Indeed, during the Middle Ages Mary Magdalene often represented this stereotype in Christian art: red hair flying, wiping the feet of Jesus with nard, crying in the garden by the tomb of Jesus, she was religious emotion pitched at the highest voltage.

However, what is especially significant for our definitional interests in mysticism is not the emotion that Bernard encourages but the interpersonal focus. Here he seems to suggest that direct experience of ultimate reality on the model that Christian faith finds best is a passionate marital union between creature and Creator, sinner and Savior.

Does this suggestion canonize passionate personal love as an experiental focus more profound, central, or otherwise desirable than the focus on being, the link through being, that to this point we have found prevailing along the spectrum of mystical experience, especially that of the Eastern traditions? Bernard does not deal with this issue speculatively to let us know through arguments what his position would be, and we can think of Eastern sages, such as Ramanuja, who think more highly of love than of reason. However, granted these and no doubt many other qualifications, we can say that a Christian romantic such as Bernard does underscore the incarnational coloring of mystical experience possible when a tradition thinks that divinity has made itself reliably, if not fully, available in given times, places, and people. We can say that while Bernard would affirm the impossibility of knowing what God (either the Trinity or the Logos alone) is, the saint's practical advice would be to love what one cannot know with all the fullness of human sexual attraction.

Whatever in the biblical portrait of Jesus could inspire or nourish such a love would be good spiritual reading or meditation. If devotion to the Christ child, to the Blessed Virgin, to Magdalene and the other passionate saints gave support, to them one ought to go enthusiastically. The monastic liturgy that monks like Bernard celebrated regularly provided a full, sensuous context that also encouraged emotional approaches to God. The incense, candle wax, light, music, bows, feastings, and fastings kept the participants aware that they were citizens of two worlds. They could, indeed had to, sense the presence of the divine through their eyes and ears, but their souls took them into the heavenly courts. Indeed, their souls took them into rooms of the heavenly mansion set aside for making love—complete fulfillment of all the potential that God had breathed into them at their outset.

Zagging back to the negative theology that kept on developing during the medieval period of Christian mysticism, we come to the highly influential (and somewhat suspect) later writer Meister Eckhart (ca. 1260–1327). Eckhart speaks an ontological language with which we should now be quite familiar, but on his tongue it gains several new intonations:

> Where the creature stops, there God begins to be. Now God wants no more from you than that you should in creaturely fashion go out of yourself, and let God be God in you. The smallest creaturely image that ever forms in you is as great as God is great. Why? Because it comes between you and the whole of God. As soon as the images come in, God and all his divinity have to give way. But as the image goes out, God goes in. God wants you to go out of yourself in creaturely fashion as much as if all his blessedness consisted in it. O my dear man, what harm does it do you to allow God to be God in you? Go completely out of yourself for God's love, and God comes completely out of himself for love of you. And when these two have gone out, what remains there is a simplified One. In this One the Father brings his Son and then there springs up in God a will that belongs to the soul. So long as the will remains untouched by all created things and by all creation, it is free. Christ says, "No one comes into heaven except him who has come from heaven" (Jn. 3:13). All things are created from nothing; their true origin is nothing, and so far as this noble will inclines toward created things, it flows off with created things toward their nothing.[10]

The general intent is relatively plain, but the individual phrases invite misunderstanding, which they received abundantly during Eckhart's own lifetime. Eckhart is exploring the border between divinity and creaturehood, where God gives all things their reality. The key instinct is that things have no reality apart from God, from the being that God alone possesses independently; but God does not give being, existence, creation extrinsically. It is not a lump of dough tossed in an oven, a pot tossed off a potter's wheel. God gives, becomes, is the being of the creature, who otherwise is nothing. The creature is more divine than limited because, in the final analysis, for the creature to be, God must be in it.

This would be merely interesting ontological speculation were Eckhart not fashioning from it a call to mystical prayer. When he interprets the use of images as a hindrance to appreciating the presence of God in the creature, he places himself on the negative side of the Christian crossroad, for the moment bracketing the question of what the Incarnation of the Logos ought to imply for Christian contemplative prayer.

The going out of the creature from images is simple enough to understand, but what are we to make of the going out of God? The Neoplatonists thought of a divine action of exit and return. Here, Eckhart seems to be thinking rather of the ecstasis that love entails. For love of the creature, God goes out of the divine sufficiency. From love of the creature, gone out, God makes with the creature a "simplified One." What this new reality is, is hard to determine, but it comes from the union of creature and Creator that love creates through their mutual outgoing.

Eckhart further complicates the picture by bringing in the divine Son, to whom the Father gives eternal birth of a sort by speaking him forth constantly. Eckhart would have the soul of the Christian contemplative attune itself to this speaking because he thinks that mystical love brings the birth of the divine Word to bear on the soul of the individual contemplative. The contemplative is being born into a divine life, a *divinization,* and the form of this birth, according to Christian faith, comes from the relationship between the Father and the Son. The freedom to which contemplation, all of lawful human desire, aspires depends on holding to the divine action of the process making the contemplative a simplified one with God.

In effect, then, Eckhart is saying that the crux of mystical experience and progress is making real, letting become consciously real, in the substance of one's being the priority of God that is always there but usually not appreciated, usually blocked from achieving its full efficacy. The successful mystic, like Jesus, lives in heaven with God. Mystics may move on earth more or less gracefully, but they define themselves by heaven, the abode, the personal being of God. Their elevating contemplative experiences admit of interpretation as moments when they see or feel that they are more divine than human.

Correlatively, their contemplative experiences also tend to confirm the Christian teaching that creation is *ex nihilo,* from nothing, making them feel that when they let their will run to creatures they lose badly. In themselves (which they never are simply), creatures are nothing, not-God. To preoccupy oneself with what is not God when one could become absorbed with what is God (everything, in its being) is to make a significant mistake, to miss a fine opportunity.

Eckhart was a Dominican, a member of the Order of Preachers founded by the Spaniard Saint Dominic (1170–1221). Along with the Franciscans, the Order of *Fratres Minores* founded by the Italian Saint Francis of Assisi (1181–1226), the Dominicans brought a new spirituality to Christian Europe. Whereas the Benedictine monastic rule had been the prevailing form of Western spiritual formation for hundreds of years, the new political and economic conditions that were moving Europe away from feudalism suggested that the Church needed more flexible pathways. Various medieval heresies and countercultural movements (for example, that of the puritanical Albigensians) called for preachers to be reformers, who might combat wrong thinking with solid sermons. The quickening pace of city life suggested that priests needed to be more flexible and less tied to the agricultural rhythms of the stable monastic liturgy. One result was an effort of the *mendicant orders,* as the Dominicans and Franciscans sometimes called themselves (because they professed a strict poverty and so had to beg for financial support), to develop a semimonastic spirituality. The mendicants would still domicile in monastic houses and celebrate the daily office, but they would move out of the monastery, performing works of active ministry (especially preaching) as the pastoral needs of a given place and time required.

Such a move out of the monastery meant a simultaneous move toward greater self-sufficiency, even individuality, in spiritual matters. The mendicant needed a formation in prayer for a life on the move, a life engaging him (at first the equivalents of mendicant groups for women permitted them only limited freedom to work outside the convent) with the mercantile world more fully than had been the case for the Benedictines. Certainly, the ordinary clergy, to say nothing of the ordinary Christian laity, had always struggled to pray in the world and do the good works that kept the village prospering, but now there was a pressure to rethink the spiritual life itself in more active terms. Now the example of Jesus walking the byways of Galilee, sitting himself down beside the Samaritan woman's well fatigued, became more relevant to mystical prayer. By modernity, this movement had gained the full outline that we find in Counter-Reformational spiritualities such as that developed by Ignatius Loyola (1491–1556) for the Jesuits. In Eckhart we catch a transitional moment, a quite personal translation of the long-standing negative tradition for a time of intense preaching, an effort to present as profoundly as possible the overwhelming reality of God.

The Late Medieval Period

Two classical mystical texts from fourteenth-century England stand at the threshold that we are calling "late medieval." Together, they demonstrate that the tension between the negative and affirmative moments or dimensions of Christian mysticism continued to be strong. The first has become a matter of great interest recently because of its highly positive view of creation as well as its feminine imagery. The second is an anonymous work, set self-consciously in the line of Dionysius the Areopagite.

The *Showings* of Julian of Norwich (ca. 1342–1423) begin with a recollection of her deathbed, where she received a vision that determined the rest of her life:

> And when I was thirty and a half years old, God sent me a bodily sickness in which I lay for three days and three nights; and on the fourth night I received all the rites of Holy Church, and did not expect to live until day. . . .
>
> The parson set the cross before my face and said: Daughter, I have brought you the image of your saviour. Look at it and take comfort from it, in reverence of him who died for you and for me. . . .
>
> After this my sight began to fail, and it was all dark around me in the room, dark as night, except that there was ordinary light trained upon the image of the cross. I never knew how. Everything around the cross was ugly to me, as if it were occupied by a great crowd of devils. . . .
>
> After that I felt as if the upper part of my body were beginning to die. My hands fell down on either side, and I was so weak that my head lolled to one side. The greatest pain that I felt was my shortness of breath and the ebbing of my life. Then truly I believed that I was at the point of death. And suddenly in that moment all my pain left me, and I was as sound, particularly in the upper part of my body, as ever I was before or have been since. I was astonished by this change, for it seemed to me that it was by God's secret doing and not natural. . . .
>
> I desired to suffer with him [Jesus], living in my mortal body, as God would give me grace. And at this, suddenly I saw the red blood trickling down from under the crown, all hot, flowing freely and copiously, a living stream, just as it seemed to me that it was at the time when the crown of thorns was thrust down upon his blessed head. Just so did he, both God and man, suffer for me.[11]

Julian recovers from this deathbed experience and spends the rest of her life examining it, drawing out the implications. The result is a profound spirituality that stresses the goodness of God, God's willingness even to suffer for us. Matched to the specificity, the vividness, of the New Testament images of Christ's passion, this appreciation of the willingness of God to suffer for us takes the mysticism of Julian some distance from the rhapsodic aloneness beyond being to which Dionysius the Areopagite aspires. Julian comes to think of God, more specifically of Jesus, as her nourishing mother, feeding her at the breast. She takes the most intense human love that she knows, that of a mother for her child, and ties it to the love by which God saved sinful humanity. The willingness of Christ to shed his blood for sinners is but the most dramatic instance of the goodness that God displays constantly, like a completely selfless mother. Compared to this goodness, which grounds all of creation, the evils and sins of the world retract considerably. They are nothing positive. Indeed, it is mysterious that they should be at all.

We can suffer through them because the positive love of God is so much greater.

Sometimes Christian masters, as masters of other mystical traditions, take up at length the question of how best to relate the imagination to the intellect and the intellect to the will. Thus we heard Meister Eckhart urge his listeners to stay free of images. Julian has none of this. She becomes highly literate, theologically, but the inspiration for her thought continues to be the powerful experience at her deathbed. The cross of Christ and the goodness displayed in the running of the red blood mean immensely more to her than any insight into where God begins in her and her own nothingness ends. Certainly, she is medieval in being more ontological than psychological, but the focus of her contemplation never wanders far from the icon of God's sufferings, the cross of Christ her savior.

This reminds us that Christianity came on the scene as a way to salvation—health, healing, restoration, forgiveness. It has remained a way of salvation, as well as a path to wisdom. The transformation that both saints such as Julian and common sinners have sought tends to involve tears of both sorrow and joy. It does not proceed in Neoplatonic detachment, Hellenistic *apatheia.* Julian was in heavy pain but glad enough to be for love of her suffering savior. The oneness in suffering that she found with Christ on her deathbed gave her pain a new meaning. The cross of Christ blotted out all other concerns, inasmuch as she found in it and through it a comprehensive economy of salvation. God was offering to all people the saving, suffering love that she felt transform her back to health.

The inmost reality of human experiences, passions, suffering, and delights was the provision of love that God was making through them. This reality was ineffable, not so much because of the divine infinity as because of the divine goodness. How could it be that God would be so good as freely to choose to suffer for human beings—ontological nothings, moral failures? Only God could explain God. For Julian the explanation preached from the cross was utterly eloquent. One could not exaggerate the goodness of God. One could not praise too much the love of God our Mother.

The second fourteenth-century English text that we treat, *The Cloud of Unknowing,* is a masterpiece in the tradition of the Areopagite, the tradition of negative mystical theology. By the time of this text, contemplatives had developed the typology of Martha and Mary (Luke 10) to place the life of contemplative prayer above the life of active, practical works. The life of active, practical works is good, but the life of contemplative prayer is better. The anonymous author of the *Cloud* imports this distinction into the exercise of Christian spirituality itself. The best portion of Christian prayer is where the contemplative rests in a cloud of unknowing, unconcerned with images or practical thoughts, concerned only to abide with God, the ungraspable divine mystery, in love. The following paragraphs illustrate this conviction:

> The lower part of the active life consists in good and honest corporal works of mercy and of charity. The higher part of the active life, and the lower part of the contemplative, consists in good spiritual meditations and earnest consideration of a man's own wretched state with sorrow and contrition, of the passion of Christ and of his servants with pity and compassion, and of the wonderful gifts, kindness, and works of God in all his creatures, corporeal and spiritual, with thanksgiving and praise. But the higher part of contemplation, insofar as it is possible to possess it here below, consists entirely in this darkness and in this cloud of unknowing, with a loving impulse and a dark gazing into the simple being of God himself alone.
>
> In the lower part of the active life, a man is outside himself and beneath himself. In the higher part of the active life, and the lower part of the contemplative life, a man is within himself and on a par with himself. But in the higher part of the contemplative life, a man is above himself, because he makes it his purpose to arrive by grace whither he cannot come by nature: that is to say, to be knit to God in spirit, in oneness of love and union of wills.[12]

In the background of this analysis is the development of the mendicant orders during the prior century and a half. Their mixed spirituality had challenged the purely contemplative spirituality of monks not working in the world pastorally. The author of the *Cloud* finds both works and prayers good, but they have an order, a hierarchy. Experientially, but also theologically, he prefers the purely contemplative preoccupations that focus the human spirit on God directly.

The most advantageous way to approach God is to put aside everything that is not God. Indeed, it is useful to shut down your mind or let the obnubilating presence of God shut it down for you. When God comes to you as God, no images carry him. Your mind is dealing with something too thin, too pure, for it to grasp. You do best to give up trying to grasp it. You do best simply to enter the cloud where the divine presence presents itself, let it drift over you, let go of all your thoughts, images, hopes, and fears. The excitement in your heart—the darts of love, responses of passion, to the beloved presence of God—is your best guide to what you should do. The more ardently, fully, deeply you can simply abide with God in the very vagueness that is all you can make of God, the better you accomplish the contemplative work.

The contemplative work is nothing but God's transforming you through love into union with divinity. The action of God in the cloud, the significance of your response through the darts, movements, desires of love, is to make you one with God, to give you the fulfillment for which you were created. The Christian symbols that root the desire of God for union with human beings in the historical, biblical order of salvation that Jesus continued and consummated are not dispensable. The author of the *Cloud* is not a Neoplatonist simply tacking evangelical warrants onto a mysticism already worked out of the natural dynamics of the human soul. Rather, the negative theology of the *Cloud* is a Greek elaboration of biblical Christian convictions about the desire of God to give us salvation, to take us into the divine deathlessness, and to love us back from sin to full health.

The stimulus for this elaboration is contemplative experience itself. People like the author of the *Cloud* tend to experience what they consider the depths of their prayer as a moment, a passion (in the sense of an undergoing), in which the godness of God appears in the freedom of God from all images, the pure spirituality of God that keeps God unlimited by any material borders or containers. In the measure that the Spirit of God calls a person to this sort of prayer and appreciation (people should not presume to call themselves, they should move by attractions from the Spirit that they cannot deny), that person can feel confident that a negative exercise of love, an energetic abiding with God in the clouds of unknowing (turning aside from any conceptual or imaginary definitions of God) and forgetting (turning aside from any concern for creatures), is the pathway along which God is drawing them.

The fourteenth and fifteenth centuries were a difficult time in Europe. Various corruptions in the church, changes in social patterns, and plagues contributed to a strong sense of depression, melancholy, bleakness. The *Devotio Moderna* that developed in the low countries reflects this depression. *The Imitation of Christ,* attributed to Thomas à Kempis, invites the reader to warm but lugubrious contemplations of the suffering Christ. The Christ of glory has largely vacated the scene. Although Eastern Orthodoxy continued to venerate the Christ of glory, the *Pantokrator,* late medieval Europe focused on the man of sorrows—despised, rejected, acquainted with grief.

The movements for reform that we associate with the names Hus, Wycliffe, Erasmus, Luther, Calvin, Zwingli, Loyola, and many others took aim at both cleansing the church of its corruptions and creating a new spirituality for what became modern times. Generally, the reformers began with an attack on the material wealth of the church, which they saw as the prime source of its decay. Additionally, most of the reformers wanted a return from the elaborate embellishments of high medieval Christian culture to the plain message of the gospel. So they strove for both a simplified approach to Scripture and a simplified sacramental, liturgical prayer. They wanted to demystify church life inasmuch as mystification seemed to keep the common people in bondage. The Protestant reformers rejected the notion that monastic life is superior to lay life. They translated the Bible into the vernacular, put the liturgy into the vernacular, and established the primacy of individual conscience over external church authorities. None of this religious change transpired smoothly, but *en masse,* as a clear vector, all of it produced a new era in Christian faith, a new stress in Christian mysticism on psychology, the self-awareness of the individual contemplative.

Luther, Calvin, and some other leading reformers turned away from mysticism insofar as they deplored the departure of monastic spirituality from the apparently simple piety of the evangelical Christ. Indeed, Protestant theology and spirituality as a whole have remained leery of mysticism, though sometimes individual Protestant writers have contemplated the presence of God in nature with enthusiasm. The Catholic contemplatives who took up the challenge of reform during the sixteenth century did not turn their backs on the prior mystical traditions. Such outstanding new teachers as the Carmelites Teresa of Avila and John of the Cross illustrate this well.

Teresa (1515–1582) created a new form of the old bridal mysticism that we saw in Bernard of Clairvaux. In addition, she described in considerable detail the various stages of progress in the mystical life, from initial contemplation to the highest, nearly continual union of the individual soul with God. Her writings brim with descriptions of ecstasies, visions, sufferings, revelations. She stands on the affirmative side of the Christian see-saw that we have illustrated, more favorable toward images of Christ, feelings toward God, than negative authors would be. Still, she realizes the impossibility of capturing God through any images, and she is more than willing to surrender the control of the mystical process to God, her espoused.

The following text suggests Teresa's interest in the different stages of mystical evolution toward complete union with God:

> Now then, when His Majesty is pleased to grant the soul this divine marriage . . . He first brings it into His own dwelling place. He desires that the favor be different from what it was at other times when he gave the soul raptures. I really believe that in raptures He unites it with Himself, as well as in the prayer of union. . . . But it doesn't seem to the soul that is called to enter into its center, as it is here in this dwelling place, but called to the superior part. These things matter little; whether the experience comes in one way or another, the Lord joins the soul to Himself. But He does so by making it blind and deaf. . . . Yet when He joins it to Himself, it doesn't understand anything; for all the faculties are lost.[13]

On the one hand, the language suggests romantic ecstasy, a mystical connection to God shaped by the Song of Songs. On the other hand, Teresa is conscious of the unknowing, the blindness and deafness, that proximity to God entails. The initiative in the process toward mystical union lies with God. The raptures that the soul experiences come when and as God chooses. Certainly, the soul has the responsibility of keeping faith with God, showing up for prayer, staying attentive, but the intense, one-to-one character of the encounter makes the soul the blank check of God. God is responsible, even tender, yet also ruthless. Union means that the soul does not belong to itself. It has lost itself to God in love, a great finding.

This psychological interest continues in the mystical theology of John of the Cross (1542–1591), who collaborated with Teresa in the reform of the Carmelite order. John also worked with the symbolism of the Song of Songs, adding a solid understanding of scholastic theology. The result was a severely negative understanding of the mystical life, according to which the safest way is the way of darkest faith. We owe to John the most famous Christian descriptions of the dark nights of the senses and soul, the mystical experiences of being stripped of all sensual and spiritual satisfaction, comfort, support. *Nada* is the watchword John favors—nothing. We ought to rely on nothing but God, desire nothing but God, cling to no will of our own. We ought to realize that God's stripping away all consolation from us is for our good. The more nakedly we reach out to and meet God, the better. In God's own good time, we may feel our souls again take fire, enter upon a spiritual marriage with God, become living flames of love, but when and how that happens is completely up to God.

John was concerned that many good people might be missing the favors God wanted to bestow on them because of inept spiritual direction. He thought that most of the priests of his day knew relatively little about the mystical life and so could easily thwart the progress of devout people called to higher prayer. Thus he reformed the traditional counsels on how to discern the spirits at work in a person's soul, trying to determine which trials came from God and which probably did not:

> Because the origin of these aridities may not be the sensory night and purgation, but sin and imperfection or weakness and lukewarmness or some bad humor or bodily indisposition, I will give some signs here for discerning whether the dryness is the result of this purgation or of one of these other defects. I find there are three principal signs for knowing this. The first is that as these souls do not get satisfaction or consolation from the things of God, they do not get any out of creatures either. Since God puts a soul in this dark night in order to dry up and purge the sensory appetites, He does not allow it to find sweetness or delight in anything. . . . The second sign for the discernment of this purgation is that the memory ordinarily turns to God solicitously and with painful care, and the soul thinks it is not serving God but turning back, because it is aware of this distaste for the things of God. . . . The third sign for the discernment of this purgation of the senses is the powerlessness, in spite of one's efforts, to meditate and make use of the imagination, the interior sense, as was one's previous custom. At this time, God does not communicate Himself through the senses as He did before, by means of the discursive analysis

> and synthesis of ideas, but begins to communicate Himself through pure spirit by an act of simple contemplation, in which there is no discursive succession of thought.[14]

John the scholastic has related the simplicity and the wholeness of God to the character of human intelligence. We cannot deal with God discursively, step by step, because all our steps are partial. God takes over our minds and hearts, our souls, to come to us precisely as God. Initially, we find this strange, even painful. It is like the withdrawal of a drug to which we had become accustomed. We cannot lean on creatures any longer, not even on images or feelings of them. God is leading us out of the sensual life into the desert wastes of spirituality. In the desert God will espouse us, as he espoused Israel, but the pilgrimage, the march, is painful. Unless we realize that it is right for us to leave meditation and take up contemplation, we are likely to abandon the effort to find God in prayer.

Wise spiritual directors realize what is happening in the souls of those they are directing. Wise directors read the signs of transition. Most interesting in the signs that John gives is the blend of disgust and attraction. Nothing pleases the soul, yet it keeps yearning for God. Eventually, John will say that the dark nights show the soul what it is on its own, apart from God. Apart from God it is worse than nothing, a horror of frustration. With God, however, suffering God's leading it out into the divine darkness, the soul can become fit for eternal union in love.

The Modern Period

We have concentrated on the historical developments demarcating the evolution of Christian mysticism in the Western European orbit. In the East, piety and mysticism certainly changed in response to political and cultural changes but perhaps less dramatically than in the West. The closer union in the East between crozier and crown, and then the influence of Islam in many Eastern areas, which tended to isolate Christians from the cultural mainstream and lump them together as subjects of a homogeneous political control worked through the hierarchs of the local Christian communities, produced a view of Christian holiness long on tradition, stability, and resistance to change.

Certainly, the iconoclastic controversy, the impact of the European crusaders, the cultural differences with Roman Christianity, the evangelization of the Slavic peoples, and other complex, highly influential historical developments kept Orthodox mysticism from stagnating. Nonetheless, one does not find in Eastern Christian mysticism the rapid production of different schools that one finds in the West. One finds rather a patient persistence in a deposit of faith secured through the first seven ecumenical councils and a splendid liturgical life that varied relatively little through the medieval centuries.

The Orthodox church has often thought of itself as participating in the eternal realities of heaven. It has not felt the full attraction to ongoing historical change that has occasionally seized the West. The pneumatic spread of divinity through the cosmos, like the drift of the incense at the liturgy, has taken the edge from ephemeral activities. Neither material good works nor political concerns have challenged radically the contemplative, monastic spirituality at the heart of Eastern faith. Orthodox bishops have been monks. Monks have been the great spiritual directors, the masters. Few nonmonastic orders or movements have competed for spiritual authority or influence over the mystical life.

Perhaps the most interesting modern revival in Orthodoxy has come through a return to *hesychast* sources. Let us describe such sources and then suggest how the use of the *Philokalia,* a prime Eastern compilation of texts, and of the Jesus Prayer, perhaps the most characteristic hesychast devotion, have oriented modern orthodox mysticism.

Of these matters Philip Sherrard has written recently:

> The hesychasts are God's dead, self-effaced so that they may unveil the countenance of the Invisible from whom they and the world they have renounced receive their new life. One can consequently only read the history of this tradition in the signs it leaves of its presence in various other forms, in icons, in the lives of the saints, in the writing and translation of spiritual texts. Thus it has been possible to attest its convergence on Mount Athos in the final centuries of the Byzantine period and above all in the fourteenth century, when its crucial doctrinal thesis—that although God in his ineffable transcendence is absolute and imparticipable, he nonetheless enters into deifying communion with created being through his uncreated energies—was affirmed with such magisterial authority by Gregory Palamas.[15]

The uncreated energies of God are reminiscent of the sparks that the Jewish Hasidim found working throughout the world to sanctify it. However, the work of the uncreated energies that interested the Christian hesychasts most was the transformation of individual human beings, their divinization. To abet this work, the Hesychasts chose the road of silence, abnegation, poverty both exterior and interior. They wanted to minimize creaturely concerns and thereby maximize the operation of God in their minds, hearts, and souls. Following a largely Greek patristic approach to scripture, they lived within the biblical text as both the source of their daily liturgy (for example, the psalms) and the treasury of images that nourished their interior lives. However, the strongly negative Neoplatonic traditions of Greek spirituality inclined them to play down the imaginative factors, the feelings, the visions that an intense biblical spirituality could produce. They had in the Orthodox tradition of venerating icons of Jesus, Mary, and the saints a sufficient anchor for both the religious imagination and the Christian incarnationalism that their Eastern heritage had long sponsored. So, for the deeper work of *divinization* (being made into the substance of God), the hesychast way headed into unknowing, darkness, penance—a *nada* John of the Cross might have loved.

The *Philokalia* has appeared in various versions through the centuries but always as a collection of stories about eminent spiritual masters, texts from mystical authorities, prayers to nourish contemplative devotions (for example, *centuries,* or sets of a hundred sentences, opinions, bits of advice). The title itself is instructive, meaning "love of the beautiful." The Orthodox sense of divinity, ultimate reality, makes God surpassingly beautiful. God is *kalos* as well as *hagios*—lovely as well as holy. The beauty of the Eastern liturgy is not accidental, merely aesthetic. Rather, it reflects the beauty of the heavenly courts, where the angels celebrate its archetype.

The Jesus Prayer dovetails with the *Philokalia* and hesychasm in that since the hesyschast revival the three have been inseparable. Taken on its own, the Jesus Prayer amounts to a Christian *mantra,* a way of trying to pray always and make Jesus sovereign in one's heart. "Lord Jesus, have mercy on me, a sinner" is one form of the prayer. Other forms, shorter ("Jesus") and longer ("Lord Jesus Christ, Son of the Living God, have mercy on me a sinner"), have also thrived. The object of the mantric exercise has been to focus one's whole being on Jesus through the prayer. Usually this effort has involved coordinating the recitation of the prayer with one's breathing. With practice, this coordination could become habitual, automatic, unthinking. The Jesus Prayer could become a cardiac rhythm, a systole and diastole orchestrating one's blood.

On a simple level, the Jesus Prayer met the challenge that the apostle Paul had set before his disciples: Pray always (I Thess. 5:17). On a profound level, it met the desire of mystics to find ways to give themselves completely to God. The Jesus Prayer has been a remarkably incarnational way for such giving. With or without images of Jesus, the person using this prayer has made a physical identification. The Greek doctrine of grace that underscores deification, God's giving the believer divine life, fits well into this physical identification. As the Jesus Prayer took over a person's metabolism and then psychology, it could take over the spirit as well. It could be transforming a mortal creature into the deathless splendor of the Creator.

Concluding a study of mysticism in the Orthodox church, Nicholas Arseniev has emphasized precisely this theme of deathlessness and the triumph of eternal life:

> For Christian Mysticism the glorification of the world is inseparably bound up with the incarnation, crucifixion, and resurrection of the Son of God. Indeed, it is precisely through His death, Passion, and humiliation that the victory over death and the glorification of life have been accomplished! Here we have the deepest foundation of all Christianity, the essence of Paul's preaching, the inmost meaning of the liturgies, the "philosophy" of the Fathers. Athanasius, for instance, writes of the joy that has burst forth from His death: "Yea, verily, it is a thing rich in joy, this triumphant victory over death, and our incorruptibility (won) through that body of the Lord." "God united Himself with our nature," says Gregory of Nyssa, "that our nature by union with God may become divine, as loosed from death and freed from subservience to the enemy; for His resurrection from the dead is for mortal mankind a beginning of the resurrection into eternal life." [16]

Neither modern Catholicism nor modern Protestantism has matched this intense Orthodox commitment to a glorious faith centered in the resurrection. The spiritual schools that have most shaped modern Catholicism (for example, the Ignatian) have offered rich counsel, along with practical exercises, about prayer and mystical union, but they have stayed under the Western umbrella raised by Augustine, which thinks first of grace as God's remedy for our sin. Related to this understanding has been a relative predilection for the cross of Christ, rather than for his resurrection. The world has seemed less redeemed than what the Orthodox vision of Christ's triumph implied it ought to be. The strong debt of Luther and Calvin to Augustine made them pessimistic about human nature, and in the main their followers through the modern centuries have retained this pessimism. Regularly this has led to a suspicion of contemplative prayer, mysticism, and the imaginative, aesthetic use of icons, incense, and other sacramental symbols.

The modern Protestant mystique has flowed from the Word, both scriptural and preached. Protestants have not been keen on the sacramentality flowing from the eucharistic liturgy. Modern Catholics have reacted by neglecting scripture and concentrating on religious rituals, many of them paraliturgical—rosaries, novenas, retreats. Until Vatican Council II (1962–1965) Catholics placed considerable emphasis on confession, the sacrament of penance through which a priest can work out with a penitent a customized absolution for sins. The Catholic genius has run to art, visual representation, creating such liturgies as that for penance. The Protestant genius has run to music as space for the Word. Neither tradition has neglected the genius of the other completely, but throughout the modern centuries they have clung to quite diverse religious styles.

From the middle of the eighteenth century, the rise of atheism, agnosticism, secularism, and other outlooks inimical to traditional Christian faith has put the European intelligentsia at odds with orthodox

Christian faith, both Protestant and Catholic. The assertion of the rights of human intelligence to autonomy that came with the Enlightenment appeared initially as a direct rebellion against faith. The strongest grounds for this assertion were the successes that critical inquiry had produced in the natural sciences. Galileo, Newton, Mendel, and other pioneers proved the case that careful empirical inquiry, combined with creative thinking about theoretical overviews and underpinnings, could lead to new and fruitful hypotheses. Speculatively and practically, for both the better understanding of the physical and biological processes of nature and the improvement of human life (medical applications, new engineering, eventually the Industrial Revolution), the evidence seemed obvious that supporting free, nondogmatic research was the straightest way to progress.

In this context, calls to surrender reason to faith and personal autonomy to church authority rang discordantly. The parallel in the political dimension was the call for freedom that the American and French revolutions had expressed at the end of the eighteenth century and the churches' flight into political conservatism. Everywhere educated human beings felt themselves to be on the verge of a new era. Only as the exposition of European ideals to the rest of the world during the colonialist expansions of the nineteenth century and then the carnage of the two world wars of the first half of the twentieth century made their full impact did the wisdom of trying to improve the world through autonomous human reason become highly dubious. The names Hitler, Stalin, and Mao stand for some of the worst abuses of human autonomy and some of the craziest flights into madness. Nowadays, as we found at the end of our chapter on Jewish mysticism in the contrast between Wyschogrod and Green, the question is how to deal honestly with both the new, postmodern situation to which recent history has led us and the perennial wisdom of the mystical traditions.

Obviously, this is a huge question, though one offering rich rewards to those who even begin to appreciate its centrality to current Western, indeed global, culture. One can note interesting experiments of the Western Christian churches, both Protestant and Catholic, as they have tried to reach accommodations with the secularism of recent higher culture. One can also study the recent history of Orthodox Christianity, which includes fascinating chapters on the survival of faith in such unpromising cultures as the now defunct Soviet Union. But the heart of the matter will remain the task of admitting two sets of data and finding them to be equally imperative.

One set collects recent historical happenings—what people have actually experienced, felt, thought, said, created since the end of World War II. The other set collects constant, universal experiences and needs that have existed since human beings first gained consciousness, that operate today, and that are certain to survive the evolution of history from its present postmodern phase to whatever succeeds the phase. A development such as thermonuclear weaponry ought to color how all people concerned with ultimate reality, all lovers of wisdom and mystics trying to reform themselves by contact with their divine source, go about this task. So ought this task itself, this effort at reform, this present version of the long-standing mystical quest, to color how all serious people think about thermonuclear weapons, genocide, AIDS, and all the other particular influences powerfully shaping their actual minds and hearts, their actual Christian faith or secularist agnosticism.

Many observers speak up for the rights of recent history. Fewer speak up knowledgeably for the rights of the perennial mystical quest. Regardless, the human condition itself forces upon many reflective people the pursuit of ultimate reality, existence that does not fail. Death, suffering, injustice, and ignorance are primal wonders, essential to any more than biological definition of *Homo sapiens,* as are creativity, beauty, generosity, and love. Mysticism is healthiest when it arises honestly, naturally, wholly from the intersection of present and perennial concerns, the intersection of negative realities such as death and positive realities such as creativity.

Summary

We have followed the historical currents of Christian mysticism from the New Testament to the present. We have also discussed the major Christian doctrines and symbols, noting especially the play of the negative (apophatic) and positive (cataphatic) predilections of representative Christian masters. In review, two features of this history of Christian mysticism beg commentary, as does the matter of how intrinsically the Christian doctrine of the Incarnation has shaped the mainstream of Christian mysticism.

The two features of the history of Christian mysticism that may strike the detached observer as peculiar relate closely. The first is the Jewish piety of Jesus, which the later Christian mystical tradition did not appropriate, indeed at times seems to have ignored. The second is the great influence of Greek ideas about God, Greek experiences of the spirit's meeting the darkness, or divinity beyond all categorization.

Let us offer a thesis on the first historical feature, the neglect of the Jewish piety of Jesus, and then qualify it. Jesus appears in the New Testament as a Jew of his times. For the early Christian exegetes, he appears out of the Hebrew Bible, the Old Testament, as the messianic prophet bringing the new, promised era of fulfillment. Jesus prays to a God he calls Father, dealing intimately with the Maker of heaven and earth. According to the evangelists composing his portrait, he defines himself by reference to his Father; his meat and drink come only from the Father's will. He dies according to the symbolic template of Isaiah's suffering servant. His resurrection inaugurates a new eon in salvation history, replacing the dysfunctional history begun with Adam.

All of these symbols, concepts, ingredients in the piety of Jesus come from his Jewish background. The whole matrix of the New Testament, in fact, is Jewish. Certainly, the peculiarity of the first century C.E., when Christianity was evolving its constitutive self-understanding and Judaism was evolving into a rabbinic, Talmudic form, explains much of the reaction of the first Christians. The Jews who rejected the Christian claims that Jesus was the Messiah went their own way, to their own new elaborations of Torah, and those Christians bruised in encounters with them bid them good-riddance. However, a fateful thing happened. Out of sight, Judaism went out of mind. The Jewish piety of Jesus became only his immersion in the Hebrew Bible, his religious formation as a child of Moses and the prophets. This formation remained valid, something for Christians to appropriate, inasmuch as Christians claimed that from the beginning (whether Adam or Abraham), God had planned to bring the covenant to consummation through Jesus. But this formation became passé, inasmuch as Christianity became the new carrier of the covenant and the rabbinic Judaism developing outside of Jerusalem became a side track.

The qualification that limits the historical significance of this thesis is that Christian exegetes still had to deal with what the Bible gave them, and what the Bible gave them remained thoroughly Jewish in both testaments. In fact, therefore, much of Jewish faith and practice entered into Christianity in earliest times and never left. For example, the Christian liturgy arose from Jewish forms, the Christian priesthood owed much to Leviticus, and the key roles attributed to Jesus—prophet, priest, king—carried Jewish overtones. Still, most of these developments were quite murky. Few Christian theologians were champions of Jewish religion, even that of the Old Testament, and few dealt positively with the cultural riches at the base of the piety of Jesus. On occasion Christian masters appreciated the benefits, as well as the liabilities, of Jewish law, but the more common reaction followed Paul in contrasting the law with the gospel as slavery with freedom.

The second feature in the history of Christian mysticism that we underscore, the influence of a Greek outlook, seems natural if we stress that the cultural milieu into which Christianity expanded was Hellenistic. In other words, the partner in the dialogue in which Christian missionaries to the Gentiles engaged inevitably was the Greek sense of reality prevailing in the Greco-Roman world of the first century C.E. During the next three or four centuries, when the fathers laid the foundations of Christian theology and mysticism, this Greek sense of reality continued to prevail. Without steady or significant dialogue with Jews, the early Christians moved into a Greek intellectual climate, perhaps not fully realizing how different it could be from the biblical climate in which Jesus had lived.

The Greek climate had no covenant, no passionate commitment to history, perhaps not even a personal God. It had no Mosaic law, no line of prophets ardent for justice, no wild symbols of messiahship, apocalyptic rescue, eschatological consummation of the world. It did not know what to make of an incarnate deity because the idea seemed self-contradictory. It celebrated no cross on which sin and death had died, no resurrection into divine deathlessness. Yet it came to furnish the basic language of Christian mysticism, from the concerns of the desert fathers for virtues and vices to the flights into the divine darkness dared by Gregory, Denis, and the author of the *Cloud*. Finally, the Greek sense of reality sponsored an enormously influential monastic life, taking Christian mysticism apart from that of Judaism, which held monasticism to be suspect.

Concerning the influence on Christian mysticism of the central doctrine of the Incarnation, the data appear to be complicated. Whenever Christian mystics dealt intellectually, in formal theological terms, with the God of their contemplative prayer, they had to contend with the creeds that made their God a Trinity. Similarly, whenever they dealt with the history of salvation, they had to contend with the Jesus who had announced the reign of God, suffered, died, and rose. The Christian church recalled Jesus at virtually every liturgy. There was no avoiding the symbols stemming from the Incarnation and no desire to avoid them.

When it came to explaining the experiences crucial in mystical prayer, however, many Christian masters, as we have seen, preferred negative language, indeed negative feelings. The way to God was a way apart from creatures, speech, and ideas, a way into plain being, silence, and unknowing. Certainly, other masters, such as Julian, took other, more affirmative tacks, and, equally certainly, the most revered masters were orthodox in not denying the trinitarian substance of the ineffable godhead. As we saw, Meister Eckhart labored to relate to the birth of the Word in the soul the mystic's letting God be God. Nonetheless, more of Christian mystical theology is aniconic than one might expect in a faith committed to the historical incarnation of its divinity. For an influential line of church fathers, Jesus was the image of the Image, the icon of the Eternal Word. To drop this double layer of sacramental mediation when it came to union with God, Christian mystics had to move well apart from their biblical matrix.

Once again, qualifications arise. The Bible notes the hiddenness of God. Perhaps all human theology, conception of God, has to be significantly negative because limited beings can never match the unlimited positive reality of God. Always we are forced to the stance of Augustine, the stance of countless Asian mystics: "not this, not that." Regularly we are bound to sympathize with John of the Cross and say *nada*. Yet when the Christian mystics did directly encounter ultimate reality, what happened to their Christological confessions? From the small amount of evidence that we have seen, those confessions seem to have carried less formative impact than one might have expected. While it was possible intellectually to speak of all things holding together in the Incarnate Word, the mystic's experience of going out from all creatures, along with the Hellenistic intuition that being is the reality of all beings and so reveals their nothingness, often made the language in which mystics spoke of their realizations of ultimate reality ontological and negative rather than historical and incarnational. The result was less personal fusion with Jesus the Christ than what Christian theology could have warranted.

Two exceptions complete this assessment, with one concluding speculation. First, some Christian mystics did develop a sense of ultimate reality that was fully iconic, finding God exactly in and through the flesh of the Incarnate Word. In the New Testament, Paul and John admit of this interpretation. The *Spiritual Exercises* of Ignatius Loyola admits of it in stretches, as does much of the mysticism expressed as devotion to the sacred heart of Christ the suffering redeemer. Nonetheless, iconic, sacramental mysticism is not the Christian standard. The flesh of the Word is not imposed. At least as often as Jesus functions at the center, the divine no-thing-ness counterprevails.

Second, the theology, liturgical prayer, and mysticism of Eastern Christianity come closer to an iconic, incarnational point of application than do the Western mystical schools. A keener appreciation of the freedom of the risen Christ from human constraints allows Orthodoxy to pray to "Christ our God" more comfortably than many Western mystics have. This is a matter of degree, not of kind. Few Western Christian mystics would have hesitated to follow Thomas in exclaiming to Jesus, "My Lord and my God" (John 20:28). Still, the Eastern Christian mystics show a keener sense of the pneumatic, pantokratic Christ than what one finds in the West. If one couples this sense with the Jesus Prayer, interesting possibilities open.

Christian mysticism has been most distinctive, most idiosyncratic and creative in its formation, reception, and understanding of direct experiences of ultimate reality, when it has aligned its faith most exactly along the axis of the Johannine/Chalcedonian Christology that not only insisted on full humanity and full divinity but made the flesh of the Word, the entire reality of Jesus the Christ, the privileged revelation and presence of God. That Christology is the most iconic and sacramental in the Christian repertoire. Yet, it is not something that could fashion mystical experience from the outside, as though God had to move in people's spirits according to precepts of given scriptural texts or church councils. Rather, it is something that, now and then, made the transformation of individual Christian mystics a function of their personal love of their incarnate divinity, a friendship or romance with Jesus the Christ.

No other religion has an incarnate divinity as Christianity does. After one has entered many necessary qualifications (shown how the Buddha or Muhammad have and have not been incarnations or how Christianity has and has not distinguished divinity from the incarnate Word), this thesis remains credible phenomenologically, in terms of how the different believers have actually believed.

Still, the history of Christian mysticism suggests that many mystics have not been so exactly incarnational, so distinctively Christian, as an *a priori* approach worked out of Christian doctrine might suggest they would have been. That is a salutary reminder that God may not be a Christian, any more than a Hindu, a Buddhist, a Muslim, or a Jew.

NOTES

1. For general treatments of Christianity, see Jaroslav Pelikan, "Christianity: An Overview," in *The Encyclopedia of Religion,* ed. Mircea Eliade (New York: Macmillan, 1987), 3:348–362, and Denise Lardner Carmody and John Tully Carmody, *Christianity: An Introduction,* 3rd ed. (Belmont, Calif.: Wadsworth, 1994).
2. Bernard Cooke, *God's Beloved* (New York: Trinity, 1991), 81; also John Meier, *A Marginal Jew* (New York: Doubleday, 1991).
3. Harvey Egan, S.J., *An Anthology of Christian Mysticism* (Collegeville, Minn.: Liturgical Press/Pueblo, 1991), 29–30.
4. Ibid., 41–42.
5. Bernard McGinn, *The Foundations of Mysticism* (New York: Crossroad, 1991), 240–241.
6. Ibid., 241.
7. Paul Rorem, "The Uplifting Spirituality of Pseudo-Dionysius," in *Christian Spirituality,* vol. 1, ed. Bernard McGinn, John Meyendorff, and Jean Leclercq (New York: Crossroad, 1985), 132–133.
8. Egan, *Anthology of Christian Mysticism,* 96–97.
9. Ibid, 177.
10. Meister Eckhart, "Selected Sermons," in *Meister Eckhart,* ed. Edmund Colledge and Bernard McGinn (New York: Paulist, 1981), 184.
11. Julian of Norwich, *Showings,* ed. James Walsh (New York: Paulist, 1978), 127–129.
12. *The Cloud of Unknowing,* ed. James Walsh (New York: Paulist, 1981), 137–138.
13. Egan, *Anthology of Christian Mysticism,* 445.
14. Ibid., p. 454–456.
15. Philip Sherrard, "The Revival of Hesychast Spirituality," in *Christian Spirituality,* vol. 3, ed. Louis Dupre and Don E. Saliers (New York: Crossroad, 1989), 419.
16. Nicholas Arseniev, *Mysticism & the Eastern Church* (Crestwood, N.Y.: St. Vladimir's Seminary Press, 1979), 148. See also Andrew Louth, *The Origins of the Christian Mystical Tradition* (Oxford: Clarendon Press, 1981).

7

Muslim Traditions

General Orientation

If we place the beginning of Islam with God, because from the beginning of their creation God wanted human beings to be Muslims ("submitters"), then Islam is as old as humanity. If we take the view of secular historians that the Arab monotheistic tradition begun by Muhammad early in the seventh century C.E. owed much to Judaism and Christianity, then Islam appears as the daughter and further stage of biblical religion. For our purposes, better than either of these points of view is the more strictly historical one that has Islam begin with the prophet Muhammad (570–632). As we see in the next section, Muhammad grew up in Mecca (in present-day Saudi Arabia), influenced by Bedouin and other Arab religious traditions. He received a call from God (Allah) and a series of revelations (the basis of the Koran) and became the head of a new religious community. Shortly after his death Muslims expanded their influence dramatically, first throughout the Middle East, then to Europe, India, Africa above the Sahara, East Asia, and lately to the entire world.

When following the major developments in Sufism, the main carrier of Muslim mysticism, we shall suggest how Islam developed historically. For the remainder of this section we concentrate on the central Muslim tenets, the faith that explains the practice.

The simplest summary of faith is the *Shahada:* "There is no god but God, and Muhammad is his prophet." This confession is the Muslim equivalent of the Jewish *shema.* Let us note the major subaffirmations and implications. First, God is personal, full of knowledge and will. Second, God is one. Islamic theology is radically monotheistic. Third, Islam is strongly opposed to idolatry. To say that there is no god but God is to clear the field of pretenders. Historically, such pretenders included the many spirits of Arab religion prior to Muhammad. Theoretically, they can be any numen or natural force, any value or movement inasmuch as it sets itself in competition with God: money, power, Marxism, science, whatever. Fourth, the sole God has spoken through Muhammad. In the past, God spoke through prior prophets: Abraham, Moses, Jesus. For the Arabs, in these latter days, God brought such prior prophecy to consummation through Muhammad. The revelations that God gave through Muhammad (the Koran) are the culmination of prior prophecy. The Koran, eternally with God, is the definitive revelation of Allah, his nature and will.

From this simple summary, we may expand our sense of Muslim faith by noting that the Koran itself (4:136) takes as articles of faith God, his angels, his books, his messengers, and the last day. God is the Lord of the Worlds, the Creator, who made all that is and who spoke through Muhammad. The angels are named and unnamed, biblical and nonbiblical, creating a world of spirits set to carry out the will of God, working in human events as God chooses. The books that Muslims venerate are the scriptures of Jews and Christians (who become known as "People of the Book"), which are holy but less so than the Koran. The Koran is the book of books, the revelation of revelations. In Islam the Koran holds the place that in Christianity Jesus holds. It is the divine word, the central revelation, the incarnation, and the sacrament of God.

The messengers of God are the prophets, including those prior to Muhammad, who carried God's message to their people. Muhammad is the definitive prophet, as the Koran is the definitive scripture. Finally, the last day is the day of judgment. From the beginning of his preaching, Muhammad stressed a time of reckoning, a winnowing of the evil from the good. On the last day, the good will gain the Garden, while the evil will gain the Fire. God holds all fates, yet people are responsible for where they end. If they realize the importance of the last day, they will take time seriously.

A third summary of Muslim faith, one oriented practically, is the quintet of obligations known as the "five pillars." The first pillar is the *Shahada* itself. Each day Muslims are to recite the *Shahada,* return to their wellsprings in the dual affirmation that there is no god but God and that Muhammad is the prophet of God. Devout mothers sing the *Shahada* to their infants, forming them from the cradle:

> [A] spiritual mother nurtures the soul of her child with the powerful effect of the recitation of the *Shahadah,* the oft-repeated prayer . . . and the beautiful Names or Attributes of God by singing them as a lullaby for putting the child to sleep or for comforting a wailing or a disturbed child. In doing so, the mother makes her contribution to permeating the very being of the child with the most powerful words of the Quran.[1]

Muslim tradition has put the number of the beautiful names of Allah (Koran 7:179) at ninety-nine. Devout Muslims often recite them on rosaries.

The second pillar of Islam is the obligation to pray five times a day. In Muslim areas, the call to prayer summons the faithful at each of these times. Geoffrey Parrinder has described this prayer (*Salat*) as follows:

> (Arabic) "prayer," in the sense of ritual or regular acts of prayer. Salat is one of the Pillars of Islam . . . which every adult Muslim is required to perform five times a day: at dawn, midday, afternoon, sunset, and night. The prayer need not be performed in a mosque, but at home or wherever the person may be, but he must face towards the Ka'ba in Mecca and pray on a prayer mat. It must be preceded by ritual ablutions, or washing face and head, hands and arms, and feet, and also by a declaration of pious intention (*niya*). The first chapter of the *Qur'an,* Fatiha . . . is recited in Arabic, followed by kneeling, prostration, and ascriptions, to form a section (*rak'a*) of prayer. Two or more *Rak'as* may be performed in each act of prayer.[2]

The Ka'ba in Mecca is a large cube, housing a black stone, probably physically a meteorite, that pre-Muslim tradition associated with Abraham. Muslims going on pilgrimage (*Hajj*) to Mecca perform various rituals in the neighborhood of the Ka'ba. If Mecca is the center of Islam (what comparativists call the *omphalos,* the navel connecting the living social reality to God), then the center of the center is the Ka'ba. In a mosque one can know the direction of Mecca from a niche (*mihrab*) in the wall. Facing the niche, which is often the most beautiful feature, the believer is facing Mecca.

The third pillar is the obligation to give alms (*zakat*). A representative explanation of this term presents it as follows:

> *Zakat* is an Arabic term that literally means "purification," "sweetening," and "growth." In the religious terminology of Islam it stands for obligatory charity or alms that every rich Muslim should pay to those who are poor and in need. *Zakat* is one of the five fundamental duties . . . within Islam. Like the daily prayers, fasting during the month of Ramadan, and pilgrimage to Mecca, it constitutes an act of devotion and piety. In the Qur'an, the command for *zakat* often comes together with that for *salat* ("prayers"), and one might say (with Jesus, *Mt. 22:36*) that on these two depends all Islam. *Salat* emphasizes the love of God, while *zakat* promotes the love and concern for one's neighbor.[3]

(Note, in passing, the reference to Jesus. This is more than a generous bit of ecumenism. For knowledgeable Muslims Jesus is a venerable prophet, honored by the Koran, and so worthy of having his words taken devoutly.)

The fourth pillar is the obligation to fast (*saum*) during the twenty-eight-day lunar month of Ramadan (Ramadan moves through the solar year, which means that sometimes Muslims are fasting during the long, hot days of summer, sometimes during the shorter, colder days of winter). If taken strictly, it entails abstaining from all food, drink, and sex from dawn to sunset. The basis for this pillar is Koran 2:183–85:

> O believers, prescribed for you is the Fast, even as it was prescribed for those that went before you—haply you will be godfearing—for days numbered; and if any of you be sick, or if he be out on a journey, then a number of other days; and for those who are able to fast, a redemption by feeding a poor man. Yet better it is for him who volunteers good, and that you should fast is better for you, if you but know; the month of Ramadan, wherein the Koran was sent down to be a guidance to the people, and as clear signs of the Guidance and the Salvation. So let those of you who are present at the month, fast it; and if any of you be sick, or if he be on a journey, then a number of other days; God desires ease for you, and desires not hardship for you; and that you fulfill the number, and magnify God that He has guided you, and haply you will be thankful.[4]

Let us paraphrase these verses briefly, conscious that the poetic character of both the Arabic original and the English translation render any such effort perilous:

You who believe in God, you Muslims: know that God has prescribed the *saum* for you. It is a precept from God, a requirement, as it was for those (biblical believers, perhaps also some monotheistic pre-Muslim Arabs) who preceded you in the line of prophetic faith. Ideally, you will take this precept to heart generously, in open fear of God and faith, wanting to do God's will with dispatch. The fast is only for a stipulated time, and if it is inconvenient or impossible for you to carry it out precisely on the days stipulated, you may substitute other days. You may also fulfill it, if you can afford this, by feeding a poor person. Yet it is better to feed a poor person with no strings attached, and it is also better for you to do your own fasting (because the discipline will bring you spiritual benefits). Muslims fast during Ramadan because that is the time when God sent down the Koran. The fast therefore commemorates the great mercy at the center of Islam, even as it offers you a useful annual discipline. God gave the Koran to guide you on the straight path to salvation, fulfillment, the Garden. Therefore, let those who are at home and can carry out the fast relatively easily do it during Ramadan, while those who are away can make accommodations and transfer it to other days. Your God is not legalistic about this. Still, he wants you to fulfill this precept, the days laid out for the *saum*, as a way of praising him for his guidance. May you hear all this clearly and do it generously.

The fifth pillar is the pilgrimage (*Hajj*) to Mecca. The following is a representative description of the *Hajj:*

> As we have seen, Mecca was a centre of pilgrimage long before Islam, how long none can say. The inference from the Quran, sura 22:25 f., is that Abraham initiated pilgrimage there. Sura 2:192 f. commands Muslims to perform the pilgrimage and provides a "ransom" of fasting, alms, or an offering. This passage presupposes the pilgrim rites; it does not explain them or set them out, for the obvious reason that those to whom it was addressed were perfectly familiar with them. We know that Muhammad suppressed certain customs and modified others; but unfortunately we do not know what the ceremony performed by the heathen Arabs actually was. A full account will be found in Burton's *Pilgrimage*, and since his day many others have described the ritual. . . . The pilgrims circumambulate the Ka'ba seven times, then run between the two small hills of Safa and Marwa hard by, and gather together at the hill of Arafat twelve miles away; on the way back they sacrifice sheep and camels at Mina, where the ceremonial stoning of the devil takes place. One of the most important acts in the pilgrim ceremonial is the kissing of the black stone set in the wall of the Ka'ba. The original ritual required the pilgrims to be nude, but Muhammad ordered that when the pilgrim came within the sacred territory he was to lay aside his ordinary clothes and put on two plain sheets, leaving the face and head bare. Thenceforth a state of taboo exists: he must not cut his hair or pare his nails or have sexual intercourse until after the sacrifice at Mina.[5]

Sir Richard Burton was a Westerner who, risking his life, passed as a Muslim and made the pilgrimage to Mecca in 1853. Medina is about 190 miles north of Mecca. It is the city to which Muhammad repaired in 622 (the *Hijra*, which marks the beginning of the Muslim tally of years) after the Meccans had rejected his first preaching, which he had based on the first of the Koranic revelations, around 610, and continued in Medina. In Medina Muhammad established a Muslim community, and from Medina he returned to conquer Mecca and establish Islam there as well. The presence of the Ka'ba in Mecca led to Mecca's becoming the center of Islam.

The rituals that transpire during the pilgrimage, and the symbols associated with it, suggest that the *Hajj* is a special time, what the anthropologist Victor Turner has called a *liminal* (threshold) time, observable in the experiences of pilgrims of other, non-Muslim traditions as well. Those who make the pilgrimage (*Hajjis*) step out of ordinary time and ordinary social relations. For once they can concentrate wholly on God. Their concentration facilitates the creation of what Turner has called *communitas* (an idealized bonding, religious body, presentation and experience of the House of Islam).[6]

Rarely do *Hajjis* not return from their trip to Mecca exhilarated by the feeling that they have met the true Islam, the wonder that God meant to create through the Prophet and the Koran. Generally, they feel much closer ties to other Muslims all over the world, whom they have seen, even if they have not spoken with them. Usually the *Hajjis* give edifying reports to their neighbors, increasing the desire of others to make the pilgrimage themselves. Some of the *Hajjis* return to Mecca for second, third, or even annual reinvigorations of their faith.

A final summary of Muslim doctrine appears in the opening surah of the Koran. As Parrinder noted in his description of *Salat,* this chapter, the Fatiha, is part of each daily session of prayer. In the rendering of A. J. Arberry it rings out as follows:

> In the Name of God, the Merciful, the Compassionate: Praise belongs to God, the Lord of all Being, the All-merciful, the All-compassionate, the Master of the Day of Doom. Thee only we serve; to Thee alone we pray for succour. Guide us in the straight path, the path of those whom Thou has blessed, not of those against whom Thou art wrathful, nor of those who are astray.[7]

Muhammad begins with an assurance that what he hands over comes from God, the source of his revelations. The stress on the mercy and compassion of God is impressive, mitigating the dreadful aspect that so sovereign a Creator and Lord might present. Still, God is the master of the last day, the time of judgment. Solid Muslims serve only God, go only to God for the help they need. (No other gods exist, and to depend on any other deep sources of support would be idolatry. Our gods are those to whom we take our sorest troubles.)

Islam, like Christianity, uses the figure of the path. Unlike many Eastern traditions, it is historical, understanding time to be linear rather than circular. The world began in an act of creation, and the world will end in an act of consummation. Devout Muslims know that God controls this unfolding of time and that they need God's help to stay on the Islamic path, the one that is straight, right, and holy, the one laid out in the Five Pillars. Other paths, straying away from the submission that is the glory of Islam, lead to the wrath of God because they are idolatrous. Astray from the path, from the program of Islam based on the Koran, human beings cannot please God.

These basic Muslim beliefs stand in the background of the historical factors and developments with which we deal in the next sections. They are common to Sunnis and Shiites, Sufis and lawyers working at the Muslim *Sharia.* What ought we to make of their general import for mysticism from the suggestion they offer at the outset?

First, we ought to note the nakedness of the core propositions. There is not much ambiguity: one God, one source of revelation. The oneness or soleness of God, affirmed by Jews and Christians as well, seems starker in Islam. The greatest sin in Islam is *shirk*, idolatry, more precisely "associating" anything with God.[8] Islam rejects the Christian idea that God could have a son, an associate in divinity either eternal or historical. Muhammad is not divine, not an associate, sharer, partaker with God in divinity. Yet, popular Muslim piety sometimes has treated Muhammad as popular Christian piety sometimes has treated the Virgin Mary, as a being so heavenly, so able to help with people's needs, that comparativists are tempted to speak of a practical divinity, the enjoyment of a functional status as a god or goddess. The Koran does not give much warrant for this point of view, however, and the high Muslim theological tradition condemns it roundly.

Compared to the Jewish concerns with the Talmud and with ethnoreligious identity, which in Judaism sometimes have seemed stronger than the concern with theology proper, Islam strikes a more imperative call to worship, a more imperative insistence that only God is all-important, even all-interesting.

This implies that Muslim mysticism is going to involve a lofty, pure, daunting, and extremely powerful divinity. To become lost in God is going to mean an extreme submission to the Creator, the Lord of the Worlds, the Master of the Day of Doom. Consequently, the strong theme that God is compassionate and merciful gains special significance. As a whole, Islam does not encourage intimate, romantic relations with God. It speaks of human beings as lowly creatures, submitters, servants or slaves of their divine Creator and Lord. Thus the assurance that God is compassionate and merciful keeps the demand for submission gracious, not autocratic or repressive.

The mystics accept the speech of the religious mainstream concerning submission, but their experiences soften it. Sufis such as Rabi'a and Rumi obviously are in love with God, pine for him when he is absent, like the bride in the Song of Songs pining for her absent lover. However, the Muslim mystics have to go gently with the theme of identification with God, loss of self into the divine reality, deification, as the fate of al-Hallaj shows. Agents of orthodoxy murdered him because he claimed to have become divine through his mystical union with God. He may have understood this claim weakly, along the lines of Meister Eckhart, who also got in trouble, though he was not martyred, for making it. Nonetheless, it cost al-Hallaj his life.

Second, we can expect that the majority of Muslim mystics, even the most eccentric Sufis, have felt a strong pressure from the creedal center of Islam, as well as the prevailing social systems, to stay close to the common practices. Like Jews making ethical practice more important than doctrine, Sufis could do themselves a great favor by performing the obligations laid out in the Five Pillars, following the religious law (*Sharia*), and keeping their sometimes esoteric views of the mystical pathway to union with God private, exposed only within the community or the lodge formed round the sheik who was its master. In that way, they could escape the notice of the guardians of orthodoxy and go their own way, which sometimes was similar to the Kabbalistic ways that used gematria.

Taking the lines of the Koran as a code, Sufis, like Kabbalists, could assign different letters different numerical values, in extreme cases thinking they were dealing with, even manipulating, the ultimate building blocks of material creation. This was a concern best kept from the lawyers involved in regulating the minutiae of practical Muslim life, lest it tempt them to start legislating for Muslim thought and religious contemplation. The Sufis were most persuasive when they argued that their intense interior lives led them to a fuller, rather than a lesser, observance of the *Sharia*.

Muhammad

We deal with Muhammad for several reasons. First, he serves well as a lens onto the very beginnings of Islam. Second, the Koran comes from his ecstatic experiences of revelations received from God. Third, for all Muslims, Muhammad has served as the fullest exemplar of the Muslim way, the most saintly submitter. Even the most exalted Islamic mystics have had to contend with the primary status of Muhammad as *the* Prophet.

Muhammad was born in the last third of the sixth century C.E., probably in the year 570. He was an orphan of the clan of Hashim and the tribe of Quraysh, his father, 'Abd Allah, having died before his birth. His mother, Amina, died when he was six, and his grandfather died when he was eight. From childhood, therefore, Muhammad was formed by powerful experiences of the fragility of life and the omnipresence of death.

Raised by his uncle Abu Talib, Muhammad married a wealthy widow, Khadija, when he was twenty-five. She had a prosperous caravan business, which he began to run. Around the age of forty, he received visions, heard voices, and wondered if he were going crazy. As critical historians reconstruct things, Muhammad, betaking himself to the desert at night, would enter a cave, wrap himself in a blanket, and try to fathom what was happening. Gradually, he discerned that these strange spiritual activities came from the angel Gabriel, sent to him by God to preach against the polytheism and idolatry then strong among his fellow Arabs.

Muhammad tried to obey, supported by Khadija and such early believers as Ali and Abu Bakr, but the vast majority of his fellow Meccans rejected his message, indeed opposed him bitterly, in good part because it threatened their large market in amulets, spells, and other polytheistic products. By 622 this opposition had grown so strong that the Prophet and his closest followers departed to Medina (the *hijra*), where he had an invitation to take charge of a community of believers. By 630 he was strong enough militarily to return to Mecca and conquer the city, destroying its idols. Muhammad died in Medina in 632. His followers compiled the revelations he had been receiving since 610 or so into the written Koran. They also gathered the memories circulating about Muhammad and the principles he had established for ruling the Muslim community in Medina into collections of *hadith* (traditions), which became another source of guidance and authority, second to the Koran but useful in interpreting how the Prophet himself had understood various injunctions of the Koran.

Toward the end of his succinct study of Muhammad, Michael Cook offers an admiring summary of what the Prophet achieved:

> It was a very considerable innovation to bring about the rule of God and His prophet over an independent community of believers. It was no accident that Muhammad achieved this in Arabia, with its predominantly pastoral and stateless tribal society. He had the good luck to be born into an environment which offered scope for political creativity such as is not usually open to the religious reformer. But it was clearly more than good luck that he found in this society the key to a hitherto virtually untapped reserve of power. The pastoral tribes of Arabia were necessarily mobile and warlike, but their military potential was normally dissipated in small-scale raiding and feuding. Muhammad's doctrine, and the use he put it to, brought to this society a remarkable, if transient, coherence of purpose. Without it, it is hard to see how the Arabian tribesmen could have gone on to conquer so substantial a portion of the known world.[9]

Muhammad conceived of Islam, the religion of submission to the sovereign soleness of God, who was uniting the Arabs to fulfill the covenants previously given to the Jews and Christians, as a missionary venture. Muslims had an obligation to publish the message of the Koran and expand the influence of Islam. Muhammad had been willing, as well as forced, to use military means to establish Islam in Mecca. The culture of the desert in which he lived was, as Cook indicates, mobile and warlike. Muhammad established himself as an effective general, as well as an effective head of the *Umma*, the Muslim community. He was a visionary but also a practical leader with both feet on the ground. Yet, more than Cook may appreciate, the core of his vision and energy came from his dealings with God. The basis of the new order that he saw Arabs developing was the Koranic revelations he received steadily over the last twenty-two years or so of his life.

A good example of the visionary experiences that first caused Muhammad to doubt his sanity and then became the basis for his conviction that God was calling him to be a prophet to the Arab people occurs in what is now Koran 53:1–11. In a previous study we introduced this text and commented on it as follows:

> In his thirties, Muhammad had developed the habit of seeking God in prayer and at the time that he began to receive the Qur'anic recitals he would sometimes go into the desert night in search of solitude. As reflected in the much-praised but rather flowery translation of the Qur'an by A. J. Arberry, Muhammad's visions were dramatic indeed:
>
>> By the Star when it plunges, your comrade is not astray, neither errs nor speaks he out of caprice. This is naught but a revelation revealed, taught him by one terrible in power, very strong; he stood poised, being on the higher horizon, then drew near and suspended hung, two bows-length away, or nearer, then revealed to his servant that he revealed. His heart lies not of what he saw; what, will you dispute with him what he sees?
>
> The celestial figure is an angel, probably Gabriel. Tradition says that the fact that Muhammad did not flinch from this encounter confirmed his worthiness to receive what followed after it—the later revelations. The text seems to come from a time when Muhammad approached some companions to explain what had happened to him and met with skepticism. His insistence on the veracity of his account, along with his description of the angelic being commissioning him, may have been designed to counter such skepticism.[10]

Here we may add that the description of the experience stresses its powerful, numinous quality. The angel is bearing the force of God, the divine authority. Muhammad does well not to be swept away. It makes sense that he might doubt his mental balance. What ordinary people have visions such as this? When the revelations continued and began to assume a coherent form as a commission to preach reform of religion (no idolatry) and embrace a new communal life under the soleness of God, Muhammad could feel confident that God was indeed calling him to be the prophet to the Arabs. Prior to that coherence, he had to struggle along in confusion.

Surah 24:35 is famous for speaking of Allah as light, a figure that gives us a hint of the splendor that Muhammad found in his Lord:

> Allah is the light of the heavens and the earth. The similitude of His light is as a niche wherein is a lamp. The lamp is in a glass. The glass is as it were a shining star. (This lamp is) kindled from a blessed tree, an olive neither of the East nor of the West, whose oil would almost glow forth (of itself) though no fire touched it. Light upon light, Allah guides unto His light whom he will. And Allah speaketh to mankind in allegories, for Allah is Knower of all things.[11]

For Allah to be the light of the heavens and the earth is for him to be the Creator. The reference to Genesis may not have been at the forefront of the Koranic revelation, but perhaps it lay in the background, mediated by Jewish and Christian views of the world. God had said, "Let there be light." Where had such light—the sun, the moon—come from if not the substance of God? The overtones of "light" also include "intelligibility." Allah is the source of all meaning, insight, dawning of significance. Luminous beings such as Gabriel come out of the divine light, sent on their missions to illumine and clarify human affairs.

Arabs lived close to the desert, where the light comes and goes quickly, dramatically. The only alternative to natural light was lamps burning oil, usually that of olives. The complicated figure of the niche, the lamp, and the glass like a shining star suggests the mediations of light, God speaking in allegories. The light of God passes from heaven to earth, glancing off this creature and that, using this carrier and that. Yet heavenly light has a purity, a way of clarifying each creature and bringing out the significance of its being, that shows its divine source. We human beings cannot make our way through the world without this light, both physical and spiritual. We wander in darkness, blindness, unless the light of God guides us mercifully. The Koran does not offer here a metaphysics of light. It does not ponder the relations among light, mind, being, and divinity. However, in a poetic way, moving suggestively though a traditional Arabic design for maximal rhetorical effect, it paints a picture of the pure light of God that approximates the Johannine Christian "light shining in the darkness." The light of Allah is the material and moral illumination of the world.

Light upon light, illumination abounding, God guides to his light those he will. There is a mystery about the light shining in the darkness. Why do some embrace it, love it, want to come to it so that their deeds may be seen in God, while others flee from the light, seem not to have been called to it? The Koran does not ponder this mystery in the prophetic terms of an Isaiah, who muses that God seems to have blinded the eyes of the many so that seeing the light they would not see, as hearing the Word they would not hear. Yet the passage does imply that walking the straight path illumined by the light and reaching the lamp at the end is a matter of providence or predestination. If God guides to the light whom he will and some people remain in darkness, then perhaps God does not will it that all people come to the light.

Yes, one can say that those whom God does not will to bring to the light have chosen against it. With God, one can make the divine predestined choice coincide with human free will, good or evil, because there is no time in God, no before and after. Thus God's willing to bring Muhammad to the light can coincide with Muhammad's generous choice to pursue the light, informing and empowering this choice without predating or predetermining it. God cooperates with any creature's action inasmuch as such an action flows from the creature's being and reshapes it. If the creature's being depends on God, who is the sole independent, necessary, noncontingent source of being, then so does the creature's action. In this way, the choice of God, the calling of some to the light and not others, is not a bullying of creatures, let alone an abandonment. It is rather the most intimate interaction with the creature, at the core of the creature's being, where light or meaning is simply the efflux or obverse of the creature's coming from nothingness to take a place in the fathomless creative plan of God.

Muhammad seems to have been content to speak of himself as the prophet of God. However, competition with Jews and Christians led some Muslims to pass beyond this modest speech. W. Montgomery Watt has summarized this tendency as follows:

> Some of the early developments within Islam came about as a result of the criticism, mainly from Christians, that Muhammad could not be a prophet since no miracles were worked through him. Muslim scholars began to search the Qur'an and the anecdotes about Muhammad to find incidents capable of being treated as miraculous and many were found, even miracles of healings, such as the tending of a wounded eye so that the individual later declared that this eye was better than the other. Stories like this could obviously grow in the telling. Muhammad's birth and childhood gave further opportunities for introducing miraculous elements. . . . The most spectacular of the miracles was the Mi'raj, Muhammad's night journey to the seven heavens. This is based on the verse "Praise to him who brought his servant by night from the Sacred Mosque to the Farthest Mosque . . . that he might show him of his signs" (17:1). Presumably this referred to a dream, but piety transformed it into a miraculous physical journey on the winged horse Buraq, first from Mecca to the farthest (al-Aqsa) Mosque in Jerusalem, and then to the seventh heaven, and accounts of what he saw on the way could be elaborated almost indefinitely. That this night journey and ascension were a reality was accepted by Muslims, though there were some heated discussions of whether he had seen God. Another verse which was often elaborated into a miracle is: "The hour drew near and the moon was split" (54:1). This was traditionally one of the signs of the Last Day, but the people of Mecca were said to have seen the moon in two parts, one on each side of a local mountain.
>
> While these miracles might fill popular imagination, the theologians tended to say that the one "evidentiary miracle" . . . was the Qur'an itself, because of the immutability of its style and contents. The theologians also held that Muhammad, like other prophets, was preserved from sins, though they disputed the precise nature of his infallibility. It also came to be an article of belief that on the Last Day Muhammad would have God's permission to make intercession on behalf of the sinners of his community; though intercession is mentioned several times in the Qur'an, it is not specifically stated that it is permitted to Muhammad.
>
> Most of what has been mentioned so far was accepted by Sunni Mus-

> lims. Among the Shi'ah and the Sufis, however, there were some who went much further. Muhammad was identified with the Perfect Man of late Greek speculation, in whom God's consciousness became manifest to itself, and who then became an instrument or agent of creation. This image was associated in turn with the "light of Muhammad." . . . For the great majority of Muslims, of course, such speculations are anathema.[12]

Muhammad was the definitive mouthpiece of God, the human channel of divine revelation. As such, he came closer to God than any other human being. This made him, *ipso facto,* a mystic in our sense of the term. He had the most direct experience of ultimate reality that Muslims could accredit. Beyond what the most exalted Sufi saints might claim, Muhammad had been filled with the Word of God, the Koran. The Word of God, abiding with God eternally, was a form of heaven, a speech of light. To have been filled with that Word, that light, and to have let it pour through his mind and heart made Muhammad one with God uniquely. Muslims did not find such oneness divinizing, in the strict sense of the term. Muhammad did not become an associate of God or a divine son. Indeed, Muslims attacked the claims of Christians for the divinity of Christ as idolatrous, even polytheistic, as they attacked the related Christian theology of the Trinity; but even as they insisted that Muhammad was only a creature, never more than a human being, Muslims strove to distinguish their Prophet from all other religious figures.

As Watt notes, the great proof for the uniquely exalted status of Muhammad lay in the Koran itself. Tradition, in fact, has said that Muhammad was illiterate and that the Koran had come to him directly and fully from God, with no significant "reception" on his part or linguistic work to express it. The Arabic of the Koran became the standard of eloquence and excellence. No human being could have created so profound and beautiful a book. The Koran shouted that God had dictated its contents and was its author in a very strong sense.

Mystics do not have to produce documents as evidence of their direct experiences of ultimate reality. Nonetheless, most of the mystics who become known and exert significant influence on others do produce documents. A danger attaches to the period when they turn their experience into explanatory or didactic products. The mystic has to shift consciousness from the receptivity involved in undergoing the experience (the initiative of which always belongs to God, if the experience is truly mystical, *patiens divina*) to the activity of describing it. To describe it, an author or artist has to gain a focus, choose verbal or visual images, bring the experience as best as he or she can within the mental range of the audience. All of this denatures the experience itself, which is ineffable inasmuch as it is the impression in the mystic's spirit of the unbounded, unfocused divine reality.

Whatever else one can or must say about ultimate reality as ultimate or truly divine, it is trans-human, transcendent of all created boundaries. It sets those boundaries inasmuch as it determines the degree or manner in which creature *x* is to participate in its divine, fontal being. Assuming that these general rules of mystical psychology apply to Muhammad's reception of the Koran (not an assumption that all Muslim analysts would accept easily), we can say that what we find in the Koran is in some ways more than what Muhammad experienced and in some ways less. It is less in that, unless God merely dictated the bare words of the Koran to Muhammad, who would serve then as an amanuensis or a secretary, Muhammad had overwhelming experiences of the full, limitless being of the Lord of the Worlds, which even his best imaginary and linguistic gifts could render only imperfectly. However, what we find in the Koran is more than what Muhammad experienced in that its words developed an independent life of their own, one out of Muhammad's control, among later readers of the Koran, especially those who used it for worship and the nurture of their own contemplative lives.

Lastly, we note that many of the "revelations" contained in the Koran are relatively prosaic laws for common conduct. The implication is that part of the message Muhammad received from God was a set of rules for the conduct of his community. Thereby, God gave divine sanction to guidelines concerning marriage, inheritance, the treatment of people captured in war, and much more. What this implies for the intercourse between the Prophet and God is interesting to consider. Perhaps God wanted a full say in how the *Umma* was to function. Perhaps, taking Muhammad as a human author, he became sufficiently intimate with God to feel entitled to attribute to God his own bright ideas for the governance of the *Umma*. Or perhaps, and we find this the most likely interpretation, the Prophet was so attuned to the presence of God that he could not distinguish between the light, the wisdom either poetic or practical, that directed his understanding of daily reality and the light of God.

In the latter case, perhaps all of the Prophet's thinking became an implicit dialogue with God, a reflexive openness to the light and Word of God, so that any probing of reality, any seeking of solutions to practical or speculative problems, was a calling upon God, or a waiting upon God, to reveal the solution. If so, then the Prophet achieved an habitual union with God of sorts because his thinking had become intrinsically dialogical. When he thought, God was in the picture, directing the *son et lumière*. When he felt, God was moving in his blood, his emotions. There was no mind of the Prophet, no heart, apart from God. This did not make him the son of God in the sense that Christians used the term for Jesus, but it did move him in the direction of identifying himself only through God. He was who he was; his being Muhammad signified being the Prophet because he had become a relational being, not knowing what it would be like not to be the Prophet, the man formed in the desert night in a cave dominated by the power of Gabriel on the horizon, by the light of the lamp hanging in the niche around which creation arranged itself, and by the pure luminescence of the holy olive oil brimming with liquid mercy.

The Koran

The Koran is more important than Muhammad, but there is no Koran without Muhammad, nor any Prophet without the Koran. The Koran and Muhammad go together. The Prophet is the Prophet because through him God gave the Koran. Still, Muhammad bows low before the Koran. His submission, and that of all other Muslims, is to the God known fully only through the Koran. In this section we probe the nature of the Koran regarding our own interests in mysticism. How does the mediation that the Koran provides, or imposes, color the directness of Muslims' experiences of ultimate reality?

The Koran that we now find between the covers of a book arose orally, in an oral culture. Before the revelations, the materials given to Muhammad to recite took form as words on different surfaces; they came to the Prophet as infrapsychic events—meanings occurring in his mind, heart, and senses. Before Gabriel became the terrible power appearing on the horizon, Muhammad had the full, more-than-verbal experience of the angel's coming. As we noted, the transposition of an experience such as this into written or other artifactual form involved both a loss and a gain. Lost was the preverbal wholeness and fullness. Gained was a definition, a delimited expression, that could enter the lives of readers and listeners and start a history apart from the Prophet's direct control.

The relationship between the human author or agent of a scriptural text and the supposedly divine source of that text is hard to determine precisely. The middle range of interpreters will struggle to reconcile two agencies, agreeing that (*1*) even scriptural texts bespeak a given time and place of provenance, owe much that is intrinsic to the culture and idiosyncracy of the human beings who issued them, and yet (*2*) what makes a text "scriptural," "divine," "saving," or "revelatory" in a strong sense are the indications that the human beings who issued it felt that its contents came to them from outside, from a divine power they could not control but had only to accept with an active passivity.

In the case of the Koran, Muhammad had to receive the recitals from God and pass them on. The conscious parts of his awareness and conscience told him that God gave him no choice but commanded his listening, receiving, obeying. God had chosen him, he had not chosen God. His prophetship was very little a matter of his own doing but almost completely the free, mysterious doing of God. This interpretation of Muhammad has inclined Muslim analysts to think of the Koran as much more the book of God than the book of Muhammad. The Koran is where God makes present in history, the world of space and time, all that only God can. Thus the Koran is not something that human beings can change, update, or pick and choose from. It is not even something that human beings can translate from one language to another with any claim that revelation passes across the linguistic divide without loss. Speaking strictly, the Koran is immutable and can exist only in Arabic. It is immutable because it came from God to express the divine will, which human beings have no right to alter in the slightest and which God has every right to express once and for all. The Koran can exist only in Arabic because (*1*) that is the form in which God gave it and (*2*) God gave it in the first place as the scripture of the Arab peoples, their way to dominance under the sun.

The far reaches of this prejudice that the Koran is much more the book of God than the book of Muhammad took the majority of Muslim thinkers to the conclusion that the Koran has existed with God in heaven from eternity. Here, the equivalence between the Koran and the Christian heavenly Word that took flesh from Mary becomes clearest. The Koran is not a divinity alongside God in heaven, an associate that would make God less than the sole divinity and so compromise Muslim monotheism, but the prehistoric, precreational expression of the intelligence and will of God that his dynamic, creative nature produced spontaneously. The historical book that came from the recitals imposed on Muhammad is a reflection of the eternal prototype, a participant in the archetype that puts into human language all that one can of the spontaneous self-expression of the divine. It is the best that human beings have, more than human beings could expect their sovereign Creator to place in their midst, but there is room for debate about the fit between the human form of the Koran, the words which the Prophet or God used for the recitals, and the eternal, heavenly Koran.

The majority of Muslim authorities, wanting close obedience to the letter of the earthly Koran, have made the fit close, if not hermetically sealed. Mystics, lost in the verbal building blocks of the Koran, might sense a greater distance, might even feel they were traveling into the unlimited divine self-expression of the heavenly Koran, the direct "speech" through which God spontaneously talked to himself. If they were prudent, however, such mystics would keep such feelings to themselves and discuss such possibilities only within the Sufi lodge, where others would be apt to understand them, or at least not to misunderstand them so that they became suspect of heresy.

More prosaically, we should note that the order of the 114 surahs that make up the Koran runs from longest to shortest. The canonical Koran is not arranged in the order in which the Prophet received the recitals. Modern scholars of the Koran, task on the text with what have become standard methods of textual (historical and literary) criticism, have offered suggestions about which surahs arose from revelations given while Muhammad was in Mecca and which while he was in Medina, but they have not achieved anything like a complete consensus. Generally, experts on the Koran say that the more dramatic revelations, suggesting the beginnings of the Prophet's task when the call first came to preach to his fellow Arabs and the work was to grasp what God wanted him to say, probably came during the early period at Mecca. The materials dealing with practical matters for the worship, legal affairs, and general running of the *Umma* probably came while Muhammad was at Medina, heading up the new community there and more or less consciously drawing the blueprint for all later Muslim communities.

Traditional Muslim scholarship, sophisticated in its own way, if only through an intense immersion in a Koranic text that is considered to be the perfect Word of God, has not performed the historical or literary surgeries that one finds standard in modern, Western critical scholarship. It is more concerned to reconcile passages that could seem to be discordant, even contradictory, and to draw up the lists of obligations that the text laid upon believers. The situation is much like that in Jewish scholarship concerning both the Bible and the Talmud. Until well into the twentieth century, Orthodox Jewish scholarship on both the Bible and the Talmud reflected the passion of the rabbis to find the meanings that God had strewn through the sacred text. The believer might, indeed ought to, put questions to the text and wrestle with it to get it to surrender its inner logic, but the believer would not stand outside the text, as the secular textual critic would, to assess in terms of human probabilities what the authors had in mind and how they had crafted their human artwork.

Lastly, it bears noting that the Koran reflects the literary, rhetorical, and poetic traditions of Muhammad's own culture. The desert culture that formed him placed a high premium on eloquence. Along with martial prowess and physical bravery, the ideal Arab man spoke well, was a leader led by the persuasive beauty of his words as well as by his example in battle, and brought prudence to the councils where the tribe debated its important decisions. One had to move men through their ears and reach their souls by seducing their minds with vivid images and pleasing rhythms. These are characteristics of oral cultures everywhere.

When books become the main coin of exchange, meaning is less immediate and a critical distance arises automatically. With books people can sit back and ponder, keeping their place with a finger. They can even use the book for *lectio divina,* the leisurely listening for the Word of God that they can assume is available in the pages of sacred texts.

The texts can become liturgical, heard again and again through the annual cycle of religious celebrations, in the common assembly where a new dimension can emerge. There, the text clearly forms the community, and just as clearly it gains its status as sacred, scriptural, "the book," from the assent that the community gives it.

However, when the main communication of meaning, either the horizontal meaning that ties the community together by expressing its mores or the vertical meaning that ties the members to ultimate reality, is oral, the tying is more spontaneous and labile. "Religion," which through Latin etymology we can understand as coming from "retying," an image of people's reestablishing regularly through their rituals their connections with their sacred sources, is then alive more vividly than when literacy has made the primary reference of "scripture" and "meaning" bookish, textural, written. Muhammad did not get from God a text, a collection of printed pages bound between two covers. He did not deliver to his fellow Arabs in Mecca a text or pamphlets or broadsheets. He received a living speech from God, or the angelic messengers serving God, and he delivered to his fellow Arabs in Mecca, and then Medina, speeches, sermons, instructions. He received his commands *viva voce* and he passed them on *viva voce.*

All of this explains somewhat the style of the Koran and enters into the substance of its command. The function of the Koran shapes its content, the meanings it spotlights. Form does follow function, while function does shape form. When one realizes what were the rhetorical traditions that Muhammad knew, who were the people he was addressing, and what were the goals his addresses sought to achieve, the exalted, at times rhapsodic, style of the Koran makes considerable sense. Muhammad had to move people's hearts as well as clarify their minds. He could succeed better by declaiming his message poetically than by laying it out in pellucid syllogisms. Even if he made the choices that he did deliberately (assuming that the initiatives of God left him some literary discretion) rather than purely instinctively, he would have been constrained by the realities of the situation in which he found himself, the actual parameters shaping the communication he wanted to achieve, to speak poetically, imagistically, in the allegorical style that the surah on light attributes to God.

Let us conclude this section by offering two short examples of Koranic materials and then reflecting on the implications of basing a mystical life on this scripture.

The two short surahs are the last in the Koran, 113 and 114, entitled, respectively, "The Daybreak" and "Mankind." According to Marmaduke Pickthall, both came to the Prophet at Mecca.

> In the name of Allah, the Beneficent, the Merciful. Say: I seek refuge in the Lord of Daybreak, from the evil of that which he created; from the

evil of the darkness which is intense, and from the evil of malignant witchcraft, and from the evil of the envier when he envieth. (113)

In the name of Allah, the Beneficent, the Merciful. Say: I seek refuge in the Lord of mankind, the King of mankind, the God of mankind, from the evil of the sneaking whisperer, who whispereth in the hearts of mankind, of the jinn and of mankind. (114)[13]

The first line is the standard reminder of the kindly nature of God the revealer, who works for the salvation of human beings, not their damnation. In surah 113, the *mise-en-scène* is daybreak, when the light of the Lord brings creation into clarity and shows what space and time are for, in a word offers wisdom. It is somewhat striking that the prayer arising at this fresh, pristine moment is for refuge, or sanctuary, from the evil that threatens all believers. This evil apparently runs throughout creation, though one cannot say definitively from this text that God created evil as something positive, an entity in its own right (that is, not a privation of right order, a lack of proper being). (Islam has no doctrine of original sin, as Christians understand the concept, at least in the wake of Augustine: willful if absurd rebellion against God. Islam inclines rather to think of human beings as weak, forgetful, needing constantly to be reminded of the revelation and mercy of God if they are to avoid developing into considerable agents of wrongdoing, wreckage.)

The darkness can be either physical—the dangers of life in the midst of capricious nature—or spiritual. Spiritual darkness is forgetfulness, wandering off the path that is straight, risking the wrath of God by idolatry of one sort or another. The evil of witchcraft reminds us of the polytheistic, or polydemonistic, culture of pre-Islamic Arabia. Literally, the figure used here has women blowing on knots, uttering curses over cords used to represent (the life spans of?) enemies.

The evil of the envier is the ill will, so difficult to combat or remove, of those who take offense at any prosperity of other people. In presenting himself as the *rasul* of God, the one commissioned by God to preach a message of repentance (in light of doomsday) and great opportunity (to become the bearers of God's definitive covenant), Muhammad was bound to arouse great envy. Even those who had no stomach for his message could easily have resented the prominence that his preaching it gave him. Indeed, this resentment could have so clouded their hearts and thickened their brains that they would have rejected his message while barely hearing it. (This is a central charge in the polemic of the Christian evangelists against the Pharisees. The Pharisees were so emotionally opposed to Jesus that they could not deal with him rationally. The heavy irony in John depends on this polemical judgment, but it applies to the majority of intensely prophetic situations, including those of such Hebrew prophets as Jeremiah. Muhammad joins this line to offer another instance of a prophet initially finding little honor at home, among his own people.)

Surah 114 begins more positively, honoring the king of all human beings, the (one) God of peoples everywhere. Whatever the special ties between God and the Arabs, the Koran never presents him as provincial. Always he (a patriarch) is the Lord of all Creation, the Master of all peoples. More than is true of both Judaism and Christianity in many places, God is a universal God, not a God who sets apart one people as chosen exclusively. There are qualifications to this thesis, of course, most of them stemming from the privileged place of the Koran itself, which admits no equals as definitive revelation. However, what we overhear in this text about the relationship between God and all men and women is a consistent theme elsewhere. God makes all members of the human race, as he makes all material creatures and God loves any members who accept him, live as true believers (submitters).

Soon, however, this surah too takes up fears and needs that the believer wants to lay before God, refuges that the believer needs to find in God. Here, the evil from which the believer wants to flee is the influence of the sneaking whisperer, probably the devil, the personification of the evil spirituality set against human beings and God as their worst enemy. This evil agency does its greatest damage by poisoning the hearts of human beings and *jinn* (spirits, from whirlwinds to local geniuses, forces of place, mood, eerie or numinous power). This text does not elaborate upon how such poisoning proceeds, but we may conjecture that it takes a full range of forms. Envy would be a powerful form, but so would loss of confidence, leading in the worst case to despair, inability to hope that the message of the Koran is faithful and true, that God is all that the Prophet assures believers he is.

Faith requires a daily struggle to stay on the path that is straight and maintain one's conviction that the work, the self-denial, the suffering, the exposure to being overwhelmed by the hiddenness of God, the immensity of ultimate reality, are worthwhile. The sneaking whisperer tries to make this struggle overwhelming, more than what the believer can bear. He exaggerates the difficulties and sacrifices involved, raising doubts. If he can distract believers and keep them from regular contact with God through the Koran, he can succeed. They need only immerse themselves in the Koran daily, however, to find a cotton with which to stop their ears. They need only pray the stipulated prayers, the *Salat,* five times each day to stay in touch with the promises that reveal the emptiness of the sneaking whisperer's words and so make them seem only so much foul breath.

What do surahs such as 113 and 114 suggest about basing a mystical life on the Koran?[14] First, regular preoccupation with this scripture would take a contemplative through a full range of emotions. In these two short surahs, confidence in God and praise of God alternate with fear of the evils in creation and a petition that God save the believer from all evil influences. Regularly, other surahs repeat explicitly, or create implicitly by the flow of their images, this same overall impression that reality is dual, though not dualistic. Good and evil interact everywhere. The sovereignty of God assures the believer that good is more significant than evil, but the weakness of human beings and the justice of God make evil formidable.

Second, the usual tendency in mystical development is for beginners to move by fear more than love, for middlers to prize illumination (coming to understand the inner implications of texts and symbols that previously they had dealt with only on the surface), and for the advanced to take the textual words as springboards to a simpler, quieter communion with divinity itself. This tendency appears in Muslim mysticism, though of course with angularities peculiar to the unique historical experiences and doctrinal commitments that shaped mystics based in the Koran.

The advanced Sufis do tend to see through, or feel through, the Koranic text, reaching out directly toward divinity. The closer they come to touching divinity, the less they have to say, the more the immediate fullness of divinity preoccupies them and quiets their spirits. Communion with God tends to be heart to heart, even more than mind to mind. Inasmuch as "mind" connotes something at the top of the human composite, while "heart" connotes something at the middle, closer to the point of balance, the center of gravity, mystical self-understanding favors the heart over the mind.

Certainly, the light of God connects more with the mind than the heart, though often we find a mystical knowledge that is cordial, born of religious love. However, the love of God correlates more with the heart. This love is both the ardor of the mystic for beautiful ultimate reality and the prior, deeper, more final love of ultimate reality itself—the affection and ardor of God, the Creator and Lord. Even traditions such as Islam that bend over backward to stress the sovereignty of God find in the practice of their mystics, the ones who take most seriously the calls of the tradition to dedicate oneself to God utterly, a conviction that the great happening is falling in love with God, being taken into the divine embrace.

Perhaps the most important instance of this great happening occurred in the life of the foremost Muslim religious thinker, al-Ghazzali (1058–1111). Although he could not have desired greater prominence than what his scholarship had gained him, in middle age al-Ghazzali went through a religious crisis and took himself off to join the Sufis. There he found a more profound knowledge of God than what his scholarship had given him, one deeply personal, indeed mystical. Like Pascal coming to feel in an experience of divine fire burning in his breast the difference between the God of Abraham, Isaac, and Jacob and the God of the philosophers, al-Ghazzali learned that the traditions, the laws, even the Koranic texts taken superficially were only the bare bones of Islam. To reach the living heart, to feel the awesome creative power, one had to open one's heart, rivet one's spirit, to God as such, ultimate reality sought with nothing held back, no compromises with mammon or mind or limited human intelligence. The utter primacy of God that the Koran proposes, indeed chants about in surah after surah, encouraged conversions such as al-Ghazzali's, as we see now in taking up the history of Sufism, the central movement of mystical Islam.

Early Sufism

From the first, the recitals of Muhammad carried an imperative call for conversion. Human beings could only gain the straight path by turning from crooked paths (various idolatries) and acknowledging the complete sovereignty of God in their lives. Implicitly, then, accepting the Koran meant accepting God as the first treasure, the overwhelming good. When honest, serious people asked how to do this, they ventured out of merely external conformity to the letter of the Muslim law and approached an interior, potentially mystical venture in dealing with God personally.

Peter Awn has described well the tension, perhaps the coincidence of opposites, that prevails in the relationship between God and human beings, as the Koran presents things:

> The vision of the God–man relationship in the Qur'an offers a study in contrasts. On the one hand God is the almighty creator and lord of the cosmos who sustains the universe at every moment (Qur'an 10:3 ff.); men and women are but servants—finite, vulnerable, and prone to evil (2:30 ff. and 15:26 ff.). God is both lawgiver and judge (surahs 81 and 82); whatever he wills comes to be (2:142, 3:47, 3:129, 5:40, 13:27). Servants of God are enjoined to embrace his will, not question its import, for men and women will be rewarded or punished according to their deeds. To breach the lord–servant (*rabb–abd*) relationship leads easily to the cardinal sin of *shirk*, substituting some other power for that of God.
>
> On the other hand the inaccessibility of the transcendent Lord must be understood in the context of those Qur'anic verses that speak of his abiding presence both in the world and in the hearts of the faithful. For did he not actually breathe his own spirit into Adam at creation (Qur'an 15:29, 38:72)? And is he not closer to humankind than his [its] own jugular vein (50:16)? God's presence is all-pervasive, for to him belong the East and the West, the whole of creation . . . "and wherever you turn, there is God's face. Truly God is omnipresent, omniscient" (2:115).
>
> The Qur'an enjoins on every Muslim the practice of recollecting God (33:41), for the peaceful heart is one in which the remembrance of God has become second nature (13:28–29). The most crucial Qur'anic verse

> for Sufis, however, describes the establishment of the primordial covenant between God and the souls of men and women in a time before the creation of the cosmos: "And when your Lord took from the loins of the children of Adam their seed and made them testify about themselves (by saying), 'Am I not your Lord?' They replied, 'Yes, truly, we testify!' " (7:172).
>
> This unique event, which confirms the union between God and the souls of all men and women, has become known in the Sufi literature as the "Day of *Alast*," the day when God asked "Alastu bi-rabbikum" ("Am I not your Lord?"). The goal of every Muslim mystic is to recapture this experience of loving intimacy with the Lord of the Worlds.[15]

Martin Lings, tracing the development of the Sufi line historically, begins with the Prophet himself:

> The Traditions of the Prophet abound in mystical precepts which show that Muhammad was in fact, as the Sufis insist, the first Sufi Shaykh in all but name. "All mystic paths are barred except to him who followeth in the footsteps of the Messenger," said Junayd [d. 910], and also "This our lore is anointed with the sayings of the Messenger of God." To this day, the differences between the orders are mainly differences of selection, by the founders of the orders, from the wide range of practices offered by the Prophet's own example and recommendation. But this function of Spiritual Master was knit together with his other functions in the unity of his own person; and analogously the community under him, for all the divergence and disparity of individual gifts and tendencies, was united into one whole as it was never to be again.[16]

Let us pause to comment on these first two suggestions about the origins of Sufism in the original charter of Islam. The text from the Koran on the creation of human beings (7:172) intrigued the Sufis because it established union with God as the basis of the plain coming into being of the human species. There would be no human species had God not worked in the loins of the first parents and made them fruitful. The lordship of God goes to the center of human character, nature, quiddity. One cannot have a human being without the action of the divine Lord. This means that whenever human beings come to themselves, realize who they are, revisit the gist if not the primordial imagery of their actual situation, they find themselves united to God. Therefore, the Sufi practices of remembrance (*dhikr*) go in principle to the original moment of creation, where one can see without fail the crucial truth about human existence.

The original moment holds pride of place in most traditional cosmologies. More often than not, beginnings receive more honor than developments or conclusions because in the beginning the creative process may have been purest, least tainted by human resistance. The Sufi return to the moment of primal union with the Creator, when God drew humankind forth for the first time, reenforced the general experience of mystics that their union with ultimate reality, their finding themselves to be the products of a free grant of being from the sole unlimited source of all that is, has already to be in place. Mystical illumination discovers what has been operating since first the creature came into being. Mystical illumination does not create the dependence of the creature on the Creator that makes the creature a beneficiary of the Creator willy-nilly. The dependence, and so the inalienable tie between the creature and the Creator, is a condition for the being of the creature, its bare ability to step outside the void of nonbeing. The illumination, whether through return to the pristine first moment or other means, makes the creature more aware of what it actually is, of how God must always and everywhere be present to it and acting within it.

Second, the function of the Prophet as the first Sufi is more than a useful conceit. Muhammad became the exemplar of Koranic sanctity because everything that Koranic sanctity implies and that Muslim contemplatives labor to achieve came to him in the gift of the Koran. Intimacy with God, which he enjoyed in rich measure by God's filling him with the divine Word and bidding him to recite it for all potential believers, was the necessary condition on which his prophetship depended. Muhammad had no word to speak apart from this intimacy. The various ways of responding to his mission that Muhammad demonstrated, his different means of altering or intensifying his union with God, became paradigms for later disciples. However, apart from the utility of one or another method, the bare primacy of Muhammad as *the* Prophet assured that his union with God would be the fullest, the most beautiful, in Muslim history. Thus the Sufis were bound to make Muhammad the greatest of their teachers, the richest of their success stories, and their model.

The first four successors of Muhammad as leaders (caliphs) of the community constitute a golden age. Despite deep controversy about who ought to have succeeded the Prophet and what criteria for succession ought to have prevailed, Muslims as a whole look back on the first quartet as especially faithful to the original impulse of the Koranic revolution. Abu Bakr, a disciple from the beginnings of Muhammad's venture in preaching the revelations God was giving him, ruled for only two years (632–634), but he launched the expansion of Islam that quickly made it the predominant power in the Middle East. As well, he stands high on the Sufis' list of saints as an early hero.

The fourth caliph, Ali, was the cousin and son-in-law of Muhammad, his closest male blood relative. The principal basis of the dispute that arose at the time of the death of Muhammad was whether his successor ought to be a blood relative. Arab tradition was divided on this point, some inclination to pass tribal leadership down the family line from father to son standing against another inclination simply to choose the most gifted leader, regardless of his connections to the departed sheik. The partisans of Abu Bakr prevailed, while the partisans of Ali, who finally ruled (656–661), took considerable umbrage at his rejection. This conflict led to the split between the Sunnis and the Shiites that has characterized Islam down to the present day. Nonetheless, the Sufis have honored Ali as also one of their earliest heroes, further cementing their movement to the first generations of Islam.

Lings, trying to render some of the euphoria that the reception of the Koran as a scripture and covenant that would raise the Arabs to the status of other great peoples, writes elliptically of the impression that Sufi authors took away when they looked back on the first four caliphates:

> The Qur'an [3:110] goes so far as to say addressing Muhammad and his Companions, *Ye are the best people that hath been brought forth (as a pattern) for mankind.* The miraculousness of that community is also made clear: on more than one occasion the Qur'an mentions that it was God who had attuned, or united, the hearts of the believers; and on the height, where it had been providentially placed this *best people* was held by force long enough to be indelibly impressed with certain principles. The natural downward course of the cycle which, for that community, had been arrested by the miracle was only allowed to resume its course when Islam had been firmly established; and the sharpness with which civil discord soon set in may be considered as partly due to a cosmic reaction against an excellence that was as a violation of the nature of the age in which it had been set. The new state continued nonetheless to be governed, after the death of the Prophet, by four successive saints, and though this could not stop the steady increase of worldliness and consequent troubles, it was a thing of untold significance, unparalleled elsewhere. Moreover, the first and fourth of these caliphs, Abu Bakr the Prophet's closest friend and father-in-law, and Ali the Prophet's cousin and son-in-law, are counted by the Sufis as being amongst the greatest of their spiritual ancestors.[17]

In the mid-seventh century the extremely rapid expansion of Islam through military and cultural conquest resulted in the development of the Umayyad dynasty, which moved the headquarters of Islam from Medina to Damascus. Concomitantly, Islam became very wealthy, even cosmopolitan, putting great pressure on what had been a relatively simple desert culture to adapt to sophisticated urban conditions. The Sufis coalesced in part as a protest against the worldliness that the newfound wealth and power were generating. Their instinct was that something crucially important was in danger of being lost. (The word "sufi" may come from the word for wool, the plain stuff that these ascetics preferred to the richer cloths of the Damascenes.) Geographical loyalties (against Syria) also appear to have played a part because the first centers of the Sufi movement during the eight and ninth centuries were Iraq, Khorasan, and Egypt.

The grand old man at the head of this protest movement was Hasan of Basra (Iraq) (642–728), who moved from Medina. He preached eloquently a stern asceticism, picturing the world as a smooth, deadly snake. Indeed, Hasan went so far as to depict the creation of the world as a mistake, saying that when God looked on creation it filled him with disgust. This is a marginal position in Islam, as it is in Judaism and Christianity, but it suggests the vehemence with which the early ascetics set out to save Islam from corruption.

Softening this ascetic fierceness, but only so as to focus piety on the purest of loves, Rabi'a al-Adawiyah (ca. 717–801), the earliest and most famous female Sufi, spent her early years as a slave, a skivvy. Freed because of her manifest holiness, she strove to eliminate all self-seeking from her heart that she might love God utterly and purely. She wanted to move beyond any desire either to avoid hell or to gain the joys of heaven, concentrating her gaze only on her divine beloved, delighting only in worshipful, even slavish devotion to him. She expressed these sentiments in prayers and poems that canonized an ideal of completely emptying one's spirit of self-concern.

An amusing development in Sufi lore pitted the deep piety of Rabi'a against the angry asceticism of Hasan of Basra, much to her credit. Some commentators tried to establish an ongoing rivalry between the two, as well as a semiromantic battle of the sexes, but the fact that Hasan died in 728, while Rabi'a was born about 717, makes any actual historical interaction unlikely. The two probably stood out as such different personalities, psychologically, that the Sufi imagination began to picture what it would have been like had they fought out face to face their variant interpretations of the mystical pathway.

As the Sufis gained respect for their nourishment of sanctity and penetrating appreciations of Koranic revelation, they produced a formidable line of spiritual teachers. Rabi'a had legitimated the idea that the Muslim could love God wholeheartedly, broadening considerably the inclination of some Muslims to make Koranic religion primarily a matter of the mind and will, a submission to the Lord through most dutiful observance of the rules laid out in the Koran and acceptance of the doctrines. Among the early Sufis who plunged into the experience of mystical union and articulated what he found it to involve, the most influential voice belongs to al-Hallaj (858–922), a Persian.

An effective preacher, al-Hallaj imprudently described his mystical transports, using a language of love that identified his substance with that of God ("I am the divine Truth"). This grated on the ears of the orthodox, for whom the worst sin was idolatry. Taking what were probably cries of emotional fulfillment as though they were metaphysical statements, the orthodox could only find al-Hallaj to be proclaiming heresy. Eventually, the Abbasids (the successors to the Umayyads, who moved the caliphate to Baghdad) imprisoned al-Hallaj and executed him. On

Sufi terms, al-Hallaj was himself largely responsible for his fate because it was a Sufi principle that mystics ought not disclose their experiences to the uninitiated. Those not yet trained in the Sufi path or experienced in the contemplative life were likely to misunderstand any popular communication of esoteric matters. al-Hallaj did not agree with this elitist principle, thinking that it inhibited the spread of Sufi resources for the enrichment of the masses. He paid dearly for his disagreement.

Awn has summarized well the end that al-Hallaj met and the provocative legacy that he left:

> . . . al-Hallaj was flogged, mutilated, exposed on a gibbet, and finally decapitated. The body was then burned. For al-Hallaj, however, death was not a defeat; on the contrary, he desired fervently to become a martyr of love. Al-Hallaj was convinced that it was the duty of the religious authorities to put him to death, just as it was his duty to continue to preach aloud the unique intimacy he shared with the divine:
>
> > Kill me, my trusted friends, for in my death is my life! Death for me is in living and life for me is in dying. The obliteration of my essence is the noblest of blessings. My perdurance in human attributes, the vilest of evils.
>
> The creativity of al-Hallaj's work is reflected perhaps most strikingly in his ingenious use of the science of opposites. . . . The Qur'anic text affirms on several occasions that Iblis, who was chief of the angels and the most dedicated of monotheists, was commanded by God to bow to the newly created Adam. He refused, despite God's threat to condemn him forever, and chose, like al-Hallaj, to become a martyr of love. [In al-Hallaj's interpretation, Iblis cries out]: "My refusal is the cry, 'Holy are you!' My reason is madness for you. What is Adam, other than you? And who is Iblis to set apart one from the other?" [18]

In the text al-Hallaj has been brought to the point where he has to choose between the authority of human beings and the authority of his own mystical experiences. It is barely a contest. He has so made communion with God the passion of his life that it alone matters. Indeed, death can beckon as a way to fuller union with God, to a quicker exit from the miseries of the human condition that keep him from enjoying all that he might. Like Iblis, then, he will glory in his commitment to the utter primacy of God, the sole full reality of God, even though that commitment places him outside the pale of orthodoxy and takes off his head.

Later Sufism

What we are calling "later Sufism" is the (medieval) state that mystical Islam had reached 600 years or so after the death of the Prophet.

By this time orthodox, or "ordinary," Islam and Sufism were not so contrary as they had been in earlier times, when Sufism arose as a protest against the worldliness that expansion and conquest threatened to bring into Islam, corrupting its spiritual core. Many of the later Sufis professed themselves to be solid members of the ordinary community. In their eyes the pursuits that the sheiks directed for those who wanted a more intense faith, a keener experience of union with God, moved such people closer to where the Prophet had wanted them to live than did a life of faith without much mystical intensity. The majority of the Sufis who wanted to downplay any tensions between ordinary faith and mystical pursuits made plain their acceptance of the standard articles of faith, the requirements of the religious law, and the common ritualistic practices. They kept the fasts, gave alms, prayed five times a day, refrained from drinking alcohol, and so forth, even when their deeper interest was contemplative exercises that might help them taste for themselves the goodness of their God.

For their part, the conservatives who worried about Sufi aberrations tended to be leery of such phenomena as the proliferation of saints (holy people who could become objects of veneration and so potential idols), dramatic ceremonies such as the dances of the dervishes and the horsemanship of other Sufi groups, an interest in miracles, and the power that Sufi sheiks could garner. All these phenomena carried the danger of taking people apart from the plain faith: no god but God, only one Prophet. They could also seem to threaten the authority of the ordinary community. Finally, sometimes Sufi thinkers, speculating on the implications of mystical union with God, spoke monistically (as though no gulf separated God and his creatures).

However, the influence of devout Sufis often drew many ordinary people deeper into Islam than they might otherwise have gone. Holiness, combined with a down-to-earth teaching and worship, tends to be very attractive. Like their counterparts in Judaism (for example, the Hasidic rabbis) and Christianity (for example, the medieval saints Francis and Dominic), some of the Sufis drew large numbers of people closer to the Muslim community. In this way, their service to the general community became obvious. They could be functioning as its heart or soul or most attractive aspect. They could be giving the *Umma* more life than it would have had otherwise.

Let us illustrate the mysticism of later Sufism from texts of eminent masters of the time and then describe the "brotherhoods" that had developed, to carry the Sufi traditions forward generation after generation.

Ibn al-Arabi (1165–1240), a Spanish authority, did some of his teaching in the form of commentaries on eminent biblical figures. In their lives, as God led them, Ibn al-Arabi found *bezels* (facets of gems) of wisdom. Concerning the eminence of Moses, for example, Ibn al-Arabi wrote:

> From his birth Moses was an amalgam of many spirits and active powers, the younger person acting on the older. Do you not see how the child acts on the older person in a special way, so that the older person comes down from his position of superiority, plays and chatters with him, and opens his mind to him. Thus, he is under the child's influence without realizing it. Furthermore, the child preoccupies him with its rearing and protection, the supervision of his interests and the ensuring that nothing might cause it anxiety. All this demonstrates the action of the younger on the older by virtue of the power of his [spiritual] station, since the child's contact with his Lord is fairly recent, being a new creature. The older person, on the other hand, is more distant from that contact. One who is closer to God exerts power over him who is further from Him, just as the confidants of a king wield power over those further removed from his presence. The Apostle of God would expose himself to the rain, uncovering his head to it, saying that the rain had come fresh from its Lord. Consider, then, how majestic, sublime, and clear is our Prophet's knowledge of God. Even so, the rain had power over the best of humanity by virtue of its proximity to its Lord, like a divine emissary summoning him in his essence, in a silent way.[19]

Concerning the wisdom of Muhammad himself, Ibn al-Arabi says:

> His is the wisdom of singularity, because he is the most perfect creation of this humankind, for which reason the whole affair [of creation] begins and ends with him. He was a prophet when Adam was still between the water and the clay and he is, by his elemental makeup, the Seal of the Prophets, first of the three singular ones, since all other singulars derive from it. He was the clearest of evidence for his Lord, having been given the totality of the divine words, which are those things named by Adam, so that he was the closest of clues to his own triplicity, he became a clue to himself. Since, then, his reality was marked by primal singularity and his makeup by triplicity, he said concerning love, which is the origin of all existent being, "Three things have been made beloved to me in this world of yours," because of the triplicity inherent in him. Then he mentioned women and perfume, and added that he found solace in prayer.[20]

Ibn al-Arabi shows us that during its first 600 years or so the Sufi movement took in many esoteric influences. Here, that of a mystical Neoplatonism seems most prominent, but in other authors Gnostic, Kabbalistic, perhaps even Buddhist influences may be at work. Certainly, after some centuries it had become legitimate for Sufis to ruminate on the Koran from a variety of interpretational stances or spiritual interests. The rumination here on the relations between the younger and older Moses shows both an interest in the interactions between parents and children and an intrigue with the idea that what is youngest is freshest because closest to God. The lovely picture of

Muhammad bearing his head to the rain, like Saint Francis honoring Brother Sun or Sister Moon, tells us that Ibn al-Arabi filled his mind with pictures of the Prophet and that it pleased him immensely to imagine Muhammad fully at home in the natural world, taking all natural phenomena as gifts, messages, from his Lord.

The second passage, on the Prophet, leads us into deeper waters. Muhammad becomes a cosmic principle, the reason for creation. Scholars are not confident that they have mastered all the ins and outs of Ibn al-Arabi's dialectical view of reality, in part because many of his important texts have yet to be edited into critical form. It seems clear, however, that he developed a sophisticated metaphysics in which relationships were extremely important. The oneness of God was absolute, but this did not mean God's isolation from the world. Indeed, there could be no world without the presence of God, and one could say that, ultimately, God was the true reality of the world. At different times Ibn al-Arabi speaks of three or four apparently coeval aspects of reality or correlatives to the oneness of God, and here he works them out with reference to the Prophet. The only certainty available from the two texts appears to be that, as the perfect creature, Muhammad contains within himself the primary dialectic, the basic unity amidst difference, of creation itself.

That Ibn al-Arabi makes love the origin of all existent being suggests that by his time the Sufis had taken the emphases of Rabi'a and al-Hallaj fully to heart and that many later mystics had confirmed the primacy of passionate communion with God. That the three treasures of the Prophet were perfume, women, and prayer is in our text only a provocation, a tease to read or think on, but we can suspect that Ibn al-Arabi's symbolic approach to appreciating the Koran and the various prophets would have led him to find in this trinity a pleasing invitation to keep probing the implications of both the traditional scriptural imagery and that of the *hadith* as a way of continuing to deepen his absorption with the primary mysteries of the spiritual life.

Perhaps that is the main impression that the rich lode of Sufi masters makes: generation after generation, devout Muslims kept absorbing themselves with the primary mysteries of the spiritual life. From the Koran, the *hadith,* and the burgeoning traditions of their own lodges, they drew stimuli to live in an interior world of wonders. Because God was spiritual and not material this world stretched forth endlessly. Because the crux of spirituality is meaning, both achieved and virtual, the tasks of gaining insight, clarifying vision, and enflaming love became the first imperative driving the Sufis' quest. Exterior conditions mattered far less than interior ones. Purification, prayer, instruction from a master, selfless immersion in the traditional texts—these were the labors that were crucial.

The relationship between Sufi master and disciple tended to specify how such labors would unfold. In a rightly ordered lodge or brotherhood, disciples took instruction regularly from a holy man. The holy man gave sermons, answered questions, and, perhaps even more significantly, served as a model of the spiritual life. In the following exchange reported by the early-fourteenth-century Indian master Nizam ad-Din Awliya we hear overtones of the personalized instruction that characterized mystical Islam at its best. Master Nizam is typical in teaching by telling stories:

> [A] man once presented himself to Khwaja Ajall Shirazi—may God grant him mercy and forgiveness—and the Khwaja conferred discipleship on him. The new disciple expected Khwaja Ajall to instruct him on the invocations and prayers he ought to observe. "Whatever you do not find agreeable for yourself," declared the Khwaja, "do not wish it to happen for others; wish for yourself (only) what you also wish for others." In short, that man went away and after a while returned, presenting himself again to Khwaja Ajall Shirazi—may God have mercy upon him. "On such and such a day," he submitted to the Khwaja, "I waited upon you, hoping that you might tell me a prayer or invocation (that I could repeat), but you told me nothing. Today I am also expectant." "On that day," replied the Khwaja, "what were your instructions?" The disciple was stupified; he did not answer. The Khwaja smiled and said, "On that day I told you that whatever was not pleasing to yourself was also not pleasing to another, and (that you ought to) wish for yourself the same thing that you wish for another. You did not remember that instruction. Since you have not learned the first lesson, how can I give you another?" [21]

The novice comes with romantic expectations. The master will give him secret practices, special mantric words, that will take him by short cuts to the highest virtues; but the master takes his measure and begins with the fundamentals, the golden rule. The novice can enter upon the path that is straight by sifting out what he or she really wants, what really pleases him or her, what all people have the right to want. This apparently simple exercise is actually profound: what ought human beings to desire?

As the Khwaja phrases it, the exercise is also radically egalitarian. The disciple ought to want nothing for himself that he cannot want for others. He ought to think of the moral life, the spiritual life, as the life commonly human, the life of men and women knit together solidly by God and seen by God as more alike than different. Because the master is experienced, the inability of the novice to grasp the favor given him in the imposition of the simple lesson moves the master to smile rather than frown. This smile, however, does not remove the seriousness of the situation. The novice has not paid attention, has not listened well, was not able to rid his heart of the presumptions with which he approached the master. Until he becomes more serious, he is not good material for progress along the Sufi path. No master can make fine clothing out of poor material. Until the disciple bestirs himself, he will make no significant progress.

In the wake of this story Nizam elaborates more positively what is important in the spiritual life. If no particular practices (prayers, fastings) are imperative (as beginning with the golden rule suggests), then what is imperative, or the heart of the religious matter? For Nizam, the answer is the long-standing Sufi conviction that to make progress toward God, one has to renounce the world:

> After that the master told the following story about a certain chaste saint. Many times he used to say that all virtuous deeds, such as prayers, fasting, invocations, and saying the rosary are a cauldron, but the basic staple in the cauldron is meat: Without meat you do not experience any of these virtuous deeds. Finally they asked that *pir* [elder]: "Many times you have used that analogy, but now explain it." "Meat," replied the saint, "is *Renouncing Worldliness,* while prayer, fasting, invocations, as well as repetition of the rosary—all such virtuous deeds presuppose that the one who does them has left the world and is no longer attached to any worldly thing. Whether he observes or does not observe prayers, invocations, and other practices, there is no cause for fear, but if friendship with the world lingers in his heart, he derives no benefit from supplications, invocations, and the like." After that the master observed. "If one puts ghee, pepper, garlic, and onion into a cauldron and adds only water, the end result is known as pseudo-stew. The basic staple for stew is meat; there may or may not be other ingredients. Similarly, the basis for spiritual progress is leaving the world; there may or may not be other virtuous practices." [22]

This is a marvelous text, cutting to the bone. All true masters know the difference between what is substantial and what is accidental. From this knowledge, they can adapt the path to the needs of given disciples. Without this knowledge, any religious group tends to canonize nonessentials. The formulas, routines, traditions that have grown up, usually without full design, become sacrosanct. It should not be so, Nizam says. Virtually all of our spiritual exercises ought to lie in the order of means, ways to the end. We should use them in the measure that they conduce to the end, and we should lay them down in the measure that they do not conduce. (Note that the master does not take up the question of rules or obligations imposed by the Koran. Presumably, good Muslims perform such obligations without question, and such obligations do conduce to the end of the spiritual life—breaking with the world and loving God.)

Islam is not a religion that denies the reality of the material world. It is not driven by a hatred of given existence, even by a Buddhistic sadness over *samsara.* However, its Sufis are at war with worldliness. From their historical beginnings, as we have seen, they feared that worldly prosperity would pervert the sublime message delivered through the Prophet. The will of God and the intimacy with God become available by submitting to that will. These were the great treasures. To live in the wide-open meaning and beauty of God was much more desirable to large Sufi hearts than any gains that political, military, or even cultural prosperity could entail. There is only one God. There is only one ultimate reality that can fill the human heart. To gain this, one must renounce all pretenders, idols, competitors. That is the heart of the Sufi conviction, the wellspring of Sufi asceticism. Only when the world has retracted in significance and been reduced to its properly subordinate value can the mystical regimes flourish.

Ibn Abbad of Ronda (1332–1390), a Moroccan master, frequently gave spiritual direction through letters. The following counsel to a man suffering trials amounts to a Muslim version of the common mystical call to reject all worldly props and offer God *carte blanche:*

> One of the mystics has said, "That which you worship is the first thought that comes to your mind when you are suffering anxiety." Another, commenting on the words of God Most High: "One who is in need may be sure of an answer when he calls upon Him" (27:62), said, "The needy person is the one who enters his Master's presence with his hands raised in supplications, envisioning no particular gift from God as though he had a claim to it, and says, 'My Master, give me whatever you have for me.' That is a needy person, even though he attains in this state the privilege of nearness to God and the special gift of love. Since one is able to profit thus even from indigence, the perplexity one experiences pales in significance.
>
> "If the answer to your prayer is not apparent to you and you are not steadfast in repentance while you continue in the state I have just described, namely, being in need of God's gifts and not relying on other means, then one of two things will happen. Either you will go to pieces with anxiety and be severely agitated, or you will be patient and resigned. In your case there is no reason why you should come undone with agitation. You have already become secure and have arrived at the Mystic Truth of faith, so you have no reason to fear that you will come to such a pass that you must choose between the two states. Therefore all that remains is patience and resignation. And in this state even the lack of a response is a response." [23]

Here we find sketched the inviolability of a personal faith in God that experience has made stable. In the measure that a person has actually found God to be different from the world, not variable like the world, the chance arises that nothing will make that person lose balance, capsize through anxiety. Prudent people never assume that they are invulnerable. The saints are unanimous in rushing to say that they stand only by the grace of their Lord, but the more fully devout people have thrown in their lot with God and surrendered their will to him, the less any reversal, rebuff, or ill fortune can shake their foundations. Their foundations are not what they want or propose. Their foundations are what God chooses for them, how God disposes their lives. If the disposition of God leads to their financial or social prosperity, so be it. Equally, however, if the disposition of God leads to their suffering, so be it. The operative conviction is that God is providential, guiding their lives for their own good. They do better to leave their lives in God's hands than to try to steer for themselves. They do better to trust in the goodness of God and in God's love for them than in their own shrewdness.

At the end of such a trust lies the paradox that even the lack of a response is a response. For the patient, those willing to suffer God's sending their days to them as he sees best, whatever God sends is what they need, for their good. If they are ill and seek a cure, a cure is a sign of God's care; but the lack of a cure, even the deepening of their illness, is not a sign that God does not care. The lack of a cure, for the person living by faith, is simply what God, who always remains all good, has decided thus far to parcel out. In the mysteries of God's knowing their hearts to the most intimate nook and cranny, they find the solace of thinking that this nonanswer, this demand that they keep suffering illness or spiritual darkness or vilification, is for their good.

Ultimately, there is no higher union with God than wanting only what God wants, being only as God desires one to be. The abnegation entailed is the summit of renouncing worldliness. The essence of worldliness is self-assertion, acting as though God were not the sole Creator, throwing one's weight or desire around to make a world, a paradise, apart from God, out of creatures. When a person turns over to God all that happens, good or bad, becoming indifferent to all worldly implications, the peace that surpasses worldly understanding can slip into his or her heart, protecting against all anxieties. "Let nothing disturb you," Teresa of Avila said. Two centuries before her, Ibn Abbad of Ronda had said virtually the same.

We have alluded to the social context of most Sufi mystical ventures, the brotherhood or lodge where beginners could take instruction and a community of like-minded believers would offer one another support. From the beginning, when they started to gauge the requirements of the intense spiritual life, devout Muslims intuited the value of companionship (*suhba*). Under the direction of a sheik, a religious master, disciples would gather together. Gradually, lines of connection developed between different lodges, family relationships. When groups became stable, extending over generations, they came to think of themselves as "ways," "fraternities," "orders" (*silsilah*). By the fourteenth century, such fraternities had been exerting great influence for hundreds of years, waxing and waning in purity, much as medieval Christian monastic groups had.

The Sufis were not monks, however. Most of them were married. Although they came to the lodge regularly, they tended to resemble the disciples of a Hasidic rebbe in having to provide for a family and live in a village community nearby. On the whole, the lodges included only men, though in Islam women have explored numerous paths to interior perfection. Because most women married and raised children, the solitary ways of a Rabi'a did not provide a useful blueprint. Better were the traditions about the outstanding women of Muhammad's generation, especially his wife Khadija and his daughter Fatima. As ardent disciples, they became the models for how women ought to serve and pray. A bevy of female saints, holy women of legendary prowess, lent an aura of sanctity and helpfulness at many local shrines, but on the whole female sanctity in Islam developed in the house, the domestic circle, more than in the mosque or any Sufi sisterhood.[24]

In the history of the Sufi brotherhoods, two of the most influential figures are al-Junayd (ca. 825–910), an Iraqi Sunni master, and Jalal al-Din Rumi (1207–1273), a Persian. We conclude this section by noting briefly the contributions of each.

Al-Junayd is known for a sober, restrained cast of mind. He is not a Sufi of emotional, enthusiastic, or ecstatic temperament. Indeed, he refused to accept al-Hallaj as a disciple. Quite sober, as well, is his simple teaching:

> The central point of al-Junayd's teaching is the doctrine of *tawhid* (unification), which he defines as "the isolation of the eternal . . . from the contingent." According to al-Junayd, the soul, shorn of its attachments to the world, returns to the state in which it existed prior to entering the physical body. This is the state in which the soul made a covenant . . . with God by answering "yes" to his question, "Am I not your Lord?" (surah 7:171). Thus, in a state of unification, the pristine soul is reunited with the divine. The last state of the Sufi becomes the first.[25]

This interpretation of the mystical quest is gnostic in the sense that it depends on a knowledge from out of time, a revelation of how things were at the primal moment (the *in illo tempore* [at that time] with which many aetiological myths begin), when God set the patterns that would obtain ever after. The assumption is that the normal human life follows a curve, descending from heaven, passing (sufferingly) through earthly times and trials, and ascending back to heaven. The Sufi can cut across this curve, shorten the phase of earthly trials, by returning in spirit to the primal moment when union with the Creator pulsed powerfully. To do this, the prime requisite is distinguishing the eternal, the one, the reality of the divine Creator from the contingent, the many, the created. When the spirit clings to the one, it is back at its beginning, reset in the relationship that makes it most itself.

Rumi, also known as Mawlana, was a great poet, probably the foremost Persian stylist. He became a Sufi through a dramatic, ecstatic experience that enflamed his poetry for the rest of his life:

In October 1244 he met the wandering dervish ["beggar," synonym for Sufi] Shams al-Din, "Sun of Religion," of Tabriz, and, if we believe the sources, the two mystics spent days and weeks together without eating, drinking, or experiencing any bodily needs. The discussions of Rumi and Shams, who must have been about the same age, led Jalal al-Din into the depths of mystical love but also caused anger and jealousy among his students and his family. Shams left Konya, and in the pangs of separation, Mawlana suddenly turned into a poet who sang of his love and longing while whirling around to the sound of music. He himself could not understand the secret of this transformation and expressed his feelings in ever-new verses, declaring that it was the spirit of the beloved that made him sing, not his own will.[26]

Rumi is always concerned with love, which becomes the medium for communion with God:

Love is personified under different guises—Rumi sees it as a police officer who enacts confiscation of man's goods or as a carpenter who builds a ladder to heaven, as a ragpicker who carries away everything old from the house of the heart, or as a loving mother, as a dragon or a unicorn, as an ocean of fire or a white falcon, to mention only a few of the images of this strongest power of life. God's preeternal address to the not-yet-created souls, "Alastu bi-rabbikum" ("Am I not your Lord?" Qur'an 7:171), is interpreted as the first music, which caused creation to dance out of not-being and to unfold in flowers, trees, and stars. Everything created participates in the eternal dance, of which the Mevlevi [Turkish for Mawlawi, the order that Rumi/Mawlana founded] ritual is only a "branch." In this ritual, the true mystery of love, namely "to die before dying," of sacrificing onself in order to acquire a new spiritual life, is symbolized by the dervishes casting off their black gowns to emerge in their white dancing dresses, symbols of the luminous "body of resurrection." For the idea of suffering and dying for the sake of transformation permeates all of Rumi's work, and he expresses it in ever-new images . . . for the heart must be broken in order to find in itself the "hidden treasure" which is God.[27]

The Sufi orders, lodges, or brotherhoods that claimed descent from eminent founders such as Rumi have exerted a significant influence in many parts of the Muslim world. From the thirteenth century, they proliferated, developing distinctive, individualizing characteristics. Peter Awn has followed this development with interest.[28] As he sees it, the Mawlawiyah, for instance, have taken from Rumi, and his son Sultan Walad, a distinctive interest in aesthetics. They have used the great poetic work of Rumi, the *Mathnavi,* as a privileged text, their characteristic way of reading the Koran (on which Rumi comments constantly). They have also developed a distinctive ritual combining music, poetry, and whirling (dervish) dancing. Quite a different group, the Bektashiyah, derives from an obscure Khorasan holy man of the fourteenth century, Hajji Bektash, and would never be confused with the Mawlawiyah. Analysts tend to find in Bektashiyah religious life elements of Byzantine Christianity and esoteric philosophy joined with Shiite Islamic convictions. The result is a spirituality less immediately recognizable as straightforwardly Muslim. Because the Janissaries, soldiers of Ottoman emperors from the time of Sultan Murad I (fourteenth century), took on Bektashiyah ideas and practices, this order enjoyed safe passage in the Ottoman Empire. The magical component in this spirituality has made it attractive to many uneducated layfolk, while its associations with Ottoman power could make the lodge much feared.

Medieval Iraq sponsored the Suhrawardiyah and Rifa'iyah brotherhoods. Characteristically the former has provided disciples with clear teaching and guidance along the spiritual path, and it has had considerable influence among Indian Muslims. The Rifa'iyah, somewhat in contrast, has been known as "howling dervishes" and is connected with dramatic religious rituals. In such rituals ecstatic disciples might eat fire, pierce their own flesh, bite off the heads of snakes, and perform other prodigies, none of them calculated to assuage the fears of sober guardians of Koranic tradition already leery of Sufi aberrations.

In Egypt the Shadhili order prospered under the leadership of the third in its line of sheiks, Ibn Ata Allah, whose views owed much to Ibn al-Arabi. Ibn Ata Allah strove to make Sufi life easier for laypeople to practice, downplaying the requirements of some other lodges that members largely quit the world. Still another Sufi order, the Qadiriyah, has had one of the largest memberships and owed much of its popularity to the aura of its saintly founder, Abd al-Wadir Jilani (d. 1116). The Naqshbandiyah, prominent in Central Asia and India, provides a last example of the many other Sufi orders that a full history would describe. The general point in such a history would be that these orders not only influenced the large numbers of people who enrolled in them, seeking spiritual advancement, they also influenced Islam as a whole, keeping up a steady pressure to avoid worldliness and pursue intimacy with God. The situation has been analogous to the impact of monasticism in Christianity. Even when Christian monastic groups did not enroll large numbers of disciples, they symbolized for all with eyes to see the possibilities for an intense religious life.

The Modern Period

In this section we bring the story of Islamic mysticism from the medieval period of six to seven centuries after the death of the Prophet into the twentieth century. By the time of the death of al-Ghazali (1111), for example, medieval Islam, including Sufism, had achieved a full synthesis of law and culture, thought and practice. The program given by God through the Prophet had expanded into a variety of cultural settings: Syrian, Iraqui, Persian, North African, Turkish. Muslims had entered Spain in 711 and the Indus Valley in 713. In Egypt and Jerusalem the Crusades, an entire panoply of adventure, war, and cultural development, had unfolded, coloring the Sufi lodges.

The later centuries were not so fortunate for Islam. The Mongols sacked Baghdad in 1258. In 1492 the Spaniards ousted Muslims along with Jews. By 1707 Islamic (Mughal) rule had declined in northern India. From the early nineteenth century a strict, perhaps even puritanical, Islam (*Wahhabism*) had taken hold in Mecca and Medina. Turkey adopted a secular constitution in 1924, a sign of the struggle of Muslim intellectuals to forge an accommodation with Western political and scientific thought. The violence in India after independence from the British Empire (1948) led to the creation of Pakistan, a Muslim entity nearly bound to suffer conflicts with the (officially secular but actually) Hindu India that remained. The rise of the modern state of Israel in 1948, at the expense of largely Muslim Palestinians, turned out to make Middle Eastern politics furious throughout the second half of the twentieth century. The development of Muslim fundamentalism over the same period has produced a major confrontation between tradition and modernity, pitting such values as reverence for the Prophet against artistic freedom in the affair of Salman Rushdie.

Modernity, in the sense of a knowing commitment to personal autonomy, the primacy of individual conscience, democratic politics, and an essentially secular worldview dominated by physical science, has yet to shape the majority of the world's Muslims. Certainly, until well into the twentieth century the majority of Muslims lived a traditional life, where *Sharia* and Sufi values prevailed. Piety continued to be much as it had been in medieval times, so we need only reattune our ear, by attending to a few recent voices, to find our main point.

Our main point is that from its unvarying foundation in the Koran and the traditions about the Prophet, "modern" Islam continued to challenge people to a mystical goal: seeking God's face. The main sharpening of this point, giving the modern era its distinctiveness, has been the slow realization among Muslim intellectuals that Western developments in science and politics enfold a spiritual, indeed a religious, challenge. On the one hand, they offer new vistas of creativity. On the other hand, they carry the threat of dissolving, or at least bringing lesions to, the bone and marrow of Islamic tradition. Postmodern mysticism in Islam, as well as elsewhere, may prove to be orthopedic—working to strengthen the bones that they may carry healthy flesh for new generations.

The cultural anthropologist Clifford Geertz has done wonders to make vivid how Islamic religious practice actually works. In contrasting the operations of Muslim piety in Morocco with those in Indonesia, Geertz allows us to sense the enormous impact of a given cultural setting—history, art, language, mores. Consider the following study of a Moroccan saint (*marabout*) as an example of how Muslim mysticism developed in North Africa during our modern period.

Ali al-Hassan ben Mas'ud al-Yusi, better known as Sidi Lahsen Lyusi, was a seventeenth-century Berber holy man (b. 1631). The Moroccan ideal that formed his piety was tough—the fierce will to survive in the harsh conditions of the desert. Lyusi earned his stripes as an apprentice *marabout* by nursing an aged saint whose linen had become repulsive from disease (Lyusi boiled the linen and drank the water as tea). Confirmed in the spiritual power (*baraka*) of a holy person to deal with disease and other evils without suffering harm, Lyusi became a prophetic moral force, chastising unjust Moroccan rulers. On one occasion he stood up for workers being treated unjustly, earning the wrath of the local sultan.

The sultan had entertained Lyusi hospitably (a significant honor in the Middle East), only to find the saint agitating against him outside the palace gates. When confronted, Lyusi said, maddeningly, "Tell him . . . I have left your city and I have entered God's." Geertz continues the story and draws some cultural conclusions:

> Hearing this, the Sultan was enraged and came riding out himself on his horse to the graveyard, where he found the saint praying. Interrupting him, a sacrilege in itself, he called out to him, "Why have you not left my city as I ordered?" And Lyusi replied, "I went out of your city and am in the city of God, the Great and the Holy." Now wild with fury, the Sultan advanced to attack the saint and kill him. But Lyusi took his lance and drew a line on the ground, and when the Sultan rode across it the legs of his horse began to sink slowly into the earth. Frightened, Mulay Ismail [the sultan] began to plead to God, and he said to Lyusi, "God has reformed me! God has reformed me! I am sorry! Give me pardon!" . . .
>
> The whole process, the social and cultural stabilization of Moroccan maraboutism, is usually referred to under the rubric of "Sufism," but like its most common gloss in English, "mysticism," this term suggests a specificity of belief and practice which dissolves when one looks at the range of phenomena to which it is actually applied. Sufism has been less a definite standpoint in Islam, a distinct concept of religiousness like Methodism or Swedenborgianism, than a diffuse expression of that necessity . . . for a world religion to come to terms with a variety of mentalities, a multiplicity of local forms of faith, and yet maintain the essence of its own identity. Despite the otherworldly ideas and activities so often associated with it, Sufism, as a historical reality, consists of a series of different and even contradictory experiments, most of them occurring between the

> ninth and nineteenth centuries, in bringing orthodox Islam (itself no seamless unity) into effective relationship with the world, rendering it accessible to its adherents and its adherents accessible to it. In the Middle East, this seems mainly to have meant reconciling Arabian pantheism with Koranic legalism; in Indonesia, restating Indian illuminationism [concern for enlightenment, on Hindu or Buddhist grounds] in Arabic phrases; in West Africa, defining sacrifice, possession, exorcism, and curing as Muslim rituals. In Morocco, it meant fusing the genealogical conception of sanctity [religious tradition passing down a line of blood or spiritual ancestry] with the miraculous—canonizing *les hommes fétiches* [the *marabouts,* those obsessed with *baraka*].[29]

The stories with which Geertz works to develop an understanding of Indonesian Muslim culture are as illuminating as his stories of Lyusi and Morocco. They suggest how saints have functioned as distinct paradigms of cultural prowess (one would not confuse an Indonesian Muslim saint with a Moroccan, the styles are nearly opposite), and how "saintliness" has woven its way through numerous traditions and institutions—politics, economics, sexual mores, warfare. Detached as some Sufi masters might be, the venture of mystical Islam attached itself to village squares, markets, schools, and mosques, like a growth of willful barnacles. There was no mystical Islam, in modernity or any other historical period, that floated unattached from history, prescinded long or well from actual human flesh.

More than Geertz may indicate, however, there was also no profound Sufi presence that did not confess the one God to be much more than the world, human culture, or any of the "many" characterizing creation. Sufi masters were philosophers of the divine One more than anthropologists of the cultural many. They might admit, even love, the great variety of ways in which God dealt with human beings, beckoning them to live well, but the Sufi masters loved even more the transcendence of God, the unlimited horizon, that offered freedom from the crush of this-worldliness. Only on the far side of the buzz and blitz of village life, city life, cultural here-and-nowness could the mystic find the full resonances of the divine Word. Only in the desert, interior more than exterior, could the Koran be fully the Koran.

In 1920 a French physician, Marcel Carret, began treating an Algerian holy man, Sheik Ahmad al-Alawi, head of a local Sufi lodge. Carret brought to the encounter the then modern mentality of a Western scientist. The sheik, though aware of modern cultural trends, stood firmly in centuries-old Sufi traditions. Gradually, Carret came to realize that this man enjoyed a profound interior life. Like a Kabbalist, the sheik would probe the numerical significance of various Koranic verses, and regularly he gave instruction to disciples, counseling them on how to develop a life of mystical piety:

> On several visits [to treat the sheik] Carret heard cries of disciples at prayer in the compound where the sheik presided. The sheik's own appearance, ascetical and peaceful, had struck the doctor as a nearly classical reproduction of the face of Christ. When asked about the cries of the disciples, the sheik replied that they were requests to God for help in meditation. The purpose of such meditation, he said, was "self-realization in God," and even though few achieved this goal, most obtained enough inward peace to justify their labors. The sheik himself, however, seemed to enjoy direct communion with God: mystical intimacy. The doctor, by contrast, would never be a Sufi [the sheik implied] because he lacked the desire to raise his spirit above himself—a fatal flaw.[30]

A Western secularist of the 1920s might have replied to the sheik that there is nothing above the human spirit, nothing to which to raise oneself. In those days, science remained mechanistic, the intricacies of the natural world had yet to display themselves as they would in the medicine of seventy years later. Yet the key word in the exchange probably would not have been "raise" but "desire." Give the sheik passion with which to work, and he might show you how your longing presses forward to extend you, even to raise your horizons beyond the material world.

The point is that in Algeria in the 1920s Sufi traditions continued to thrive. Holy men continued to probe the Koran and the long-standing mystical practices, convinced as their medieval predecessors had been that God was a treasure both beyond compare and (for those who wanted it badly enough) merciful to the suffering.

To conclude, we note that, writing of the foundations of Islamic spirituality, the contemporary Shiite authority Seyyed Hossein Nasr has stressed the unique influence of the Koran, implying that at all times, in all places, through all of the historical permutations of Muslim mysticism, the Koran has been the guiding beacon:

> The Quran is the source of not only the Law but also the Way or the *Tariqah*. The spiritual life of Islam as it was to crystallize later in the Sufi orders goes back to the Prophet, who is the source of the spiritual virtues found in the Muslim soul. But the soul of the Prophet was itself illuminated by the Light of God as revealed in the Quran, so that quite justly one must consider the Quranic revelation as the origin of Sufism. It is not accidental that over the ages the Sufis have been the foremost expositors and commentators upon the Quran and that some of the greatest works of Sufism, such as the *Mathnawi* of Jalal al-Din Rumi are in reality a commentary upon the Sacred Text. Only the Sufis have, in fact, been able to cast aside the veil of this celestial bride, which is the Quran, in order to reveal some of her beauty, which it hides from the eyes of those who are strangers to her.
>
> Even Islamic art, which may be called "the second revelation" of Islam, is rooted in the Quran, not in its outward form or as a result of applying

> explicit instructions contained in the Text but in its inner reality. Without the Quran there would have been no Islamic art. The rhythm created in the soul of the Muslim, his predilection for "abstract" expressions of the truth, the constant awareness of the archetypal world as the source of all earthly forms, and the consciousness of the fragility of the world and the permanence of the Spirit have been brought into being by the Quran in the mind and soul of those men and women who have created the works of Islamic art. Islamic art is the crystallization of the inner reality of the Quran and the imprint of this reality on the soul of the Prophet and, through him, on the soul of Muslims.[31]

In the past two centuries, as Islam has confronted the modernity that developed in the West after the eighteenth-century Enlightenment, the Sufis have come under new attacks. The Muslims who wanted to embrace European political or scientific ideals tended to find the ways of Sufi sheiks inimical to their desire. The sheiks often functioned as minor princes, full of political power that the reformers found an obstacle to social progress, while Sufi views could oppose faith to scientific reason. Thus the secularist program pushed through by Atatürk in Turkey in 1925 included the abolition of the Sufi orders.

On another front, strict reformers, such as the Wahhabis in Saudi Arabia, could consider the Sufis un-Koranic and so oppose their religious programs. If the standard for orthodox Islam were admitting nothing not based in the Koranic text, then numerous Sufi practices would become suspect. Suspect as well could be the Sufi goal of gaining union with God. Did that not smack of idolatry, an effort to bridge the unbridgeable chasm between Creator and creature? Moreover, the weaving of Sufism through the sometimes vulgar folk religion that developed at the tombs of Muslim saints tended to besmirch Sufism in the eyes of those appalled by such folk religion. So the recent centuries have been difficult for the Sufis: neither those wanting to embrace modernization nor those wanting a return to strictly Koranic religion have thought of the Sufis as allies.

Nonetheless, Sufi traditions have continued to be influential in some areas, such as India and North Africa. Indeed, wherever folk Islam has remained vital (most of the Muslim world), Sufism has continued to shape many features of popular religion. Both Muslim and non-Muslim students of mysticism have had to deal with Sufism, perforce, but the recent revivals of fundamentalist Islam have not made Sufism popular, all the more so when they have taken on militaristic political agendas. Sufis have generally tolerated the world more than they have loved it or thought that worldly power was significant. Thus it is in poetry, folk ritual, music, mythology, and other often nonpolitical, indeed nondoctrinal, aspects of Muslim culture that the Sufis have had their greatest influence recently.

Summary

We have offered an overview of Muslim history and doctrine, spent considerable time on the foundations of Muslim piety in the life of the Prophet and the sacred text of the Koran, and traced historically some of the developments in Islamic mysticism, largely by following the course of the Sufis. If nothing else, these efforts should impress us with the richness of Muslim spirituality. The great numbers of people influenced by Islam and the wide diversity of cultures to which it has transplanted itself have ensured that the variations on the basic themes would become almost endless.

Studying such variations, cultural anthropologists such as Clifford Geertz have described what we might call the "humanizing" of Islam, its taking flesh among given people at given times. Accepting all this, let us make it our purpose in the rest of this summary to suggest the simplifying counterpoint to cultural diversity that Koranic religion (taken as the foundational influence that Nasr has described) has provided. Our thesis will be that in the measure that a given cultural period or zone entered into mutual influence with the Koranic text the transcendent aspects of the Koranic message challenged leading local thinkers to explore the depths of their situation, indeed the depths of the human condition itself, in search of the sole Lord. This would explain how and why mystical Islam has thriven in so many different times and locales and where observers ought to look for the unity of Muslim mysticism. A subordinate claim, under this main thesis, is that the Sufi masters have been the principal carriers of the call to counterpoint, the challenge to get to the depths where the sole Lord dwells.

Muhammad was acutely conscious that the revelations God was giving him implied the recreation of the Arab people. Those revelations demanded a retreat from polytheism into the awesome mystery of the one Lord of the Worlds. They also established momentum for a new Arab identity, one based on common submission to God rather than on long-standing tribal affiliations. These re-creative energies led the Prophet forward toward what we have called the "humanization" of Islam, the incarnation of Koranic piety. People could discover in the lyrical texts a holy ultimacy closer than the pulse at their carotid. They could also discover strong demands that they care for the widow and orphan, that they join with their fellow Arabs to advance the path that was straight. Bowing to God and confessing that there is no god but God was not the enemy, the antagonist, of the movement for pan-Arab unity. It was the trigger and the fuel and the explosion. What Muhammad created through his rule of the *Umma* gave social form, public body, to the Lordship that all Muslims were to acknowledge. The house of Islam was built on the rock of the sole lordship of God. No catalogue of historical, geographical, or cultural differences among Muslims collected more of the reality of Islam than was available in the simple creed, the *Shahada*.

There are many ways of arguing for this thesis, but probably the simplest and most direct is to ask whether a dynasty or culture or religious school could ever meaningfully have called itself Muslim without ascribing to the *Shahada*. The answer is that none ever could have, any more than any group could ever meaningfully have called itself Christian without following Jesus or any group could ever meaningfully have called itself Jewish without embracing the Torah. We require certain elements if we are to say that a group or a proposition or a practice meets a given definition. Those elements then bulk large because they serve as our principal symbols for the heart of the matter denoted in the definition, the core of the reality described.

When it comes to keeping alive the heart of a religious matter, practice is a crucial factor. Those who practice the faith and morals of the religion year after year keep it vital. Certainly, they shape the definitional elements, giving them colors and smells peculiar to particular beaches and cul de sacs; but the definitional elements shape the practitioners even more, making Muslims of people who otherwise would be only Pakistanis of Lahore or Iraqis of Baghdad.

The Sufis were the great practitioners of Muslim spirituality. The preponderance of ardent efforts to realize the heart of the Muslim religious matter, to find the oneness of God, and to verify the beauty of the prophetship of Muhammad came from their ranks. Therefore, without denying the considerable diversity of the Sufi traditions, we submit that Sufism has provided Islam considerable unity. Shortly into their spiritual exercises, even Sufi novices, such as those of the Algerian sheik al-Alawi, realized that they were setting out for self-realization in God. They were not striving for wealth, power, or eminence in their local cultures. They were striving for something radically and universally human: self-realization, mystical fulfillment, in ultimate reality—the one Lordship of God. It is hard to overestimate the impact of this orientation. It is hard to imagine what Islam would have been without it. The unity of the Koranic God has stamped Islam through and through. There is no god but God, and there is no Islam, common or mystical, without God as the sole center, the only Lord before whom to bow.[32]

NOTES

1. Saadia Khawar Khan Chishti, "Female Spirituality in Islam," in *Islamic Spirituality: Foundations*, ed. Seyyed Hossein Nasr (New York: Crossroad, 1987), 205.

2. Geoffrey Parrinder, *A Dictionary of Non-Christian Religions* (Philadelphia: Westminster, 1971), 241.

3. Muzammil H. Siddiqi, "Zakat," in *The Encyclopedia of Religion*, ed. Mircea Eliade (New York: Macmillan, 1987), 15:550.

4. A. J. Arberry, *The Koran Interpreted* (New York: Macmillan, 1956). This is the most influential English version, useful for its efforts to render poetically the lyrical Arabic original.

5. Alfred Guillaume, *Islam* (Baltimore: Penguin, 1956), 70. Other serviceable overviews of Islam include Frederick Mathewson Denny, *An Introduction to Islam* (New York: Macmillan, 1985), and Fazlur Rahman, "Islam: An Overview," in *Encyclopedia of Religion*, 7:303–322.

6. See Victor and Edith Turner, *Image and Pilgrimage in Christian Culture* (Ithaca, N.Y.: Cornell University Press, 1978).

7. Arberry, *Koran Interpreted*, 29.

8. See Parrinder, *Dictionary of Non-Christian Religions*, 255.

9. Michael Cook, *Muhammad* (New York: Oxford University Press, 1983), 85–86.

10. Denise Lardner Carmody and John Tully Carmody, *In the Path of the Masters* (New York: Paragon, 1994), 140–141.

11. Mohammed Marmaduke Pickthall, *The Meaning of the Glorious Koran* (New York: Mentor, 1953), 256.

12. W. Montgomery Watt, "Muhammad," in *Encyclopedia of Religion*, 10: 145–146; see also Chishti, "Female Spirituality in Islam," 48–110.

13. Pickthall, *Meaning of the Glorious Koran*, 435.

14. See Mahmoud M. Ayoub, "Qur'an: Its Role in Muslim Piety," in *Encyclopedia of Religion*, 12:176–179.

15. Peter J. Awn, "Sufism," in *Encyclopedia of Religion*, 14:105.

16. Martin Lings, *What is Sufism?* (Berkeley: University of California Press, 1977), 101.

17. Ibid., 100–101.

18. Awn, "Sufism," 108–109. See also R. C. Zaehner, *Hindu and Muslim Mysticism* (New York: Schocken Books, 1969), 110–161.

19. Ibn al-Arabi, *The Bezels of Wisdom* (New York: Paulist, 1980), 252.

20. Ibid., 272.

21. Nizam ad-Din Awliya, *Morals for the Heart* (New York: Paulist, 1992), 88.

22. Ibid.

23. Ibn Abbad of Ronda, *Letters on the Sufi Path* (New York: Paulist, 1986), 72.

24. See Chishti, "Female Spirituality in Islam."

25. Muhammad Abdur Rabb, "Junayd, al," in *Encyclopedia of Religion*, 8:209.

26. Annemarie Schimmel, "Rumi, Jalal al-Din," in *Encyclopedia of Religion*, 12:482–483.

27. Ibid., 484. On the overall history of the Sufi orders, see Seyyed Hossein Nasr, ed., *Islamic Spirituality: Manifestations* (New York: Crossroad, 1991).

28. Awn, "Sufism."

29. Clifford Geertz, *Islam Observed* (Chicago: University of Chicago Press, 1968), 34, 48.

30. Denise Lardner Carmody and John Tully Carmody, *The Story of World Religions* (Mountain View, Calif.: Mayfield, 1988), 250. This is based on Martin Lings, *A Sufi Saint of the Twentieth Century* (Berkeley: University of California Press, 1973).

31. Seyyed Hossein Nasr, "The Quran as the Foundation of Islamic Spirituality," in *Islamic Spirituality: Foundations,* ed. Nasr, 8–9.

32. On the ecumenical situation regarding Christian and Muslim understandings of monotheism, see Annemarie Schimmel and Abdoldjavad Falaturi, eds., *We Believe in One God* (New York: Crossroad, 1979).

8

Mysticism Among Oral Peoples

General Orientation

We have considered the mystical traditions of five world religions, as students of humanity's interactions with the divine tend to describe the large, multicultural religious complexes: Hinduism, Buddhism, Judaism, Christianity, and Islam. In addition, we have studied the characteristics of the direct experiences of ultimate reality that seem to have been most important in East Asia among Chinese and Japanese people influenced by several points of view: Confucianism, Taoism, Buddhism, and (for Japan) Shinto. None of our considerations could be exhaustive. All did well if they suggested the main features of the intense spiritual life that developed in the tradition or cultural zone in question. Inasmuch as our focus has been mysticism and the world religions, however, the bulk of our work is done.

In this chapter we deal with several traditions or cultural groups that historically have not have scriptures, literate sources of guidance, or (without qualification) a linear sense of history. Specifically, we consider briefly the religious experiences that seem to have been most important to Native Americans, Aboriginal Australians, and Africans. Whether such religious experiences represent an extension of the oldest spiritual quests of humankind, in effect offering for our study a continuation of how the first true human beings tried to make their way to solid meaning, no one can say. Perhaps from the time that they gained reflective reason human beings did seek ecstatic communication with the powers they sensed were directing their lives.

Certainly, the oral peoples we meet in more recent times have prized such communication. However, the tens of thousands of years lying between, for example, the prehistoric people who created the famous art on the walls of caves in France and Spain and the relatively recent oral peoples whom European colonizers met in the Americas, Australia, and Africa suggest that we must proceed cautiously with any comparisons. We simply do not know in any detail the lines through which relatively recent oral peoples have developed their religious cultures or how much has changed since the beginnings of their histories.

Still, it is interesting to speculate that many oral peoples have been shamanic. As *yoga* has played a large role in the development of the mystical outlooks that we can trace to India and prophecy has been significant in the biblical spiritualities of Judaism, Christianity, and Islam, so shamanism has characterized many oral tribes as the bent or gift or calling of many oral religious practitioners. Certainly, many gifted oral religious leaders appear not to fit a pure typology of shamanism, any more than their counterparts in Asian religions fit a pure typology of yoga. The biblical religious figures have been priests and sages and healers, as well as prophets. The oral religious figures have offered sacrifices, divined dreams, regulated rituals; but at the core of the spiritual lives of many native North American, Latin American, Australian, and African groups we find an interest in ecstasy, traveling out of "ordinary" experience, moving into a further world of spirits, gods, and numinous forces.

Mircea Eliade's influential monograph on shamanism reveals both the wide extension of ecstatic desire that the history of the religions reveals and the problems that using the term "shamanism" to cover this width raises. Eliade is most interested in what he calls "archaic techniques of ecstasy," stressing the age-old, certainly prehistoric roots of the shaman's enterprise and skills. When he wants to become precise, however, Eliade has to ground shamanism in the particular central Asian traditions from which observers have derived its name and to which many comparative studies of shamanism refer as the prime analogue:

> Shamanism in the strict sense is pre-eminently a religious phenomenon of Siberia and Central Asia. The word comes to us through the Russian, from the Tungusic *Saman*. In the other languages of Central and North Asia the corresponding terms are: Yakut *ojuna (oyuna)*, Mongolian *baga, boga (buga, bu)*, and *udagan* (cf. also Buryat *udayan*, Yakut *udoyan:* "shamaness"), Turko-Tartar *kam* (Altaic *kam, qam*, Mongolian *kami*, etc.). It has been sought to explain the Tungusic term by the Pali [Indian] *samana*. . . . Throughout the immense area comprising Central and North Asia, the magico-religious life of society centers on the shaman. This, of

> course, does not mean that he is the one and only manipulator of the sacred, nor that religious activity is completely shaped by him. In many tribes the sacrificing priest coexists with the shaman, not to mention the fact that every head of a family is also the head of the domestic cult. Nevertheless, the shaman remains the dominating figure; for through this whole region in which ecstatic experience is considered the religious experience par excellence, the shaman, and he alone, is the great master of ecstasy. A first definition of this complex phenomenon, and perhaps the least hazardous, will be: shamanism = *technique of ecstasy.*[1]

"Ecstasy" does not mean emotional fulfillment. Eliade is using the word in the sense of its Greek roots: standing outside (oneself). Shamans step outside themselves, travel, often fly. They may go down or up or to the side, but whatever the direction or symbolism, they enter upon another reality, or another dimension of a complex reality. This is something not ordinary, something sacred, prized, associated with the gods or sacred or horrific powers that they intuit are at work in the processes of nature, keeping the world going, or that they think they have seen or heard in dreams, visions, special moments of insight or trial. It is this otherness, this standing *outside,* this motif of travel or transcendence that explains both the allure of shamanism and the practicality that the tribes in which it thrives attribute to it. Their journeys fulfill the shamans. Their abilities to step outside themselves function in their work as healers, intermediaries with the gods, guides of the dead.

Not every vision or ecstatic journey of a person living in an oral, nonwriting culture qualifies as a significant shamanic episode. People can experience vivid dreams or feel the power of evil raising bumps along their flesh without practicing the dances or trances or fastings or removals into solitude characteristic of "professional" shamans. Shamans generally function for their tribes, as well as for themselves. Generally, they fill important social roles, including serving as repositories of traditional lore—old languages, traditions about where the key ceremonies came from, theological convictions about how things were at the beginning, when the gods fashioned the world into its present shape. This social functioning has both given shamans immense influence and channeled their work. For instance, healing has been a great preoccupation. Working on sick people psychosomatically, shamans have striven to restore a holistic sense of health. They might use sand paintings or visions of bits of foreign matter contaminating the patient's system or convictions that an enemy had cursed the sick person, but they would employ any such physical or semantic tool while in trance or ecstasy, while their spirits stood outside their bodies, their selves, more sensitive to extraordinary influences than in unecstatic moments.

The following testimony of a Siberian shaman, who represents usefully the primal analogue on which Eliade has based his monographic study, suggests in a personal voice many of the elements that we have seen thus far:

> It was not the talent I inherited [he comes from a bloodline of shamans], but the shaman spirits of my clan. . . . I had been sick and I had been dreaming. In my dreams I had been taken to the ancestors and cut into pieces on a black table. They chopped me up and then threw me into the kettle and I was boiled. There were some men there: two black and two fair ones. Their chieftain was there too. He issued the orders concerning me. I saw all this. While the pieces of my body were boiled, they found a bone around the ribs, which had a hole in the middle. This was the excess-bone. This brought about my becoming a shaman. Because, only those men can become shamans in whose body such a bone can be found. One looks across the hole of this bone and begins to see all, to know all and, that is when one becomes a shaman. . . . When I came to from this state, I woke up. This meant that my soul had returned. Then the shaman declared, "You are the sort of man who may become a shaman. You should become a shaman, you must begin to shamanize!" [2]

Much in this report probably follows protocols of the man's Siberian tribe. He does not find it unusual to discover his vocation through a dream and does not dispute the validity of such a medium. He feels a strong connection with his ancestors, the previous generations from whom the tribe has derived its existence. Symbolic dismemberment occurs frequently in the initiations of shamans, as though to dramatize their freedom from ordinary limits imposed by the body, and perhaps to suggest as well that they are more capable of dealing with nonhuman reality (especially animals, but also plants and natural forces) than are ordinary people. The black and the fair men may represent the fortunes that the shaman will have to manage or the forces of good and evil.

The excess-bone is a conceit to give the shaman a physical distinction, to make shamanism a vocation marked into his body, to let it appear to be fated. The implication here is that the bone functions in the wide knowledge that the shaman requires, the ability to know all. That this candidate returns to health and wakes confirms the success of his initiatory dreaming. He shows the signs of a shaman's vocation. He must begin to take up the work (depending on the traditions of his people, that may mean apprenticing himself formally to older shamans or simply starting to learn, by doing, what his gifts imply for healing, dealing with the gods, and so forth).

What does any of this have to do with mysticism? That will be the question playing in the background of the brief studies that follow. Do the shamanic data that we bring forward alter the picture of mysticism that we have developed by studying the world religions, or can we see them as flowing toward the imaginative side of the incarnational apparatus and experience that we have been discovering in direct experiences of ultimate reality mediated by literate, scriptural traditions?

Probably our interpretations will be more suggestive than definitive, but working them out will allow us to take more oblique look at our understanding of mysticism and so be more than worthwhile.

Native North American Traditions

Native North Americans may well have descended from Asian peoples dominated by shamanism. Although there is no consensus concerning whether Native Americans arrived in the Western continents by crossing a strait at what is now the Bering Sea or when such a crossing might have begun, a Central or northern Asian origin remains the favored hypothesis. Whether it follows from this hypothesis that Native North American religious practices, many of which fit a broadly shamanic profile, derive from a Siberian or similar origin is also uncertain. Finally, the many centuries through which Native Americans have lived in the Western continents have given them ample time to develop their own distinctive traditions. Regardless of whether or not such immigrants began with Central Asian shamanic traditions, after at least ten thousand years of interacting with a different physical environment they could well have reached cultural understandings of the world quite different from those with which their ancestors crossed the Bering Strait.

We shall describe briefly the major characteristics of Native North American groups, insofar as their significant variety allows. Here, let us anticipate a question that we shall face at the conclusion of this chapter, using some reflections of Joseph Epes Brown, the foremost senior American scholar of Native North American religions, to guide us:

> What criteria can be used to affirm the spiritual authenticity of religious traditions such as those of the native Americans, and the reality of mystical experiences to which such traditions give rise? The tradition in question must have its origins in a sacred source that is transcendent to the limits of the phenomenal world. All the expressions and extensions of this tradition will then bear the imprint of the sacred, manifested in terms appropriate to the time, place, and condition of humankind. The tradition provides the means, essentially through sacred rites, for contact with and ultimately a return to the transcendent Principle, Origin, or by whatever name this is called. True and integrated progress on such an inner journey demands the means for accomplishing the progressive and accumulative integration of the following elements or spiritual dimensions: (1) purification, understood in a total sense, that is, of body, soul, and spirit; (2) spiritual expansion, by which an individual realizes his or her totality and relationship to all that is, and thus integration with, and realization of, the realm of the virtues; (3) identity, or final realization of unity, a state of oneness with the ultimate Principle of all that is. Spiritual expansion is

> impossible without the prerequisite purification, and ultimate identity is impossible outside the realm of virtue, wholeness, or spiritual expansion. These themes of purification, expansion, and identity are inherent in all the spiritual ways of the orthodox traditions of the world.[3]

Brown believes that, at their best, many Native American traditions have fulfilled these criteria for mystical authenticity. They have carried people to the transcendent, required that people develop the moral virtues, and generally both expanded people's horizons and facilitated their "identification"—their becoming themselves. The main mechanism that the Native American peoples, north and south, have used in this work, which is less a conscious project than simply the common social task of helping people become fully human, has been ritual. Their ceremonies, regular or occasional, have afforded them the means of entering wholeheartedly into the convictions supporting tribal life. The rituals have displayed the main myths by which the tribes have sketched the order of the natural world. By dancing, chanting, and identifying themselves with sacred figures, participants have moved outside themselves, into realms considered realer than what workaday consciousness encountered. Tribal rituals of any significance have required preparation—fasting, abstinence from sexual relations, meditations on the meaning of the ceremonies to come. In addition to the social coherence that the rituals have fostered, they have instructed individuals from youth in the symbols and stories that their people held to be most sacred. Thus they have been great educators.[4]

Some of the rites of Eastern Woodlands peoples suggest how Native American ritual consciousness has tended to work out in practice:

> For the Ojibwa, certain sacred spaces of human manufacture stood for the power that ran through nature and might break out at any exceptional site. In a "medicine rite" (a ceremony designed to tap the power of the cosmos), the Ojibwa would construct a lodge of arched trees. They would place a rock on the earthen floor and erect a pole in the center. The symbolism was meant to recapitulate the structure of the cosmos: earth beneath, sky above, directions (north, south, east, west) arching around the people's life. The pole stood for the people's hope that they were truly connected with the heavens above. Each of the arches of the cardinal directions could also symbolize a water spirit or snake associated with the vital power mediated by that direction (warm wind, gentle rain, cold wind, snow and ice, etc.). Up the tree or cosmic pillar ran different spirits of the heavens—layers of power. (For shamans, the tree also represented a way to ascend to the powers holding the tribe's fate.)
>
> For the Iroquoian and coastal Algonquian tribes, who tended to live in "longhouses," the ordinary tribal dwelling was a miniature of the cosmos. Thus the Delaware "big house" associated the floor with the earth, the ceiling with heaven, one side with the rising of the sun, the opposite side with the sun's setting. A "good white path" through the house stood for

> the journey that human beings make from birth to death. In ceremonial dances, the Delaware would exit from a door in the western wall and circle back to the east, as an expression of their hopes for rebirth. A post in the center of the big house stood for the axis of the world. It had twelve levels, to represent the different layers of the cosmos. At the top was the abode of the Great Manitou, the foremost power. The post was like the staff of the Great Manitou, who dispensed the power of creation down it. All around the lower levels of the big house carvings represented spirits who carried this power. The general result was a picture of Delaware life in which the people were always surrounded by numinous, creative powers.[5]

These descriptions hint at the rich, highly symbolic ways that Native Americans have constructed reality. Few material things have been only themselves, bare rocks or birds. Most have appeared within a sacred context, an archetypal set of possibilities. When people danced in the big house, they entered explicitly into the possibilities carved in the poles, but even during less ceremonial times those possibilities guided their senses of themselves. Mystical moments would have been those in which people saw directly or felt immediately or engaged passionately with the numinous possibilities that their traditions said surrounded them constantly. The final proof of the value of such moments would have been how the experients lived afterward, but the moments themselves would have carried a confirmatory value. If a person saw or felt the Great Manitou, then the conviction that the power of this greatest force ran down into the many aspects of the tribe's life would have been easy to sustain.

The great advantage of a ritual religious focus is the wholeness that it makes possible. People can feel the forces of ultimate reality, as well as deal with them conceptually. They can dance and sing their way into more than intellectual communion. Ritual alone does not make people holy or mature or mystical, but when people are serious about transforming themselves, becoming realer by closer connection to ultimate reality, ritual offers them a most useful set of disciplines. Apparently instinctively, most Native North Americans have realized this. Again and again, they have developed thorough systems of symbols, myths, and purificatory practices that have made probing the spiritual significance of nature attractive, challenging, a fine way to move outside the restrictions of the self, or even the tribe, into a horizon of creativity as such, the power by which everything rises or falls.

Perhaps we can suggest further some of the sensibility, the openness to spiritual influence, that Native Americans enjoyed traditionally by noting the impact that open space and relatively unrestricted travel on sparsely populated land tends to develop. The following description comes from *Arctic Dreams*, a provocative work by the naturalist Barry Lopez that perforce imagines the Inuit (Eskimo) closeness to nature.

Lopez begins with some reflections of the geographer Yi-Fu Tuan and then continues under his own steam:

> "In open space," writes Tuan, "one can become intensely aware of [a remembered] place, and in the solitude of a sheltered place, the vastness of space acquires a haunting presence." We turn these exhilarating and sometimes terrifying new places into geography by extending the boundaries of our old places in an effort to include them. We pursue a desire for equilibrium and harmony between our familiar places and unknown spaces. We do this to make the foreign comprehensible, or simply more acceptable.
>
> Tuan's thoughts are valid whether one is thinking about entering an unused room in a large house or of a sojourn in the Arctic. What stands out in the latter instance, and seems always part of travel in a wild landscape, is the long struggle of the mind for concordance with that mysterious entity, the earth.
>
> One more thought from Tuan: a culture's most cherished places are not necessarily visible to the eye—spots on the land one can point to. They are made visible in drama—in narrative, song, and performance. It is precisely what is *invisible* in the land, however, that makes what is merely empty space to one person a *place* to another. The feeling that a particular place is suffused with memories, the special focus of sacred and profane stories, and that the whole landscape is a congeries of such places, is what is meant by a local sense of the land. The observation that it is merely space which requires definition before it has meaning—political demarcation, an assignment of its ownership, or industrial development—betrays a colonial sensibility.[6]

Not burdened with a colonial sensibility, not pressured to "develop" the land, Native North Americans have tended to experience more mysteries than their later white counterparts and probably also to struggle more for "concordance with that mysterious entity, the earth." Yes, so abstract an expression comes from Lopez more than Native Americans, but the reality is likely valid. The earth, whether stark (here) or teeming (there), is mysterious, its ways being both. No mature people take the earth for granted, any more than they take for granted their opportunity to participate in the earth's rhythms and life. At their best, the rituals of oral peoples have freed them to explore the earth, appreciate the work of the cosmogonic spirits, moving occasionally toward their own equivalent of the open vistas of the Arctic, the purifying simplicity of the Koranic One.

Native Latin American Traditions

Although tribes such as the Delaware might speak of a supreme spiritual power, rarely if ever did they accredit a single divine agency with total power. Rarely, in other words, did they speak of a "Creator" in the all-encompassing sense that the biblical traditions used when referring to their lord. This does not mean, however, that Native American peoples did not experience the natural world as contingent or requiring explanation. Indeed, some of the most central aspects of Native Latin American oral religion have dealt with this requirement. One of the most important has been rituals to celebrate different cycles of existence. The fact that the world operates through cycles has suggested to many South American tribes the need to commemorate at least the most important of such patterns. Their commemorations provide for both aspects of a natural reality that manifestly keeps moving: its change and its continuity.

On this score Lawrence Sullivan has written:

> Cycles are fundamental temporal structures. These symbolic complexes make time apprehensible. The meaning of a cycle is enmeshed in its symbolic expression (sounds, smells, colors, textures, fruits, etc.); cycles of time are the symbolic predicaments of specific existential meanings . . . [for example] the actions of Avireri, the Campa culture hero and trickster, create cycles of time. He brings about the seasons and the night by playing music and dancing dances appropriate to them. Urged on by his curious grandson's questions, Avireri becomes very drunk and happy and begins, for the first time ever, to play the panpipes. "It was he who showed it, therefore, all festival activities also." Today the river water levels rise and fall to mark changes of the Campa seasons because Avireri once played powerful panpipe music to bring on the rains. The rainy season is called Kimohanei, from the word which means "big water." The dry season, Osarenci, serves as a way of measuring years, which are reckoned in dry seasons, or "summers." [7]

The Campa myths in which Avireri appears are storied efforts to probe the existential structures, the ontological designs, of the world that the Campa experience. Certainly, the language in which Sullivan couches his interpretation of cycles in general and the work of Avireri in particular is academic, owing more to social scientists than to South American tribal peoples, but the language can illumine what the rituals and myths of the people generate. The rituals and myths offer tools for understanding the world in which the people find themselves. The commemorations take account of all the major natural and human conditions that beg explanation.

For instance, how ought we (Campa) to make sense of the regular alternation between rainy and dry seasons? How ought we, at a deeper level, to trust a world with such fluctuations? Because the panpipes were important to the Campa, they associated the music of the panpipes with the crucial moment when Avireri brought on the rains. That he was drunk and showing off when he did this hardly mattered. Drunkenness can be a mode of ecstasis, a door to perceptions, feelings, that seem extraordinary (at least to the drunk). The point is that through the story something recurring gains its archetypal voice. A myth of explanation gives the cycle of rains and dry times enough order to take away its potential terrors.

Most oral peoples have used their explanatory myths participatively. They have entered into them, dancing and singing, not stood apart, chin on fist, to analyze the logic. The truth of the myth of Avireri's playing for the rains lies in what happens when people commemorate the story ritualistically. If their commemoration takes them out of themselves and into the primal moments and energies when the world gained the cyclical rhythms that characterize it now, the myth is true. If their commemoration wards off the potential horrors of a world that might be only discrete change, unorderable movement, the myth is precious.

Oral peoples do and do not know that they themselves have made their myths. The ritualistic setting does and does not attribute what it celebrates to the ancestors of the past. On the whole, oral peoples place brackets around this question, acting as though their stories were self-evident or as though the gods had handed them over straightforwardly. The point is that the myths and celebrations work only as long as the people using them can believe that they render, or illumine, the world credibly. Once a key myth has become questionable, no longer transparent for natural or social meaning, the religious life of the tribe accustomed to using it has fallen into crisis.

Myths never remove the mysteriousness of human existence, any more than do fundamentalist diatribes. Reality is intrinsically mysterious: the divine never surrenders its name. However, a functional religious complex allows people to deal with the mysteriousness of life equably, offering enough plausible scenarios of how the spirits have structured the cosmos to keep people from feeling threatened constantly. Yet such a functional religious complex also suggests that something deeper, simpler, darker lies beyond any set of rituals and myths, however pleasing. This suggestion gives shamans and others primed to search out deeper wonders an intuitive justification.

In the tropical forest of Brazil, among the Mehinaku, the most important belief is that many spirits shape the fortunes of the people. No beings are simply material; all are also spiritual, each carrying an insubstantial as well as a substantial form. Such insubstantial forms or spirits move like the wind, creak like the vines in a storm. These are the forces that visit the sick, that the shamans see on their journeys, that appear in every crack in ordinary material reality.

On the whole, such spirits are frightening: huge heads, long teeth, eyes that glow in the dark, horrible voices. People are most vulnerable to the influence of such spirits when they are ill, depressed, or lonely. The shaman has to fight against depression and loneliness, making sure that people do not lose their souls. One can give gifts to the spirits through ceremonies that pay them off, but a firm will to resist them is more important.[8]

Here we see the crux of the shaman's work, set in the context of the polytheism or polydaimonism that prevails among oral peoples. The shaman's job is to keep people sufficiently balanced and hopeful so that they do not succumb to the negative possibilities in their difficult situations. Sometimes the accent of the shaman's ministrations will fall on physical things: helping people regain their health, finding game for the tribe. Other times the accent will fall on spiritual things: explaining how the occult forces are working, retelling the myths of creation that show that the gods made the world good. Always, however, the revered shaman is a champion of meaning, possibility, and hope. Typically, he stands in the community as the embodiment of the capacity of human beings to deal adequately with their mysterious, dark condition. Shamans are acquainted with death (frequently, as in the case of the Siberian shaman, the initiation ritual kills the shaman symbolically), the most consequential of humanity's terrors, but from this acquaintance they can derive a profound appreciation of life.

The mystic solves the problem of possibility, significance, hope by experiencing directly a meaning that is secure, indubitable. The analogy among oral, tribal peoples tends to be a vision or other spiritual experience gained through contact with a shaman that allows the person to confront the awesome creative powers and not be destroyed. From such an experience, people can develop the willpower to resist the terrors of contingency, cyclical existence, a forest filled with fierce animals, a psyche filled with evil premonitions.

On the therapeutic or prophylactic level, then, shamanism usually provides an elementary mental health. People can weave their way through a demanding natural world, even a threatening spiritual world, if they can believe that they have adequate resources—enough light shining in the darkness to keep the darkness from comprehending it. However, when the myths and ministrations of the shamans fail, the darkness becomes deep indeed, but this is not a frequent occurrence. On the whole, oral peoples in the Americas and elsewhere have parlayed their myths, rituals, and the gifts of their shamans into a solid psychic survival.

Native Australian Traditions

We now focus on the artworks that many native Australian groups have used to express their relations with both the land and the powers believed to have formed the land at the time of creation:

> Rock incising and cave paintings abound where, in the local mythology, mythical beings have "turned themselves into" these; "they made themselves thus" . . . "they left their physical presence" in that form, and with it part of their spirit, part of their real self, *and their image.* Over and above ground paintings, the actual shapes of secret-sacred ritual grounds serve an iconographic function: they represent attributes of a creative being, if not that being himself. Ritual postulants and others decorating their bodies and wearing emblems are regarded as being *like,* as representing, the sacred beings themselves or their actions, or associative aspects. Such postulants are recognized, in many Aboriginal areas, as possessing an essential sacred essence which, through conception and/or birth, makes them part of a particular mythic being, as an extension of that being.[9]

The mentality at work here is artistic, symbolic, mythic, archetypal, participative—gather whatever other adjectives seem good. Aboriginal Australians regard the land as filled with holy presences. The land did not come into its present shapes accidentally. The present shapes are both the work and the presence of creative powers. One can tell the character of such powers from the character of the land. The ceremonies that fit a given sacred site reveal the original realities of the place. Even the participants in the rituals express the character of the place by dancing and singing out its mythical explanations, its spirits, its localizations of the universal relation of the people to the dream time, the original period of creation when the ancestral, divine spirits fashioned the natural world.

Ecstasy in the Aboriginal Australian spiritual system is less a matter of dancing and singing the self out of the body than of withdrawing to contemplate the dream time. Certainly, Aboriginal Australian ceremonies take the participants out of themselves, sometimes very much as southwestern American *kachina* dances take Native Americans out of themselves and into the doll figures that they represent, or better, that they become. However, the pleasures of contemplating the creation of the land in the dream time and their own return to this better, archetypal zone appear to be the crucial motive of the ecstasy that is more distinctively Australian. The farther along the life cycle they are, the more traditional Australian aborigines want to turn aside from profane occupations and free themselves to enter the dream time and connect their spirits with what is most real: the creative power of the forces that made the land.

The land is all-important, but the land is not uniform or generic. The land undulates and is specific. So the work that fashioned the land was particular, tailored, as is the character of the creative forces that a particular hill or valley reveals. The result is an endless map of spiritual opportunities. As Bruce Chatwin's elegant book *Songlines* suggests, the songs through which native Australians have named their local lands, mapped their local endowment from the forces of the dream time, have been the staff of Aboriginal spiritual life:

> Having established that the Perenty [large lizard] Songline followed a north–south axis, I then swivelled round and pointed to Mount Cullen. "OK," I said. "Who's this one?" "Women," Joshua whispered. "Two women." He told the story of how the Two Women had chased Perenty up and down the country until, at last, they cornered him here and attacked his head with digging sticks. Perenty, however, had dug himself into the earth, and escaped. A hole on the summit of Mount Liebler, like a meteorite crater, was all that remained of the head wound. South of Cullen the country was green after the storms. There were isolated rocks jumping out of the plain like islands. "Tell me, Joshua," I asked, "who are those rocks over there?" Joshua listed Fire, Spider, Wind, Grass, Porcupine, Snake, Old Man, Two Men and an unidentifiable animal "like a dog but a white one." His own Dreaming, the Porcupine (or echidna), came down from the direction of Arnhem Land, through Cullen itself and on towards Kalgoorlie.[10]

No doubt Aboriginal dreaming has degenerated in recent decades, as has Aboriginal culture overall. Nonetheless, Joshua is old enough, or traditional enough, to carry in his head dozens of songs that organize the land around him. The Inuit, with whom Barry Lopez worked in the Arctic, displayed an analogous mastery of their land. They could draw detailed maps of inlets, marshes, brakes of trees, and slopes of hills, even places they had visited only once or twice, dozens of miles away.

The essential orientation of oral peoples has been to the land. Generally, the divinity in their milieu has spread itself through forces of the natural world. Seldom has it located itself in a single, all-possessing god. Practically, the land has furnished hunters and gatherers, as well as cultivators, their means of survival. The better hunters knew the land, the better their hunting was likely to go; but the typical relation of a small-scale, oral tribe to its environs has been more than practical. Typically, the people have loved the land and considered it fully alive. It has been a presence—Mother Earth, if one could move out to such an abstraction. It has been replete with the spirits, the ancestral holinesses, that the dream time or the myths of creation or the annual ritual dances said had formed it.

So, we suggest that the depth and center of aboriginal, prehistoric human existence has lodged in the people's perception of the divinity of the landscape they felt moved to celebrate in songlines, the sacredness of their environs and so of themselves. Yes, tribes have also worked on their communal lives, using their rituals to rebind themselves together as such-and-such a people. Yes, it has been common to find tribes, and larger ethnic units, considering themselves to be the first instance of humanity as such and consigning all foreigners to a less than fully human status. In our opinion, however, the relationship to the living earth that has mediated the tribe's sense of creative, divine power has been more influential than even the tribe's social relations. The land, and the divinities that continually fashion it, have been more stable, more ultimately real, than the (manifestly vulnerable) tribe, even though one could trace the tribe back for numerous generations. The land has continued to turn green each spring, while the ancestors were bound to fade deeper into memory. The children that arrived in each new generation were thought to be the spirits of the ancestors returned, as witnessed by many oral tribes' practice of naming infants for grandparents or great-grandparents; but this ongoing generativity has been fragile, nowhere near so impressive, so imposing, as the annual regeneration of nature.

Among most traditional Australian Aborigines, the very conception of human beings in the womb has depended on the action of the spirits responsible for shaping the land and renewing it each year:

> Traditional Aborigines believe that no child can be conceived or born without its fetus first being animated through some action relating to a particular mythic character that also involves the child's parents and/or other consanguineous relatives. For instance, in the central desert area, an unusual event, or one that is interpreted in those terms, must be experienced in order for animation or conception to occur. A man may spear a kangaroo which behaves in a "strange" manner. Later he gives some of the meat to his wife, who vomits after eating it. She may know immediately, or after he has identified the cause in a dream, that a mythic being has either transferred part of his or her sacred essence, via an intermediary, to the fetus within her or has stimulated or activated the conception. In northeastern Arnhem Land, spirit-landings fulfill this function. Spirit children (or child spirits) are associated with particular water holes or sites and with the relevant mythic beings connected with them. They take the shape of some creature that, when caught, escapes from the hunter or otherwise behaves unusually. Then the spirit child appears to him (or to one of his close sisters or father's sisters) in a dream asking where its mother is. In all such cases, the particular area in which such an incident takes place is of significance in the social and ritual life of the child who is eventually born, and the area concerned is preferably related to the father's country.[11]

Aboriginal Australian rituals are not obviously shamanistic.[12] The appreciation of the work of the ancestral divinities on the land and in the conception of children comes less from ecstatic journeys than from regular returns to the dream time. One might call ceremonial time ecstatic, and so shamanistic, for the broad reason that in it and through it people move outside their workaday imaginations; but the majority of traditional aboriginal peoples have had no "workaday," secular imaginations, at least not as modern Westerners have had. The majority of Aboriginal peoples have moved within a physical landscape and an interior worldview that has pulsed with creative divinities. Dreaming has been a form of escape, if not ecstasis, as has ritual dancing, singing, carving of pictographs, and painting of local spirits; but little of this has gone on with the drama, the concern for magical flight and spiritual combat, that we find surrounding the Siberian shaman, who has tended to lived in demanding landscapes dominated by ice and snow. More of it has reflected the vast constancy of the Australian continent, the mile after mile of bleak openness, the day after day of burning sun.

One might call a habitual orientation toward dreaming mystical if flexibility and accommodation were the linguistic order of the day. Through their dreaming, many Aborigines apparently have reached what they have considered the ultimate zone of reality, the time and action that ground the world; but how focused such a time and action usually became and, correlatively, how intensely they focused the spirit, the self, of the dreamer is hard to say. Even if one puts off the vague, sleepy connotations of dreaming, the stories of the creative powers and the descriptions of the traditional rituals lack the passion for unity that our studies of other mystical traditions have uncovered.

To be sure, this finding or impression may result more from methodological problems—the different kinds of access to their prime religious experiences that the various mystical sources allow—than from actual Aboriginal religious life. Nonetheless, it is hard to imagine traditional Australians alongside an ecstatic al-Hallaj or an al-Hallaj dreaming his way to God or a Patanjali agreeing to lodge at the level of the *atman*, where dreams occur. For the likes of al-Hallaj and Patanjali, an intense desire for ultimacy, which they correlated with soleness and beyondness, kept driving the beat. The plurality of the forces that have populated the native Australian landscape seems to have diluted or obviated such a drive, as has the looseness of the mythical heritage that has directed the dreamers' searches. This may often have produced a more relaxed, humane, incarnational relationship to (a landed) divinity than what one finds in Islam or *yoga*. It does not appear to have sponsored a mysticism anything like their equal in energy or in passion to lose (and find) one's being either in being itself or through love of a sole lord.

African Traditions

The traditional Australian reference of current times and current features of the natural world to archetypal events and forces has its parallel in many traditional African societies. Consider, for example, the opening paragraphs of a recent description of the function of masks in traditional West African rituals:

> Even today, in the traditional societies of West Africa, the institution of masks is closely tied to the agrarian rituals, funeral rituals, and rituals of initiation that are so significant to village communities. The ceremonies in which masks play a central role almost always have as their purpose the recalling of the mythical events that took place at the time of origins and that lead to the organization of the universe in its present form.
>
> That is why when today one tries to describe a mask within the framework of the cosmogony of which it is a part, the description cannot be limited to the part of the mask that covers the head of the person wearing it. In the great societies of initiation among the Bambara, when an initiate speaks of the "head of the Komo," he means all the elements that constitute the mask: the head that borrows its morphological elements [shapes] from the swollen skull of the old hyena (associated with profound understanding), from the "mouth" of the crocodile who was the first to put the ark of creation in the pond, and from the horns of the antelope that symbolize by their pointed ends the first lightning of creation; the tunic, made of strips of cotton on which the feathers of vultures have been affixed, bearing the 266 signs of creation; the "elephant's feet" tied to the dancer's ankle and symbolizing the pillars, beams, and supports of the universe; the iron or copper whistle that evokes by its strident sound the first whistling of creation; the stiletto of thaumaturgy [miracle-working], the foremost instrument of ritual executions; etc. Finally, "the head of the Komo," called *komo ku,* also designates the wearer of all these objects and the dance that the wearer performs.[13]

The mental world in which we find ourselves here lies at the farthest possible distance from the union that beckons to the mystics of many of the world religions. The dominant faculty at work in the construction of the costume of the dancer in the *komo ku* is an archetypal, religious imagination. This squares with the general shamanic use of imagination rather than pure spirit, as well as with the particular traditional Australian tendency to picture and sing of both the original dream time and the present dreamer's contemplative way back to it. The ecstasy involved in the Australian case is more a transport to the forces and time of the original creation than to a present "being" responsible for the wonder that there is something rather than nothing, and the same seems true for the ceremonies of the West African mask societies.

Still, the forces and time of original creation that one contacts, or enters upon, through traditional masked dances are not only long ago and far away (distant, transcendent, enough to seem attractively other, divine at least partially in the mode of being more than human) but also contemporary, present here and now, inasmuch as the energies of the original powers continue to form the world; the original *theriomorphic* (animal-shaped) creative forces continue to determine the reality, natural and social alike, in which the Bambara participate each day because they are links on the chain of being that the gods have created between themselves and humankind. The ceremonies obviously trade in "both/and" rather than "either/or."

When it comes to estimating the value of knowing the divine characters that shape the traditional dances and appear through the traditional masks, our authority opines:

> All of this is evidence of the highly sacred character and the deep significance of the wearing of the mask of Komo, which appears to be a microcosm, a dynamic summary of the universe. For this reason, it is called "the burden of the universe" and, by analogy with this name, "the deep knowledge of the universe." Originally, the infallible knowledge for which the mask is here the support was given to Faro, whom the Bambara and the Malinke alike regard as the helper of creation and the instructor of the universe. This knowledge, like all that exists, has two natures, one visible, palpable, and concrete, the other hidden, secret, inward, and profound. Faro taught these two aspects of knowledge to two of his chosen ones, the vulture and the hyena, the future patrons of the societies of masks and the most assiduous, attentive, and respectful listeners to their instructor's commentaries on creation. Under the patronage of these two carnivorous scavengers, the vulture and the hyena, were placed the societies of the masks that represent them among the Bambara, the Malinke, and the Minyanka. As a result, these societies are regarded as the keepers of the "true values," the "true signs," and the "totality of knowledge." [14]

African tribal religions often make an important place for the diviner, who ideally has access to the totality of knowledge. Using a scheme such as the 266 signs of creation mentioned in the mask ceremony, a diviner may sift a basket of 266 chits to determine what the future is likely to be or why a patient is suffering an illness. (This recalls the use of the *I Ching* for divination: the patterns of a toss of yarrow sticks suggest how the future will unfold.)

Dreams are important to most African tribes frequently because they bring the dreamer into contact with an ancestor who is functioning as a protector and source of counsel. Tribal kings wield sacred power, serving as links with the archetypal realms of the gods. What goes on below, on earth, in the human kingdom, ideally would be in harmony with what goes on above, among the gods. The great gift of the African gods and the great interest of African tribal religions overall is vitality—fertility, well-being, health, and strength to enjoy natural and human beauty and to resist illness, debility, and sadness.

Many of the herding tribes, such as the Nuer, focus on sacrifices of blood, for which priests become important figures. (Aboriginal Australians also employ blood in some of their rituals, as did the classical Aztecs to horrifying proportions, suggesting that blood can become a basic currency for dealing with the gods, an elementary exchange of a commodity essential to life.) The relations between the Nuer and their cattle take on rich, symbiotic overtones. This holds true generally for hunters and herders, people who live by their relations with animals. The coastal Eskimo, who depend heavily on the seal, like the Plains Indians, who depended heavily on the buffalo, never held these animals at arm's length. While pragmatically and unemotionally they killed as their survival required, spiritually they lamented the necessity of taking life to preserve life, usually apologizing to the animals they had to slay and often preserving a part of the animal (for example, the liver of the seal) and returning it to the animal's habitat (the sea). Such a ritual was self-serving in that the return was supposed to ensure a new generation of game, but it was also reverent toward the animals on whom the tribe depended, a recognition that they were brother and sister creatures, perhaps as much like human beings as unlike them.

What in any of this, either the ceremonial use of masks or the generalizations one tends to find about African tribal religions overall, is shamanic or mystical? In our impression, shamanism according to the Siberian prototype does not fit African religious complexes well, at least not as the linchpin or central arch. A better image pictures a variety of different religious functionaries: diviners, kings, priests, prophets, mediums, witches. What binds them together is a concern for vitality. Further, many of them use trance (for example, diviners or mediums [people through whom spirits, often of the dead, can speak]) or heightened spiritual awareness in their work (for example, an elder handing on the tribe's inner understandings of its myths and rituals). However, in Africa we seldom see the activities characteristic of the prototypical shaman: magical flight, highly advertised personal incarnation of the duality of the human condition (the two worlds, earthly and spiritual, in which human beings dwell), dramatic ritual initiation involving death and resurrection.

Certainly, healing is as important in traditional African religion as in shamanism, and in both cases healing is psychosomatic. Noxious spirits, perhaps unleashed by an enemy, are the source of sickness, so fighting them, or counterattacking by sending woe upon the enemy, can become a passion for both shamans and African religious functionaries. African witches (incantators calling upon deities to become their familiars) can dabble in black magic, solicitation of the forces of evil, and so can evil shamans; but in African tribal religions there is no shaman (or witch or priest or diviner) holding the pride of place that, for example, a Native American shaman, to say nothing of a Central Asian one, usually would hold.

Concerning what may be mystical in traditional African religions, the main problem, as frequently is the case elsewhere as well, is semantic. If the mystics of the world religions have accustomed us to think of mystical experience as a negation of the many of the world so as to approach, or let oneself be taken over by, the one ultimate reality, whether a Brahmanic being or a biblical lord far beyond the world (though also intimate to it as its Creator), then little in African religion, or in the religions of oral peoples overall, is strictly "mystical." As Joseph Epes Brown hinted, however, many rituals may be transcendent in that they are efforts to deal with powers more ultimate, holier, than those of human beings and to be transformed by them to mature virtue.

Moreover, from the side of the world religions, we ought to note that many mystics (not all) have felt a call to return to the world, to find ultimacy in the marketplace, and to preach the word of communion, love, fulfillment, and transformation far and wide because they have sensed that at least a few other people might be hungry for this word, apt for this venture.[15]

Lastly, we note as perhaps distinctively African the imaginative creativity suggested in the description of the use of masks among the Bambara, the correlation of the masks with a twofold wisdom potentially profound, and the analogous imaginative creativity of a holy man such as Ogotemmeli, the Dogon elder who initiated the French anthropologist Marcel Griaule into the mythology of the Dogon people.[16] Inasmuch as Ogotemmeli can represent traditional African wisdom, this continental area, like Central Asia and the Americas, has nourished a nearly endless imaginative fecundity. Like a Rumi generating poetic image after poetic image of love, the African holy functionaries especially, but their shamanic counterparts in other religious complexes as well, have kept on exploring the world imaginatively. What they have lacked in critical control, submission of their images to hard reason to make clear the difference between what might be and what has been verified actually to exist, they have in part compensated for by the vividness and the vitality of their images as at least plausibly grounded in the creative forces that have organized the natural world.

Summary

We have considered, briefly, some aspects of native North American, Latin American, Australian, and African traditional, oral, tribal religious cultures that bear on our study of mysticism and the world religions. Starting with the hypothesis that the shaman has been the religious functionary most important in oral religion overall, we followed the description (specializing in archaic techniques of ecstasy) that Mircea Eliade pinned to the (Central Asian) shaman, finding it useful, though not exhaustive, for clarifying the sensibility of Native Americans. Such an understanding of shamanism has been less useful for understanding Aboriginal Australians and Africans, though we have found it stimulating methodologically: if not shamanism, what has been central? How better to characterize the transcendence, the at least quasi-mystical move out of the ordinary and into the sacred dream time, or the realm of the traditional gods, where one might find healing and fulfillment?

Dreaming came to the fore in the Australian case, along with creating the song lines through which many Aborigines have constructed maps of their environs. A many-sided, many-ritualed concern for vitality, fertility, vigor, health rather than sickness, and strength rather than weakness seemed the best analogue in Africa. Both dreaming and the vital side of many African religious activities (for example, the trance of the diviner) bear similarities to the ecstatic interests and methods of the classical shaman, but in neither Australia nor Africa has the ecstasis been as intense, so much a matter of flying to the gods and combatting death, as it has been for the prototypical shaman.

The greater likeness among the overall religious profiles of traditional natives in the four cultural areas that we have described probably resides in the excitation of religious imagination that we found in them all. Oral religion *en bloc* has involved dazzling exercises in picturing how the gods originally fashioned the world, what the different shapes or sounds or smells of given creatures say about their origins and rightful uses, and how human beings gained their present social relations. In contrast to the concentration on reason, will, understanding, and love that we found among the mystics of the world religions (who tend to distrust the imagination, the senses, and the feelings, even though of course they have to use them), the leading figures among oral religionists have tended to be poets, artists, masters of masked dances, and sages such as Ogotemmeli, who have gone in the night to star after star, possibility after possibility, ultimately because, after years of practice, they have come to love such work, finding in it both aesthetic fulfillment and transcendence. That probably has been their greatest kinship with the more classical mystics. The religious functionaries through whom it is easiest to interpret oral religion have been like the classical mystics in following a path of maturation that led them more and more to identify their treasure, what they had been made for, as a reality richer than themselves, more ultimate, with which they might, if the gods favored, gain blessed union.

NOTES

1. Mircea Eliade, *Shamanism* (Princeton, N.J.: Princeton University Press/Bollingen, 1972), 4.
2. Joan Halifax, *Shamanic Voices* (New York: Dutton, 1979), 50.
3. Joseph Epes Brown, *The Spiritual Legacy of the American Indian* (New York: Crossroad, 1982), 113.
4. See Catherine Bell, *Ritual Theory, Ritual Practice* (New York: Oxford University Press, 1992).

5. Denise Lardner Carmody and John Tully Carmody, *Native American Religions* (New York: Paulist, 1993).

6. Barry Lopez, *Arctic Dreams* (New York: Scribner's, 1986), 278–279.

7. Lawrence E. Sullivan, *Icanchu's Drum* (New York: Macmillan, 1988), 159–160. See also Gary H. Gossen, ed., *South and Meso-American Native Spirituality* (New York: Crossroad, 1993).

8. See Carmody and Carmody, *Native American Religions,* 194–195.

9. Ronald M. Berndt, *Australian Aboriginal Religion* (London: Brill, 1974), fasc. I, 18.

10. Bruce Chatwin, *The Songlines* (New York: Penguin Books, 1987), 153.

11. Ronald M. Berndt, "Australian Religions: An Overview," in *The Encyclopedia of Religion,* ed. Mircea Eliade (New York: Macmillan, 1987), 1:532.

12. An exception might be the initiations of holy healers, who usually underwent ritual death and obtained tokens of their ordeal, such as spiritual rock crystals that were set in their bodies to facilitate their medicinal work after their resurrection. See Mircea Eliade, *Australian Religions* (Ithaca, N.Y.: Cornell University Press, 1973).

13. Germaine Dieterlen, "Masks in West African Traditional Societies," in *Mythologies,* vol. 1, ed. Yves Bonnefoy and Wendy Doniger (Chicago: University of Chicago Press, 1991), 45.

14. Ibid.

15. See the explanation of the traditional Zen Buddhist pictures ("Herding an Ox") of the progression of enlightenment in Philip Kapleau, *The Three Pillars of Zen* (Boston: Beacon, 1968), 295–313.

16. See Marcel Griaule, *Conversations with Ogotemmeli* (New York: Oxford University Press, 1972).

9

Conclusion

Direct Experience

Virtually no commentators on mysticism deny that it grows from roots in vivid religious experiences. Usually you do not become a mystic by studying academically (though studying academically is certainly an experience, and now and then such study has mediated living encounters with ultimate reality). Usually you start on the path into mysticism through an experience of transport—being taken out of yourself through a response to beauty or pain. The experience is so good, so distinctive, that it turns you at least half-way around, "converts" you at least partially to the proposition that reality is much more than what you thought it was before this happened to you.

Charmed and challenged, you set out to learn more about what happened to you and, if you are fortunate, you find guidance into regular, habitual disciplines calculated to make constant your dealings with the reality that is much more than what you appreciated initially. These disciplines are important, but it is axiomatic in the lives of the mystics that experience got them going and made them persevere. Thus after witnessing sickness, old age, and death, the Buddha was so moved that he left his cushy life in the palace and determined to gain an enlightenment that might solve the problem of suffering. The parallel for Jesus was probably his compound experience of being baptized by John the Baptist and led into the desert to be tempted by Satan. After this, he never wavered from his work of preaching that the Kingdom of God was dawning. Muhammad became the Prophet by following his hunger for contemplation into the desert night. Wrapped in his mantle against the cold, he began receiving the revelations that soon transformed both his own soul and the entire Arab people.

Therefore, the problematic word in the first part of our working definition of mysticism is not "experience" but "direct." On more than one occasion, we have had to introduce qualifications, even doubts that anything in human experience is unmediated by the culture in which people find themselves, regardless of whether they are mystics or utter worldlings. From the moment of birth, if not of conception, human beings start to acquire experiences that shape their consciousnesses, the ways that they process both sensory and spiritual experiences. There is no *tabula rasa* in normal human beings. There are no heads empty of cultural formation, formation from prior biographical experience, influence from sensual perceptions, from current bodily health, from current emotional concerns, and so on.

This means that we ought to take with large grains of salt any claims of mystics that they leave ordinary human existence ecstatically or enstatically (by going within, through a yogic stripping, in search of *samadhi,* a state of pure self-possession below any determination by sensation, intellection, or volition, where you rest on or in the pure being of reality itself) to meet ultimacy purely, unmediatedly, with no preformation, and come away made utterly new.

This does not mean, however, that mystics do not leave more of such ordinary preformations behind than nonmystics can appreciate. It does not mean that vivid encounters with ultimate reality do not show mystics the fragility and contingency of everything social and personal, indeed of everything created, and so distinctively lessen the hold on their selves of anything that seems less than ultimate, other than the sole and full reality of the divine.

This mystical movement out of contingency toward a unique necessity is a warrant for retaining the word "direct," as long as we do not canonize it or misunderstand it to entail so strong a denotation that we imagine the mystic to leave the created world behind completely. Speaking absolutely, that does not happen. Speaking relatively, it may. The mystic may well come significantly, perhaps even qualitatively, closer to ultimate reality than the nonmystic. This degree of closeness or directness or this lack of mediation may well be a prime mark distinguishing mystics from nonmystics. It has none of the precision that needle-nosed, scientific researchers desire, but when one steps back from the full *gestalt* of mystical activity it presents itself as potentially significant.

What is the source of any such distinctive directness? On the testimony of mystics themselves, the most important source is the action of ultimacy on their inner depths. The reality upon which they verge, with which in peak moments they unite, is not inert, static, lifeless Hindu mystics on the order of the Vedantists may describe *Brahman* as beyond passion or action, pure in its self-sufficiency, and this may lead them to seek a *samadhi* that seems more stonelike than human; but enough visionary activity and desire to understand remain in the Vedantic mystic's nontrance moments to suggest that *Brahman* is appealing, that its ultimacy makes a winning impact.

Certainly the *Upanishads* present the quest for the simple unity that might explain the continuance in being of the many complex items in the world as exciting, an adventure, a potential discovery of a truth that human beings either were made to know or find brings them an unexpected fulfillment. In the noetic fulfillment of enlightenment, where they realize the import of the Upanishadic sages' saying "That Thou Art," the monistic Hindu mystics show us why their quest has been worthwhile. The light coming from the presence in the mystic of the divine ultimacy, uniting mystic and *Brahman* in a participative communion, is ineffable if we speak exactly. If we speak symbolically, we can say that such light from the divine gives a glow to the mystic's mind that supports a hope that living in communion with *Brahman* is the highest of estates and graces.

Does it matter greatly whether mysticism is direct or indirect, unmediated or mediated? No, perhaps not. The average mystic, especially the average theistic mystic, is more interested in reaching the goal than staying with one method purely. So if images, icons, or symbols prove helpful, at least for some stretches of the journey, many mystics will use them gladly. Increasingly, the enlightened realize that all such aids are provisional and must in the final stages fall away, or at least be chastened considerably. But in themselves they are not bad, not necessarily idols trying to substitute for ultimacy. Indeed, they may be inevitable inasmuch as the Bible or the Koran or the Hindu scriptures or the Buddhist scriptures or the ceremonies that fill all the world religions or myriad other material things serve almost always and everywhere to mediate the divine.

The most puritanical efforts to strip away idols, icons, sacraments, and other material embodiments of faith still depend on words, which create pictures in the mind. Puritans may denude their churches or mosques of images and so think that they are pursuing a pure God uncorrupted by any desire to bring divinity under human control, but if such puritans use biblical or Koranic texts or music or even deliberately plain architecture, they are still enmeshed in symbols.

The fact is that all human communication, both external and internal, is symbolic and as such is tied to our bodies and imaginations. Even the mystic's sense of darkness and silence is symbolic, worked out in (negative) images. As long as reason is operating, people are trying to make sense through images. No human beings ever become angels, pure spirits with no debts to matter. All mystics come out of religious communities, use inherited languages and rituals, face problems of diet, sleep, sex, thought, social responsibility, and clothing. The "oneness" of the monk (*monos*), the solitude of the desert anchorite, the uniqueness of the most accomplished contemplative are all relative, and none of them is completely free of the body, completely unmediated by memories and instructions stored in the brain.

The large number of mystics who have favored negative paths, pushing off from limited, created realities to deal with ultimacy as "not this, not that," suggests, however, that mystics themselves have often wanted an ultimacy radically different from what they experienced fully imaginatively, with large debts owed to their bodies. The soleness of the biblical God can seem a welcome retreat from the "manyness" of chaotic life in the world. The stability associated with Buddha nature, Suchness, can seem a happy relief from the mutability of *samsara*. With his father, Jesus found a reliability, a sureness of love, that he found nowhere else. With Allah, countless Sufis have found the rock on which to build their lives, the counter to all the sand of human institutions.

The negative, apophatic strain so strong in the East, and also prominent in Christianity and Islam through their debts to Hellenistic, Neoplatonic thought, would have the mystic prefer darkness to light, faith to vision, even suffering to prosperity. The *nada* of John of the Cross is a hymn to the otherness of God. However, due to the paradoxes of dealing with *ultimate* reality, this *nada* is more like the triumphant "Te Deum" of the imperial liturgy, gilded and full-choired, than most readers realize initially.[1] For it says that you, God, are so magnificent that nothing human, nothing that we can experience comfortably, can do you justice. If we want to do you justice, offer you condign praise, we have to suffer your otherness, let your actual unlimited light be for our impure limited minds thick darkness. Certainly, if you choose to be with us, come to us, through historical events and sacramental forms, we should use these gladly, obediently. However, when we seek you and find only no-thing-ness, or nonconsolation, we should accept such apparent failure with equal gladness.

For John, worry does no good. Fighting the *nada* that can become the closest presence of God is senseless. The genuine believer, the best candidate for mysticism, wants God to be all and self to be insignificant. Remembering the fate of Christ, such a Christian believer wants to surrender the spirit into God's keeping and offer the Lord *carte blanche*.

We have seen that Sufi masters, such as Ibn Abbad of Ronda, have felt the same way. The closer they have come to God, the less significant they have thought their own desires. God encourages people to pray to him with confidence, but it is their submission that makes Muslims

Muslims. They have to believe that God will provide all that they need, to the point that an apparent lack of an answer to their petitions becomes a definite answer. The apparent refusal of God to grant the cure or the fortunate outcome to the business deal or the favorable wedding of the daughter for which people have prayed becomes a word from God that such an outcome is not for their good.

Inshallah ("as God wishes") often enters the speech of pious Muslims, as "blessed be He" often enters the speech of Jews pious enough to mention their Lord often. If faith in a living God is the actual wellspring of a people's life, then the will of that God, the dispositions of that ultimate reality, take center stage. Truly religious human beings do not presume to say what is best for them. They presume only to say that obedience to God, keeping God's law, praising God without ceasing (the Hesychast ideal) is the highest thing that their kind can do.

Mystics tend to be the believers who try hardest to realize such ideals and who succeed best. They tend to be the ones who receive the contemplative inclinations and the graces of special willpower necessary to persevere on the often trying, often negative journey of seeking God's face (all the while knowing that God has no face, only a name that is no name and a will for them that is bound to be surprising). They are heroes of the spiritual life because they brave the divine awesomeness most courageously. Even when they say that they have little choice, that God has chosen them more than they have chosen God, we can use them as models. We should not imitate them slavishly, but we are wise to study their experiences and distill what they suggest about ultimate reality. To the degree that ultimate reality is our own passion, the mystics can become our beloved friends, perhaps our best benefactors.

Ultimate Reality

If "direct experience" captures much of the subjective side of the mystical moment, the living encounter that mystics prize most, "ultimate reality" captures much of the objective side, standing for the other, the sacred partner in the mystical moment, the one with whom or which the mystics become joined. Just as "direct experience" limps, is a faltering analogy, so does "ultimate reality." Can human beings, who are always proximate to themselves and somewhat intermediate to others (our bodies are barriers as well as forms through which we can meet), ever reach anything ultimate? Can we ever deal with divinity as it is in its own right? John Calvin said that the human mind is a factory of idols. The mystics strive to go below their minds and to chasten their imaginations because they intuit that everything finite, defined, or specified misrepresents the sacredness, the utterly simple reality, that draws them forward, at times working palpably in their spirits, toward the light, or darkness, of their best encounters. The mystics prefer to speak of "spiritual" events, even though they may sing hymns to inner sensations, because "spiritual" seems a purer, more adequate metaphor for their own most central or ultimate or real selves.

The mystics want to meet, deal with, praise, and be transformed by the best, the brightest, the most ultimate (necessary, independent, non-contingent) reality there is. If there is a predominant imagery or language in which mystics the world over have tended to work out their desire and articulate what they have experienced and want to urge upon others, it is the imagery or language of being. For the Indo-European world religions, those indebted to India and Greece, thinking of ultimacy in terms of being, is-ness, seems to come naturally. Thus Hindus, Buddhists, and many Jews, Christians, and Muslims have worked ontologically, with a strong inclination to stress (*1*) the relation between the source or ground of the world and the world itself and (*2*) the dependence of the second, limited partner on a grant of being from the first, unlimited partner.

Brahman, Buddha nature, and God have all been first, unlimited partners in a relationship with the world, the ecological whole of all creatures, because all have been the source of the reality of the world, the font of its existence. The Indo-European intuition has been that what we experience bodily does not explain itself or ground its own existence. What we experience bodily, the world through which our rational animality or embodied spirituality moves every day, is changing because it is mortal and, therefore, it is nothing upon which we can rely absolutely. In Indian terms, this world is *samsara,* mottled, enslaved to desire, imperfect, a realm of great suffering. In European terms, this world is the work of a Creator, a pale reflection of his divine splendor, to its core a debtor, always dependent on his grant of being for its reality, existence, actuality.

This ontological slant has shaped large portions of the world's mystical literature. Whether mystics spoke of illumination or love, they were reporting and probing a being-to-being relationship, a mind-to-mind or heart-to-heart communion that linked all that they, the limited partners, were with their unlimited source. That unlimited source could be a totality of light (often experienced as too bright for them and so as darkness) or a totality of love and goodness or a totality of being with the power to diffuse itself and share its limitless reality. Distinctively, the Indo-European thinkers worked with the conviction, or at least the hypothesis, that being is convertible with truth, goodness, oneness, and beauty. These are European Christian scholastic terms (known as the "transcendentals," the qualities that apply to beings always and everywhere, in virtue of being itself), but they have their equivalents in Hinduism and Buddhism, as well as in Greek-influenced Judaism and Islam.

The mystical mind coming from this Indo-European cultural foundation has assumed that reality is intrinsically intelligible. Perhaps Mahayana Buddhism has made this point or relied on this assumption most strongly, finally reaching the point where *bodhi* (wisdom, light, intelligibility) so characterized all beings that they seemed to the enlightened, buddhas and *bodhisattvas,* perfect. The problem or wonder, then, was why human beings were ignorant of such splendor, which was obvious to the Buddhas.

With equal force, though usually more fears of abuse, the mystical traditions rooted in Indo-European ontology could sponsor unions of love between limited beings and their unlimited source. Being itself was convertible with goodness. Reality itself was desirable, not to incite grasping but to elicit movement toward union, to stimulate longing to join oneself to what was truly, ultimately real. Having felt such a great longing, the searcher could easily find lesser longings—for wealth, sexual pleasure, the praise of other people, even good health and long life—considerably less important or worthy. On occasion, such lesser longings might even appear ignoble, embarrassingly limited, provincial, or narcissistic in comparison with the great longing for union with ultimacy.

Implicit in such a great longing has been a desire to worship the ultimate. The ultimate has, on the whole, beckoned as something magnificent, a crystal palace of light, a maternal goodness, a lovingness symbolized well by the Bodhisattva Kuan-yin. The spontaneous response of the human mystical spirit to even slight glimpses of such splendor, such *kavod Yahweh,* has been to sing psalms of praise. Worship is the purest of religious acts, not because human beings cannot contaminate it with their own impurities, but because in itself it represents a straight acclamation by the creature that the Creator, the source of all that the creature is, deserves all honor and glory and attribution of power and might.

In the mystical moment the person drawn into union with ultimacy realizes that everything less than ultimacy exists only on sufferance, through the generosity of the Creator, as an utterly gratuitous grace (*gratia gratis data*). Nothing in the world of finite beings need be. Their very finitude proves that creatures are compounded of being and nothingness. The mystery is what the nothingness is and how the ultimate can compound it with being to make limited beings, forms that stop divinity at definite borders, is-ness that is bound to particular whatnesses.

For the correlation that the Indo-Europeans have made between being and mind, the limitation of being in creatures and what in the ultimate is unbounded, makes them significantly unintelligible. The non-being responsible for finitude offers the mind no light. Similarly, the limited goodness, to say nothing of evil (lack of even the order and being that one expects), of finite beings creates problems with loving them. Inasmuch as the human spirit wants to love without restriction, wholeheartedly, it finds nothing in its ordinary experience to slake its want, and so it finds frustration. The whole world lets it down, and that can become an electric prod to its moving out of the world, denying to the world the lordship that, irrationally, the world seems to ask of it. Although limited, both individual creatures and the entire "system" of the world can appear to be presenting themselves as all there is, the *de facto* divinity before which human beings ought to bow. When the mystics sense this, they speak of an idolatry more profound than any bowing before little gods arrayed on altars of stone and wood. Their wrestlings are with the constant inclination of a (fallen or radically ignorant) human nature to flee from the absolute, potentially terrifying demands of a truly ultimate holiness into a more manageable religion, a covenant making easier demands.

The two words "ultimate" and "reality," therefore, carry immense force. What is ultimate, the end in the chain of contingency, where necessity, a qualitatively different is-ness, occurs, threatens to make creatures insignificant. Whenever the mystics speak of the unreality of the world or the nondivinity of creatures, they are saying, in complementary ways, that from an ultimate point of view only *nirvana* or the godhead can claim to have reality unconditionally. All other beings exist only in the measure that *nirvana* or the godhead lets them be, gives them existence participatively, makes itself the final being of their being. This insight or conviction leads Mahayana Buddhists to say that *nirvana* is in the midst of *samsara*. It leads a Meister Eckhart to speak of letting God be God in your soul, divinizing you, using you to speak his word. It leads an al-Hallaj to say that he is the truth because he realizes in his mystical moments that God is not only closer to him than the pulse at his throat but is also the light of any light in him, the being of any existence he enjoys.

We could keep traveling this ontological path virtually endlessly, so important has it been in the majority of the mysticisms and speculations about ultimacy that we find among the mystics of the world religions. With a strong inclination to speak negatively, both Buddhists and Christian Neoplatonists, such as Pseudo-Dionysius, have tried to take the ontological quest into a silence beyond any human representation, perhaps even any human appreciation, of ultimate reality, of being so completely holy that any limited sense of its existence is almost insulting. This has been the stimulus to prizing ineffability. Those who know do not speak, those who speak do not know.

Certainly, the requirements of human social life, the entailments of embodiment, the dependence of human knowing and loving on sensation, imagination, virtue, and emotion, have made silence, crucifixion of the imagination, and the other negative, apophatic mystical ways of trying to live with ultimacy unsustainable for more than relatively brief periods and small fractions of the human population. The desert fathers and mothers have been relatively few, and even they have had to sleep, eat, move their bowels, plait their grass mats, take care of their bare feet and vulnerable skin. Nonetheless, as they wove together the ontological orientation of their native Indo-European culture and an ascetic program geared to letting this orientation free their spirits to commune with reality at its fullest, its most unveiled, the desert fathers and mothers, with their Hindu, Buddhist, and Sufi equivalents, have tended to evaluate the spirit as more significant than the body and to prize the operations in which the spirit seemed least indebted to the body as the best, the most appropriate, ways of dealing with, being transformed by, God, or ultimacy.

In addition to what we are calling an Indo-European ontological cultural formation, two other major cultural formations beg attention for their influence in the world religions and their different interpretations of ultimate reality or divinity: the Confucian social orientation that came to dominate the mores of most of East Asia and the Semitic, biblical wilfullness deriving from the personal monotheistic divinity at the source of the revelational, prophetic dimensions of Judaism, Christianity, and Islam.

The Confucian social outlook mingled with Taoist, Buddhist, Shinto, and native shamanistic traditions throughout East Asia, after the missionary advent of Buddhism from India (overall, in the early centuries of the Common Era), dominating political life. That social outlook, as we have seen, stressed the Way of the ancient sages, which the Taoists saw running through nature as well as society. The Confucians were the great codifiers of traditional ritual practices, the great champions of a rather formal, hierarchical social structure. The most important feature of this structure was reverence for elders—parents, in the first place; elder siblings, rulers, and teachers, in the second place. The Confucian system was patriarchal, if not misogynistic. It did its business through the proprieties of *li*—protocol, customary etiquette. At its best, this social formation from the behavioral outside formed the minds and hearts of those receiving it toward *jen*—warm, supple fellow-feeling, humaneness, love.

However, in our view the Confucian outlook as a whole was functional and horizontal, rather than mystical and vertical. The distinction between nature and divinity was not clear in either Confucianism or Taoism, both traditions tending to run the two zones or dimensions of reality together into one. No personal divinity stood outside the natural world as its Creator, and the ultimacy of the Tao was blurrier than the ultimacy of *nirvana* or Buddha nature or *Brahman* or the Western God. So Buddhism and Taoism were the great nurturers of mysticism in East Asia, not Confucianism. Buddhism was usually the Mahayana variety that we have seen reach its deepest achievements in the Prajnaparamita literature that finally equated *nirvana* and *samsara*. Taoism was most mystical in the *Tao Te Ching* and the *Chuang-tzu*, where ruminations on the Way produced a poetic, paradoxical, witty, quirky naturalism. The leading Taoists knew that the *Tao* moved through them, inviting them to flow along with the natural grain, but few of them speculated about the relations between this prime mover and the secondary, more limited beings that it moved.

The Semitic, biblical traditions also were not ontological as the Indo-European traditions were, though many Jews, Christians, and Muslims studied Hellenistic philosophy and took its ontological cast of mind deeply to heart. Philo, Maimonides, Augustine, Gregory of Nyssa, Thomas Aquinas, Avicenna, and Averroes all followed Plato or Aristotle as their conceptual master. However, neither the Bible nor the Koran is ontological in the Indo-European sense that we have been giving to the term, and it is easy to argue that both the Bible and the Koran have shaped the Western prophetic cultures more than Hellenistic philosophy. The personal God who emerges in the Bible and the Koran is willful in the extreme, the Lord of the Worlds before whom all creatures must bow. That this Lord is compassionate and merciful keeps him from being a tyrant, but all power and glory and sovereignty belong to him, so much so that Islam often charges the Muslim to find glory in being the slave of God.

This Lord has fashioned convenants with the biblical peoples, stimulating them to think that they have been elected to the uniquely important status of his "chosen." From this thought they have interpreted their histories as privileged sojournings with the Creator, the Lord of their people as well as of the natural world. They have believed that God has guided them providentially, through what has amounted to a history of salvation. Against this background, many Jewish, Christian, and Muslim mystics not educated in Hellenistic philosophy or drawn to a language of being have had their experiences of ultimate reality mediated by study of Torah or sacramental worship of God through union with a Christ taken to be the resurrected Lord of the Christian community or fidelity to the *Sharia*, the law developed from the Koran.

Perhaps more distinctively, the Jewish mystics formed through immersion in the *Tanak* and the Talmud, the Christian mystics formed through immersion in the New Testament, and the Sufis formed through immersion in the Koran have prized imagination, fantasia, more than have their more ontological brothers and sisters. The *merkavah* mystics and the Kabbalists, the Protestant divines fixated on the biblical Word, the Catholics and Eastern Orthodox in love with God through icons, and the esoteric Sufis lost in the letters of the Koranic text have all flown to God on the wings of elevated, sometimes free-floating, imagination.

Despite the prohibitions on making images that one finds in Judaism and Islam, the scriptural Word, full of vibrant images, has been the great locus of all piety, including that of the Jewish and Muslim mystics. De facto the Torah and the Koran have been "incarnations" of divinity nearly as influential as the Incarnation that Christians have accorded to Jesus, the divine Word become flesh. Even though considerable numbers of Jewish, Christian, and Muslim mystics have followed either Indo-European ontology or a desire to deal with the divine otherness by developing the ineffability of the divine into a negative theology, the affirmative nature of scriptural religion, its status as a covenantal relationship revealed to be the will of the Creator of the world, the redeemer of the people from waywardness into holiness, has ensured that the biblical, prophetic religions would give a high priority to historical paradigms—to remembering as definitively revelatory such moments as the Exodus from Egypt, the crucifixion and resurrection of Jesus, and the departure of Muhammad from Mecca to Medina.

Certainly, all the biblical traditions fostered sages as well as prophets—rabbis, priests, mullahs—but the Scriptures that formed the biblical communities and nourished their mystics from the cradle keep the willful, passionate, loving nature of the revelatory Lord at the front and center of their religious lives. At that center, the comparatively apathetic quality of the Indo-European ontological tradition, its efforts to move beyond passion and febrile imagination, could appear awkward, giving the mystics and deeper theologians a wide chasm to bridge. Indeed, the struggle of these groups to integrate being and creation, light and salvation, has provided them considerable frictional energy.

Lastly, we note in passing that all the world religions have housed significant "little traditions"—peasant customs and beliefs, ways of coping with an apparently enspirited if not bedeviled natural, agricultural world—that have sometimes brought them close to the shamanic imagination of the oral traditions we studied in the last chapter. In much popular Judaism, Christianity, and Islam, warding off an evil eye or enlisting the power of a charismatic saint or petitioning divinity nearly magically for good health has flowed out of a rich imagination that populated the world with the spirits of the departed, fallen angels, mythical figures such as Lilith, the first wife of Adam, and *jinn* howling through desert nights.

Although none of these traditions canonized shamans as such or put archaic techniques for achieving ecstasy at the center of their religious programs, all enrolled many simple people with labile consciousnesses susceptible to superstition, magic, and credulity. Moreover, such people tended to be biased toward a psychosomatic view of healing and to exhibit both a rootedness in the natural world and a loyalty to an extended family circle of blood relatives that put them at considerable odds with the ascetic, spirit-oriented mystics of the "high," orthodox strata of their traditions. Thus the "ultimate reality" obtaining at the peasant level of the world religions was usually considerably more proximate than what the mystics of those traditions, intent on trampling on worldliness, had in mind and heart when they contemplated their lord.

Eastern and Western Mysticisms

Here, two principal matters come to mind. First, Eastern mysticism initially appears to be involved with an impersonal deity more than a personal one, with an ultimate reality based more on people's experiences of the natural world than on their dealings with their selves or other human beings. In contrast, the mystics of Judaism, Christianity, and Islam initially appear to be concerned with a highly personal, demanding God, who wants their full obedience. Related to this is an initial impression that the sage, detached and concerned with the wisdom that comes from placing everything in an ultimate perspective based on an analysis of (*1*) the cycles of nature and (*2*) the desires of the human spirit, is the primary religious type in the East, even for mystics. In contrast, the primary religious type in the West is the prophet, who is formed through the revelatory word of a personal God, though sages soon arise to ponder the ("legal") expressions of that word thoroughly.

However, on further analysis things become more complicated, and the two halves of the mystical whole of the world religions, the Eastern and the Western, take up positions close to those of mirror images. Most of the primary concerns of the East appear in the profiles of the Western mystics, though in different proportions, and vice versa.

For example, we find a strong Western tradition of detachment from passion (*apatheia*), and we also find highly personal, passionate elements in the *bhakti* traditions of India, more prominently among Hindu than Buddhist mystics, but to an important accommodated degree in Buddhism as well (for example, the Japanese saint Shinran [1173–1263]). We find rabbis, priests, and mullahs who love wisdom, as their scriptures mediate the mind of God to them, and we find many Christian monks, as well as some Jewish and Muslim saints, who travel a mystical path of unknowing into the mysteries of being, coming to deal with divinity as an ultimate reality sovereignly free, uniquely care-less.

Second, the Eastern concern with *karma* places its mystics apart from those of the West, where the closest correlative category is probably sin. Let us develop the apparent implications of this second difference for comparing the two halves of the mystical world and then return to the first issue, the impersonal and personal aspects of the ultimacy or divinity with which Easterners and Westerners have both had to deal.

When we analyze the impact of *karma* on Eastern mystics, in contrast to the impact of sin (or fallenness or weakness or forgetfulness) on the Western mystics rooted in the Bible or the Koran, we find a different impulse to get out of the world. Mystics in both hemispheres want to escape from the world inasmuch as "the world" stands for a profanity or insanity or enslavement that frustrates their pursuit of fulfillment. The world drives them to seek something, someone, unworldly, unlimited, unconditioned, able to give complete satisfaction. Everything in the world fails them, so they are primed to hear a word from beyond the world or to seek a reality not material, changing, and flawed.

Eastern mystics indebted to Indian traditions grow up thinking of the fetters of the world as karmic bonds. They learn that the current shackles on their freedom derive from their actions in past lives. A tight if not rigid, indeed inevitable, moral causality conditions strongly their dealings with all of the realities in their milieu: inanimate objects, plants, animals, other human beings, the many gods of the different celestial realms. The great suffering that comes from being trapped in this karmic, samsaric realm casts a pall over Indian estimates of the human condition. People stay in thrall through death and rebirth after death and rebirth because they continue to desire pleasures, prosperities, satisfactions. Only the very few who rid themselves of desire draw near to *moksha, nirvana:* the states of release, the being–awareness–bliss of fully free, unconditioned existence, where the flames of desire have blown out.

Enlightenment, the act of realizing how to construe the world without desire, is the gateway to liberation, because enlightenment overcomes a noxious ignorance. Gautama is the paradigmatic sage in the East because in his own biography he presents most dramatically the path to enlightenment and then becomes a great teacher, articulating cogently how others can follow the path. Enlightenment transforms much more than the mind. It floods the saint's entire being, driving desire and karmic indebtedness out. Like a bone marrow transplant, it gives the sage a new system, in this case one psychosomatic. Detached and compassionate, the buddhas and *bodhisattvas,* and their equivalents among the Hindu saints, exhibit consummate freedom. They fear nothing because death and rebirth have no hold on them. They injure nothing because they want nothing, hate nothing. Their communion is with the being, the lightsomeness, that dances in everything they meet. Their mysticism is a direct experience of a reality beyond *karma,* ultimate because too full, too holy, to countenance desire.

The Western mystics formed by theories of sin move through many of the same psychological states that the Eastern mystics traverse but the overtones, weights, colors tend to be different. Sin encompasses the mind, but it spotlights the will. Sin entails a darkening of the mind, one that increases as the sinner becomes more habitually sinful, but the core of sin is the refusal of the creature to submit to the Creator. All human beings ought to be Muslims, submitters to the will of Allah. The account of the sin of the first human beings, Adam and Eve, that we find in the first chapters of Genesis stresses their disobedience. They knew, and did not know, that they were dismissing the will of the Lord who had made them.

Why they did this, what prompted their revolt, remains unclear. The symbol of the seductive serpent hints at a natural desire for independence, but more provocatively than lucidly. The first parents did not trust that God knew best for them. Whether distrust such as theirs is necessary for moral maturation, realism about the twistedness of personal conscience and social ethics, is a matter for debate. In their imperfect, "fallen" state, human beings certainly are ambiguous morally. Whether this ambiguity is necessary, natural, and unavoidable entails a large question about the graciousness with which God desires human beings to live.

At any rate, the Western mystics labor ascetically to cleanse themselves of sin. This is probably clearest in the Christian denizens of the desert, who explicitly wanted to remake human nature by purging themselves of the classical vices and developing a new being empowered by the classical virtues. Jewish and Muslim mystics have often pursued analogous purifications, but in their traditions monasticism has been suspect, especially the celibacy that Christian monks have embraced. Less influenced than Christian ascetics by Hellenistic dualism, which pitted the body against the spirit, Jewish and Muslim mystics appear usually to have found material creation more congenial than did the Christian denizens of the desert. How the radical incarnationalism that orthodox Christian theology has preached by confessing the strict divinity of Jesus the Christ has fitted into this whole psychological complex is a knotted issue. Suffice it to say that frequently the negative Christian mystics appear to have bracketed their doctrinal commitments to the Incarnation, in practice working and praying as though the flesh of Christ were not the primal sacrament of the presence of God.

Many qualifications attend such a saying, prominent among them being the importance of the Christian eucharistic liturgy, where the preeminence of the Word made flesh imposed itself upon Christian consciousness regularly. Nonetheless, the desire of many Christian mystics to rid themselves of sin led them into views of the material world, including their own bodies, that verged on puritanism: a deep distrust of the way that God had made them less than the angels.

The correlation between the willful revelatory God of the Western scriptural mysticisms and the willful rebellion called sin has meant that much of Western piety has entailed repentance, conversion or reconversion, weeping over infidelity, and renewing pledges to keep the covenant better. The accent has been personal, at times the passionate charges and countercharges of lovers struggling to accommodate divergent selves. God has been the dominant partner, but the human partner could complain, beg, cajole. Thus the great Hasidic rabbis would go down on their knees before the Master of the Universe, trying to protect their people against another pogrom. The nuptial symbolism of numerous Christian mystics would take the relations of sin and grace into the chambers of romance, as would the passions of Sufis such as Rabi'a and Rumi.

On this note we can return to our first topic, the matter of the impersonal and personal aspects of divinity, as the Eastern and Western mystics have highlighted them. A figure that may not distort the profiles of the two hemispheres unconscionably would line up in the Eastern column the items (*1*) impersonal ultimate reality, (*2*) attraction to the cycles of the natural world, and (*3*) conviction that the key to liberation (eventually, to mystical union) is the elimination of desire through a profound experience of enlightenment. Parallel to these items, in the Western column, would run the following counterparts: (*1*) personal ultimate reality (God), (*2*) attraction to the ruling word of a powerful monarch (God), and (*3*) conviction that the key to liberation (eventually, to mystical union) is the fusion of the creature with the Creator (the subject with the monarch) through a heart-to-heart communion that blends intimate knowledge and love.

The Eastern triplex is not absolute, as we hinted when noting a personal phenomenon such as *bhakti,* which has led many Indians to intense relations with personified gods such as Krishna, Rama, Kali, and Shiva. Nonetheless, when we characterize the physiognomy of the Indian and, to a lesser yet still recognizable degree, of the East Asian mystics, we find a concern with being or *nirvana* or the *Tao* that places the human seeker in a largely impersonal, ultimate frame of reference or construal of reality. Physical nature inspires this conception for the mystical imagination, more than individual or social humanity, and the distinction of divinity from physical nature is often weak at best. The work to drop desire, step away from *karma,* and gain dramatic release through enlightenment aims at the fulfillment of the individual largely by breaking through the limits previously thought intrinsic to the human spirit, in a show of "That thou art." The *Brahman* grounding the world is the *atman* grounding the individual human being. The *nirvana* offering the Buddha complete freedom glows in the midst of *samsara.* The *Tao* moves the ten thousand things, even as it flows along before them, apart from them, on its own.

The Western triplex begins with a sense that the world, or reality as a whole, stems from the free choice of a powerful, eminently personal God to share divine being in limited ways by making other beings. Why God chose to do this remains mysterious, though most of the mystics attribute creation to the divine goodness or intrinsic bent to make reality afresh, to bring forth life as an overflow of its own vitality. On the whole, the dominant image for this personal Creator likens him to a lord, a king, a monarch. He speaks and expects his subjects, whether angels or human beings, to obey. His speech, revelation, expresses himself and becomes in his subjects law. The Torah, the Gospel, and the Koran are books of divine instruction, forms of divine revelation, inscriptions of divine will.

These books, the revelations behind them, and the will of the monarchical Creator that they express form the scriptural peoples—Jews, Christians, Muslims—into communities considering themselves elect. Jewish mystics assume that they dwell among a chosen people whose laws exist to make them holy as their God is holy. Christian mystics assume that they live in a church, a gathering of people called by the father of Jesus Christ to be his body, the branches of his vine, and to share intimately the divine life of the Trinity. Muslim mystics assume that they live in the House of Islam, the community gathered by the *Rasul*, the seal of all the prophets, the spokesperson celebrated in the consummation of "biblical" faith: "There is no god but God, and Muhammad is God's Prophet."

Lastly, the mystical way that relations between the Western God and the people brought up in such elect communities tend to stimulate a union from the heart that blends intimate knowledge and love. The temperature is warmer than what one finds in the East. Certainly, many of the Western mystics become well aware that God is not personal, as human beings are, and that whatever they say about God is more unlike the divine reality than like it. However, the word of the monarchical Lord remains surprisingly gentle, an invitation to trust, and so it remains significantly personal, despite the infinity, the endlessness, the deathlessness that separates him from all creatures.

In the *Bhagavad-Gita* the word of Krishna to Arjuna that he loves him comes relatively late, after lengthy and quite impersonal yogic instructions. In the Western scriptures, the personal interest of the Creator, if not always the Creator's love, rings out from the beginning. There is nothing invidious in this comparison. Any ranking of one style above the other depends on judgments of value that one would have to defend through many twists and turns, and as we have noted, when one studies the full profiles of the Eastern and Western mysticisms many of their initial apparent differences smooth away considerably, lessening grounds for making them antagonists. Nonetheless, in our opinion the Western mystics remain to the end more personal than the Eastern, and we find this difference a useful reminder that ultimate reality, the true divinity creating the world, is susceptible of both designations, though captured by neither.[2]

Personal Implications

We noted at the outset that mysticism is not the whole of any religion, let alone the epitome of human culture at large. Nonetheless, we find among the mystics of the world religions, in the minds and hearts of the men and women who appear to have experienced ultimate reality more directly than the rest of us, such interesting, at times compelling, and now and then supremely edifying specimens of humanity that studying them has become as much personal pleasure, in a high sense, as professional obligation. Indeed, we find that studying the mystics has become the staff of our own spiritual lives, *lectio divina,* that we undertake apart from any academic responsibilities. This finding inspires our final section. What might a balanced, nonproselytizing review of the experiences of the mystics of the world religions suggest about the potential in any alert reader's life for doing likewise?

All human beings suffer and are ignorant. The wisdom of the Buddha is incontrovertible: all life is suffering. All human beings die, and it is not clear that not dying would solve our basic unhappiness, for as long as we live in the body we do not find in this present (*samsaric,* sinful) form of existence a lasting peace, happiness, or fulfillment. Therefore, many of us intuit with Augustine that our hearts will be restless until they rest in reality of another order, one out of time and change, suffering and death, ignorance and desire, the sins of pride, covetousness, lust, anger, gluttony, envy, and sloth. Many of us intuit that *nirvana* is our true home, the Garden our proper destiny.

Such an intuition marks us as religious people. Our humanity shows itself again and again bound to a goal, which turns out to be a source different from that of our less troubled, more secular neighbors. Whether or not they really are less troubled and actually more satisfied with worldly goals and achievements, we can seldom know because seldom do we neighbors, secular or religious, bare our souls to one another, make plain the thoughts we suffer at four in the morning. On the surface it seems that the majority of our American neighbors have spirits tamped down by television, busyness at work or with the kids, practical concerns of a dozen major kinds that limit their contemplative undertakings. Our American culture is rich and broad. Seldom is it deep. It is pragmatic, generous in good deeds. Seldom is it ontological or mystical. To set off on a contemplative, mystical path is therefore an idiosyncratic choice. The madding crowd will not know what to make of it. The solitary ways of the monk or artist or research scientist tend to draw as much suspicion as admiration.

So you should not follow lightly an attraction to developing the mystical potential in your spirit. You should test thoroughly the comings and goings of your urges to drop out of the busyness of the American mainstream and contemplate deeper things as a solitary. Unless you are so fortunate as to find a congenial community of likeminded seekers, contemplative ventures will probably be lonely treks demanding significant intestinal fortitude. You will have to battle pride, the temptation to think that you may be elect, chosen. If you persevere through the initial, beginner's phase of delight in entering upon truly serious questions, you will find that solid enlightenment usually requires dark nights that strip your senses of consolation and plunge your spirit into darkness to test your understanding of faith.

Steadily it will develop that you are engaged with you know not what. The full reality that you seek, the holiness that is a foundation without spot or wrinkle, will never come under your control. The Lord does not give his name to Moses. The Muslim never becomes the one submitted to. The mature Buddhist has no self, realizes that emptiness is his or her birthright. The mature Taoist moves freely on the current of the Way, quite careless of little needs or desires.

Yet, all this is curiously, strangely, sustainingly nourishing. Even in the dark nights, the mystics sense that they have found a better place, are being put to a better task, than would be so had they remained content with worldliness. They never know what God is, how ultimate reality can be ultimate, yet in the center of themselves, too primitive for anything like adequate articulation, they are content. What they are commingles with what everything is, with who or what everything comes from, and this commingling, this communion, gives them rest. Unanimously, the mystics, small and great alike, say that a day in the courts of the Lord is worth years of labor, even pleasure, outside those courts. An hour in the Temple is worth days in the marketplace. Certainly, mystics whom the East might call fully realized become at home in the marketplace, at ease in the world. Before enlightenment trees are trees and rocks are rocks. After enlightenment, trees are again trees and rocks are again rocks. The difference is the middle that the mystic has gone through. The result is trees and rocks that are fully themselves because we sense them to be in, transparent to, what makes everything be.

The lure of such an end point, such a comprehensive vision of the world, will vary from individual to individual, as will the lure of a personal, loving communion with a mysteriously personal divinity, a covenantal Lord compassionate and merciful, long-suffering and abounding in steadfast love. Without a strong attraction, there is little likelihood that a person will persevere on the mystical path. Equally, without the support of a religious community, the deep immersion in the traditions of the *sangha* or the Torah or the Church or the *Umma* that only ongoing participation in the worship, study, and spiritual direction of an established religious tradition can provide, one's chances of staying on the path that is straight are small.

Therefore, Rabbi Hillel told his disciples not to separate themselves from the community. The sober Sufis insisted on keeping the *Sharia,* not speaking or acting ecstatically or idiosyncratically so as to draw wrongful attention to oneself. Certainly, the "crazy gurus" of Tibet and the Tantrists of many traditions have made the important point that love of ultimacy makes one eccentric, perhaps usefully shocking to the stolid hoi polloi. However, many mystics have preferred the argument that their contemplations, their direct experiences of ultimate reality, are simply moments of full humanity potentially available to all their sisters and brothers and nothing that they themselves ought to trumpet as exceptional.

Related to this argument is that the best experiences of the majority of mystics have been gifts, not personal achievements. The language of mysticism features grace, not duty or the gains of work. Mystics do not pull themselves up by their own bootstraps. They work hard, but the best of their labors is *wu-wei,* the Taoist not-doing that is an active passivity, an attentive seconding of the actions of the primary mover, the *Tao* or Godhead drawing them on. The drama in the soul of the mystic is a dialogue, a cooperation, a synergy. Ultimate reality reaches into the proximate spirit of the mystic and reveals how its ultimacy habitually grounds all that the creature always is. The accent in the mystic's inner recording of this divine action may be ontological or personal on a biblical model, but either way it establishes that no mystic, indeed none of us creatures, is alone or belongs to the self except provisionally. We belong to who or what makes us be and gives us existence as a participation in its sole fullness of reality, to that which is always closer to us than the pulse at our throats.

Speech such as this, poetic, oblique, and paradoxical, is the best that mystics can do. It may also be a useful criterion for determining one's aptness for the mystical life. One can argue that all human beings have the obligation to understand their situations as best they can and that the unexamined life is not worth living. One can argue that all human beings have the obligation to become mature morally in their choices of what is truly good, honest, and loving. It is less clear that one can argue that all people ought to strive to become mystics. We can conclude by probing this lack of clarity through a reflection on mystical diction.

Once we were answering questions after a lecture and got into a bind with a young woman. There seemed no meeting of minds, a complete missing of one another's intentions. Finally, we asked her what she did for a living. "I study cement," she said. A light dawned. We were talking about movements of the contemplative spirit. Her world, with which she seemed fully content, focused on cement. We were students of religion, ultimate reality. She was an engineer, interested in laying down roads, putting in the foundations of office buildings. We found the situation amusing, but humor did not appear to play a large role in the study of cement. So, we bailed out of the exchange, judging it had no future.

You can work cement into many useful forms without venturing into paradox or poetry. If your view of the world, what you are brimming to say, forces you to speak paradoxically, you probably ought to stay out of civil engineering. The language of the mystic is sacramental, metaphorical, intent on showing how the revelations of ultimate reality come dazzling like light through a kaleidoscope. The mystic sees the world as a congeries of types and antitypes, archetypes in heaven and miniature versions on earth. Moreover, beyond this iconography lies darkness, silence, unlimitedness that sometimes seems more pregnant than any of the forms it inspires. Sometimes it seems like nothing less than a cosmic womb, the Buddhist *Tathagatagarbha,* or a cornucopia of beings streaming down, the illuminations of a great Being of lights.

If speech such as this means little to you, you have experienced little about mysticism. More significantly, if speech such as this puts you off, leaves you cold, then at the moment you have little aptitude for mystical studies. Mysticism aims at clarity but only through difficult unknowing. Mystics find themselves using speech to defeat speech, so that silence, wordless love, heart-to-heart communion, from the fine point of the soul through the pointless place where ultimacy infuses being, takes center stage and determines the play.

In the end, everything exits into mystery, as, mystics begin to suspect, everything entered mysteriously in the beginning. We are engaged with much more than we know or can appreciate. If such an engagement intrigues you, you may have a calling to try the mystical path.[3] Otherwise, go in peace, contenting yourself with doing justice, and loving kindness, and walking humbly with your God.

NOTES

1. See John Carmody, "On Writing About God," *Studies in Formative Spirituality* 13, no. 1 (February 1992): 85–91.
2. See Frederick Copleston, *Religion and the One* (New York: Crossroad, 1982).
3. On the mystical outlook overall, see Huston Smith, *Forgotten Truth* (New York: Harper & Row, 1976); Denise Lardner Carmody and John Tully Carmody, *In the Path of the Masters* (New York: Paragon, 1994).

Index